Agricultural and Kitchen Waste

Apart from being termed as a pollution source, agriculture and kitchen waste is also a rich source of carbohydrates, minerals, antioxidants and vitamins, and can be utilized to develop value-added products and for energy production, which is the main theme of this book. It also focuses on the minimization of this waste via different routes like conversion into bio-fertilizers, organic acids, other industrial products, and efficient energy production. It comprises different topics and concepts related to waste utilization contributed by recognized researchers and experts.

Features:

- Covers all the technical aspects of utilization of agricultural and kitchen waste.
- Discusses the quality characteristics of value-added products.
- Provides overview of different options for processing of organic wastes.
- Includes production of acids and enzymes from agriculture/kitchen wastes.
- Reviews effects of kitchen/agricultural waste on environment and its role in pollution control.

This book is aimed at researchers and graduate students in chemical and environmental engineering.

Novel Biotechnological Applications for Waste to Value Conversion

Series Editors:
Neha Srivastava, IIT BHU Varanasi, Uttar Pradesh, India and *Manish Srivastava*, IIT BHU Varanasi, Uttar Pradesh, India

Solid waste and its sustainable management is considered as one of the major global issue due to industrialization and economic growth. Effective solid waste management (SWM) is a major challenge in the areas with high population density, and despite significant development in social, economic and environmental areas, SWM systems is still increasing the environmental pollution day by day. Thus, there is an urgent need to attend to this issue for green & sustainable environment. Therefore, the proposed book series is a sustainable attempt to cover waste management and their conversion into value added products.

Utilization of Waste Biomass in Energy, Environment and Catalysis
Dan Bahadur Pal and Pardeep Singh

Nanobiotechnology for Safe Bioactive Nanobiomaterials
Poushpi Dwivedi, Shahid S. Narvi, Ravi Prakash Tewari and Dhanesh Tiwary

Sustainable Microbial Technologies for Valorization of Agro-Industrial Wastes
Jitendra Kumar Saini, Surender Singh and Lata Nain

Enzymes in Valorization of Waste
Enzymatic Pre-treatment of Waste for Development of Enzyme based Biorefinery (Vol I)
Pradeep Verma

Enzymes in Valorization of Waste
Enzymatic Hydrolysis of Waste for Development of Value-added Products (Vol II)
Pradeep Verma

Enzymes in Valorization of Waste
Next-Gen Technological Advances for Sustainable Development of Enzyme based Biorefinery (Vol III)
Pradeep Verma

Biotechnological Approaches in Waste Management
Rangabhashiyam S, Ponnusami V and Pardeep Singh

Agricultural and Kitchen Waste
Energy and Environmental Aspects
Edited by Dan Bahadur Pal and Amit Kumar Tiwari

For more information about this series, please visit: www.routledge.com/Novel-Biotechnological-Applications-for-Waste-to-Value-Conversion/book-series/NVAWVC

Agricultural and Kitchen Waste

Energy and Environmental Aspects

Edited by Dan Bahadur Pal
and Amit Kumar Tiwari

CRC Press
Taylor & Francis Group
Boca Raton London New York

CRC Press is an imprint of the
Taylor & Francis Group, an **informa** business

Cover images: Shutterstock

First edition published 2023
by CRC Press
6000 Broken Sound Parkway NW, Suite 300, Boca Raton, FL 33487–2742

and by CRC Press
4 Park Square, Milton Park, Abingdon, Oxon, OX14 4RN

CRC Press is an imprint of Taylor & Francis Group, LLC

ISBN: 978-1-032-15812-9 (hbk)
ISBN: 978-1-032-15813-6 (pbk)
ISBN: 978-1-003-24577-3 (ebk)

DOI: 10.1201/9781003245773

Typeset in Times
by Apex CoVantage, LLC

Contents

Figures

Editors' Biographies

Dr. Dan Bahadur Pal
B. Tech, M. Tech, Ph.D.
Assistant Professor, Department of Chemical Engineering, Harcourt Butler Technical University, Nawabganj Kanpur-208002, Uttar Pradesh India

Dr. Pal is currently working as Assistant Professor in the Department of Chemical Engineering at Harcourt Butler Technical University, Nawabganj Kanpur-208002, Uttar Pradesh, India. He has received his M. Tech and Ph.D. in the field of Chemical Engineering from Indian Institute of Technology (BHU) Varanasi, Uttar Pradesh, India. Before that, he completed his B. Tech in Chemical Engineering from UPTU, Lucknow. Dr. Pal's research interest is nano-technology, catalysis, energy and environment and waste management with a special focus in developing process and materials by using waste as raw materials. He also prefers to work on bio-waste processing and value addition. Dr. Pal has published more than 65 publications in reputed journals and books, along with 22 book chapters and four book edits.

Dr. Amit Kumar Tiwari
B.Sc, M.Sc, Ph.D.
Assistant Professor, Department of Chemical Engineering, Birla Institute of Technology, Mesra, Ranchi-835215, Jharkhand, India

Dr. Tiwari is currently working as Assistant Professor in the Department of Chemical Engineering at Birla Institute of Technology Mesra, Ranchi, Jharkhand. He received his Ph.D. in the field of Food Engineering from Birla Institute of Technology Mesra. Before that, he completed his M.Sc in Fruit and Vegetable Technology and B.Sc in Agriculture from Kanpur University (Chhatrapati Shahu Ji Maharaj University), Kanpur. Dr. Tiwari's research interest is development of new and novel food products with a special focus on developing process techniques for fruits, vegetables, food grains etc. with enhanced nutritional and edible qualities. Dr. Tiwari has published more than nine publications in reputed journals and 10 book chapters. Dr. Tiwari is also working on agricultural waste utilization through value addition, energy and environment and food waste management with a special focus on development of process and value-added products by using agricultural and food waste as raw materials. Dr. Tiwari also prefers to work on biomass application.

Preface

This book deals with the agriculture and kitchen-based value-added products and their application in various areas. The world's population is continuously growing, and this growth is responsible for global changes like increases in temperature, air pollution, water pollution and soil pollution etc. These changes are directly associated with various anthropogenic activities such as agricultural practices, urbanization, uncontrolled use of available resources, evulsion of huge kitchen waste and agricultural residues etc. Due to irresponsible behaviour of humans, these wastes are not properly disposed of; therefore, these wastes are a big issue for our environment. In the last few decades, food scientists and agriculture and environmental researchers from various countries proved that food and agricultural waste have potential as raw materials for the production of different value-added byproducts like bioethanol, bioactive compounds, water purifiers and adsorbents, and so on. During the 19th and 20th centuries, the focus was mainly on production of bioethanol from different agro-biomass, and significantly less attention was given to development of value-added commercial products. But now, the scenario has changed drastically after realizing that product development industries can get more benefited if they will engage them in the production of other value-added byproducts from the processing waste. Proper utilization of these waste materials by which we can control environmental pollution is discussed in this book.

Chapter 1 discusses Current Scenarios of Forest Residues and Its Utilization in Various Sectors; Chapter 2 discusses Commercial Production of Antioxidants from Fresh Agricultural Waste; Chapter 3 discusses Utilization of Agricultural Residues for Energy Production; Chapter 4 discusses Conversion of Agricultural and Forest Wastes into Water Purifying Materials; Chapter 5 discusses Sustainable Production of Nanomaterials from Agricultural Wastes; Chapter 6 discusses Domestic and Agricultural Wastes: Environmental Impact and Their Economic Utilization; Chapter 7 discusses Sustainable Utilization and Management of Agricultural and Kitchen Waste; Chapter 8 discusses Potential Benefits of Utilization of Kitchen and Agri Wastes; Chapter 9 discusses Utilization of Biomass from Refineries as Additional Sources of Energy; Chapter 10 discusses Effects of Agricultural Wastes on Environment and Its Control Measures; Chapter 11 discusses Kitchen and Agri Waste as Renewable, Clean and Alternative Bioenergy Resource; Chapter 12 discusses Various Value-Added Products from Agricultural and Bio-Waste.

1 Current Scenario of Forest Residues and Its Utilization in Various Sectors

Arkadiusz Stańczykiewicz, Marek Pająk and Marcin Pietrzykowski

CONTENTS

1.1 PAST USES OF FOREST RESIDUES

A report published by FAO (2020) indicated that the global production of wood biomass in the form of wood chips, low-quality wood of unspecified dimensions, and forest residues exceeded 507.8 million m^3 in 2019, which was 55.8% higher compared to 325.8 million m^3 harvested and processed in 1999. The total amount of this type of raw material harvested over the course of 20 years was over 9208.5 million m^3.

During this period, the largest portion of this type of wood biomass was harvested in the Americas (38.8%), Europe (29.1%), and Asia (26.6%), yielding 3571.33,

DOI: 10.1201/9781003245773-1

TABLE 1.1
Top 10 Countries in the Production of Wood Chips, Particles, and Residues in the Period 1999–2019 (based on FAO 2020)

Country	Production quantity [$m^3 \times 10^6$]	Share [%]
China (Mainland)	1750.42	19.0
USA	1546.41	16.8
Canada	1463.02	15.9
Sweden	409.66	4.4
Australia	394.42	4.3
Brazil	372.33	4.0
Russian Federation	363.18	3.9
France	343.37	3.7
Finland	308.07	3.3
Japan	254.57	2.8
Total	**7205.44**	**78.1**

2681.72, and 2446.61 million m^3, respectively. The remaining 5.5% (508.84 million m^3) was harvested in Oceania and Africa. The countries that produced and utilized the majority of wood biomass include China, the USA, and Canada, where over 4.7 billion m^3 is harvested and processed, accounting for more than half of the wood biomass harvested globally between 1999 and 2019 (Table 1.1).

According to data published by FAO (2020), more than 78% of wood chips, particles, and residues were harvested and processed by the top 10 countries alone. In Asia, China harvested and processed more than 71% of forest biomass. Within both American continents, more than 84% of biomass was harvested from forests in the USA and Canada. In Europe, the largest amount of biomass (over 28%) was obtained and processed in Sweden and Finland, while in the Oceania region Australia accounts for the highest amount of harvested and processed forest biomass (over 97%).

1.2 AVAILABLE FOREST RESIDUE RESOURCES

The global available resources of forest residues are expected to be in the range of 24–43 EJ (exajoules) per year in 2030 (IRENA 2014). If all of this material is processed into woodchips (1 ton of woodchips = 10.4 GJ (gigajoules)), then the available resources would range between 2300 and 4500 Tg (million tons).

Taking into account the current guidelines for sustainable forest management, the potential availability of forest biomass in the form of logging residues and stumps in Europe is estimated at 12% of the total forest biomass harvested (Verkerk et al. 2019). This is equivalent to approximately 48 Tg (million tons) of dry matter per year ($Tg \times yr^{-1}$). The potential availability of forest biomass within the 28 member states of the European Union is estimated to represent 84% of the total potential for all 39 European countries. Sweden, Germany, France, Finland,

and Poland account for about half of the total European forest biomass potential. Assuming the possibility of increasing the use of forest biomass primarily for energy production, about 26% of the potentially available forest biomass (i.e., about 122–125 Tg × yr^{-1}) may be harvested and processed. On the other hand, assuming that biomass utilization can be possibly increased with absolutely minimal technical constraints, over 190–195 Tg × yr^{-1} would be harvested and processed, which would represent about 36% of the potentially available forest biomass. However, the recent actions by European Union bodies targeted at improving the conservation of biodiversity may limit the availability of forest biomass to only logging residues, which would account for around 11% of total forest biomass harvested (Verkerk et al. 2019).

1.3 TECHNOLOGICAL SYSTEMS FOR HARVESTING AND PROCESSING FOREST RESIDUES

Logging residues, which are formed when solid wood is logged for industrial purposes, are wood raw materials harvested by different technical means such as machinery used for collecting and transporting and machinery used for chipping the gathered material. Standard logging, including both thinning and clearcutting, utilizes motor-manual as well as fully mechanized operations. In the absence of sudden weather changes resulting from natural disasters (mainly windthrows or snowfalls) or forest fires, it is recommended to use machines as much as possible, which allow the fastest and safest harvesting of timber for industrial purposes. However, this will lead to the accumulation of large amounts of non-merchantable wood in the form of low-quality wood of undefined dimensions, root stumps, or parts thereof. During breeding operations, noncommercial timber harvesting is also carried out as part of precommercial thinning. Regardless of the origin of the raw material to be processed, wood chips are the final product used for energy production (Kallio and Leinonen 2005).

The most common technology used in countries where most forest residues are harvested for energy purposes is whole-tree chip harvesting, which is performed with chainsaws or small harvesters equipped with packer heads in the youngest stands, from which the entire raw material will be used for energy production. The obtained young tree bundles are extracted by forwarding to forestry depots, where they are chipped using chippers. The chipped raw material is then transported to the combined heat and power (CHP) plant by containers (Leinonen 2004).

According to Leinonen (2004), the following methods are used for converting forest residues into woodchips in older thinning and clearcutting stands:

- comminution at the roadside landing involves year-round, continuous shredding of residues in the form of branches and treetops or small-diameter trees, using truck-mounted chippers for transferring the shredded material into disposal containers;
- comminution at the terminal is similar to the previous one, but it additionally deposits larger amounts of various forest residues (including uprooted stumps and parts of low-quality trees) on huge stockpiles for drying them,

sometimes cleans the residues of impurities, and then shreds them using more efficient machines;

- comminution at the stand involves chipping with the use of self-propelled terrain chippers that transport the chips stored in a container to a depot where they are loaded into disposal containers; and
- comminution at the power plant, which is indicated as the most optimal solution, involves bundling forest residues and shredding them after transportation to storage yards that are in the immediate vicinity of the CHP plants.

1.4 DIRECTIONS OF USE OF FOREST RESIDUES AND RAW MATERIALS PRODUCED FROM WOOD-BASED COMPOSITE MATERIALS

According to Braghiroli and Passarini (2020), forest residues are used for producing wood-based composites (e.g., plywood, oriented strand board, particleboard, and fiberboard) as they are characterized by better physical properties that enable reducing the effects of moisture and water (less absorption and swelling) in comparison to boards that are made entirely from solid wood for industrial processing. The study of Iwakiri et al. (2017) showed that the use of wood chips obtained from forest residues (up to 50%) in the preparation of raw material for wood-based panel production results in final products with improved or acceptable mechanical properties. Hence, wood-based composites are used as both nonstructural and structural materials for interior and exterior applications. Forest residues are not the only source of raw material used for manufacturing wood-based composites. Construction waste and waste derived from roundwood processing in sawmills (shavings, sawdust, wood fibers) are also used for this purpose (Ross 2010). In some wood-based composite panels, waste-derived raw material makes up to 50% of the volume, and thus replaces shredded solid wood, the share of which is limited mainly for economic reasons (Shmulsky and Jones 2011).

Wood-plastic composites (WPC) have been receiving increasing interest for several years, primarily in the aerospace and automotive industries. They are also widely used in the construction and packaging industries due to increasingly stringent regulations aimed at limiting the production and use of petrochemical-based materials for environmental protection (Ashori 2008). It has been reported by Kim and Pal (2011) that WPCs exhibit high durability and favorable physical and mechanical properties compared to the basic raw materials (thermoplastic polymers and wood fibers applied for filling) used in their production. Therefore, it has been projected that the global WPC market would account for over US$8.5 billion by 2022, and an average annual growth rate of more than 12% has been observed in the past 5 years (Zion Market Research 2017). However, it should be noted that pure chips and sawdust generated during industrial solid wood processing (e.g., sawing into lumber or turning) are better suited for the production of WPCs (Ashori 2008) compared to bark-rich forest residues (Migneault et al. 2014). This is due to the fact that the fibers of forest residues are more difficult to process than wood fibers, as they differ in chemical composition and fiber density (Saddem et al. 2019).

Moriana et al. (2015) suggested that processing of forest residues into several forms (wood chips, thinner branches, or needles) allows optimal exploitation of their properties in the production of specific cellulosic polymeric materials. Cellulose fibers of different structures and morphologies can be isolated from forest waste forms that are more abundant (e.g., wood chips) and used in the production of cellulose-reinforced composites. On the other hand, forest waste is characterized by higher calorific values due to significant content of lignin in branches and needles, and hence can be considered as a biomass waste with high potential for use as solid fuel.

Singh et al. (2020) emphasized the importance of polymer composites made from natural fibers derived from agroforestry waste. Their results indicated that composites can serve as alternative building materials to wood in low-cost housing construction. Necessary treatments can be applied to polymer composites with the use of chemical additives for improving their resistance to, among others, accelerated aging under the influence of water, and attack by insects, pests, and fire. This may increase the strength of composites to qualify them as an acceptable material for various applications including shuttering boards, terrace, and roofing materials or floor panels. The economic viability of polymer composites makes them competitive building materials, and the Life Cycle Assessment confirmed their superiority over glass fiber-based composites based on their environmental benefits. However, as agroforestry waste-derived natural fibers have been recently shown to exhibit variations in properties due to their origin, age, growing conditions, extraction methods, and degree of susceptibility to various external factors, there is a need to develop an up-to-date database on the properties of natural fibers and their composites in order to exploit their potential in the composites industry.

1.5 FOREST RESIDUES AS RAW MATERIAL IN WOOD FUEL PRODUCTION PROCESSES

Forest residues, especially wood waste in the form of chips, are ideal raw materials for producing heat and electricity. These are prepared as briquettes or pellets for combustion in appropriate boilers. Braghiroli and Passarini (2020) indicated that the biggest exporters of pellets to the European Union are the USA and Canada and the global consumption of wood pellets has been increasing and is currently estimated at more than 20 million tons. Mechanical processing of wood chips into pellets, which is the more desirable form of biofuel, allows meeting increasingly stringent environmental standards for biofuels (Pellet Fuels Institute 2020). Therefore, the type of raw material used in the production of such compacted wood fuel plays a major role. Rhen et al. (2007) analyzed pellets made from logging residues and from industrial (sawdust) waste and showed that the type of raw material had a greater influence on the combustion properties than the density of pellets. In addition, they found that pellets containing bark were characterized by almost 50% longer burning times compared to those made from sawdust, while pellets made only from branches showed about 15% longer burning times than those prepared from tree trunk-derived wood. The pelletization process is influenced by factors such as the content of lignin and ash in the wood used for pellet production. Nguyen et al. (2015, 2016) pointed out that higher content of lignin and ash promotes increased frictional and compressive strength, which is related to the lower longevity of trees whose wood was used for producing pellets with almost the same calorific value as pellets produced using wood from more vigorous trees.

As the demand for forest residues for energy purposes has been steadily increasing in many parts of Europe (Routa et al. 2013) and around the world (International Energy Agency 2017) and the costs of collection and transport of these raw materials are rising (Kizha and Han 2016), there is a need to search for new technological solutions involving the use of forest residues, which will increase their availability, reduce processing costs, and enhance environmental and social benefits. One of the solutions is to directly process the residues on cutting plots or in their immediate vicinity (Han et al. 2018) into biochar and briquettes and torrefied wood based on mobile biomass conversion technologies. According to the results of the "Waste to Wisdom" project, production of torrefied pellets and briquettes allows enriching renewable fuel-based energy sources used in bio- and coal-fired power plants. An additional advantage of the solution is that it can reduce the likelihood of forest fires by disposing of large amounts of flammable material lying on cutting surfaces after harvesting. Moreover, conversion of forest residues into biochar can be an effective strategy to sequester carbon and improve the productivity of forest soils.

1.6 FOREST RESIDUES AS RAW MATERIAL IN THERMOCHEMICAL PROCESSES

Multistep thermochemical processes such as hydrothermal carbonization, torrefaction, pyrolysis of different intensities, and gasification can convert biomass in the form of forest residues into gases, chemical compounds, and energy (Braghiroli and Passarini 2020). These processes are carried out in reactors of different designs and with different physical parameters (duration, temperature and rate of its increase, reaction of the surrounding atmosphere), and produce various forms of final products such as gases (syngas), liquids (bio-oils), and solids (hydrochar, torrefied biomass, or biochar).

Syngases (formed by gasification or flash pyrolysis) are used in the production of liquid fuels. These fuels include methanol or other multiatom hydrocarbons which are produced through chemical reactions using the Fischer-Tropsch synthesis process (Speight 2019). Syngas can be used in hydrogen fuel cells to produce electricity, which will facilitate the development of some hydrogen industries based on, among others, forest biomass (Capodaglio and Bolognesi 2019).

Bio-oils obtained by liquefaction or by intermediate and fast pyrolysis differ from petroleum-based fuels in their physicochemical properties. However, due to their high water content, viscosity, ash content, and acidity, as well as low calorific values due to high oxygen content, crude bio-oils are undesirable for fuel applications. Hence, they should be upgraded by techniques such as emulsification or chemical extraction or by the addition of diluents (Xiu and Shahbazi 2012). These techniques will enable the use of bio-oils as fuels for combustion in boilers and gas turbines, and for the production of chemicals (Pattiya 2018). Pinheiro Pires et al. (2019) emphasized that it is possible to develop a market for bio-oils produced from forest biomass by applying, among others, state-of-the-art analytical techniques and instrumentation, latest realistic models and procedures for identifying their physicochemical properties, and new alternative ways of fractionation and purification. Subsequently, Braghiroli and Passarini (2020) pointed out that intensive research should be performed on the development of novel techno-economic analyses, life cycles, and environmental consequences of bio-oil production.

The origin of biochar dates back to anthropogenic dark earth of Amazonia, known as terra preta. Studies proved that these soils were highly fertile as they possessed a high level of stable soil organic matter, biochar, and nutrients derived from inorganic and organic substances applied to soils (Glaser and Birk 2012). In the 19th century, biochar was used as a soil conditioner to improve crop production in Europe and North America (Barrow 2011).

Biochar is a carbon-solid product obtained from various types of biomass by pyrolysis under restricted oxygen conditions at 300-1000°C (Verheijen et al. 2010; Bis 2012). Shackely et al. (2012) described that biochar is a porous carbonaceous solid formed through thermochemical conversion of organic materials in an oxygen-depleted atmosphere and exhibits favorable physicochemical properties for safe and long-term storage of carbon in the environment. According to the International Biochar Initiative, biochar is fine-grained charcoal which is rich in organic carbon and largely resistant to decomposition. It is produced during the pyrolysis of plant and waste feedstock. As a soil amendment, biochar creates a recalcitrant soil carbon pool that is carbon-negative, and allows net withdrawal of atmospheric carbon dioxide stored in highly recalcitrant soil carbon stocks. Biochar-amended soil displays enhanced nutrient retention capacity which not only reduces the total fertilizer requirements but also diminishes the climate and environmental impact on croplands. The European Biochar Certificate (EBC) standards define biochar as a charcoal-like substance pyrolyzed from sustainably obtained biomass under controlled conditions, which can be used for any purposes that do not involve its rapid mineralization to CO_2. Thus, biochars are specific pyrolysis chars that are environmentally sustainable, possess high quality, and can be used for various applications. However, the EBC does not recommend the use of biochar products obtained through other processes such as hydrothermal carbonization, torrefaction, and coking. Essentially, biochar and charcoal are the same material and show similar structural and physicochemical properties, but they are distinguished by their applications. The term biochar is new and was coined to indicate the fact that this substance can be applied to soils, and be produced from biological materials, whereas charcoal is a renewable energy source. Other terms used for biochar include agrichar, agricarbon, biocarbon, black carbon, or pyrogenic carbon. Biochar and charcoal are similar to hydrochar, which is produced as slurry in water by hydrothermal carbonization of biomass under pressure. However, biochar and hydrochar differ in their physical and chemical properties (Bargmann et al. 2013).

Biochar can be produced from two groups of feedstock materials: (1) biomass which was primarily produced and (2) waste biomass (Brick 2010). These groups include various materials such as energy crops (e.g., miscanthus, willow, pine), agricultural biomass (e.g., rape, sunflower, corn stalks), agroforestry residues, urban biodegradable waste (e.g., green waste, food waste, used cardboard), agri-food processing

TABLE 1.2
Pyrolysis of Biomass—Typical Yields of Products Obtained with Different Modes (Mohan et al. 2006; Brown 2009; Verhejien et al. 2010)

Mode	Temperature	Residence time	Liquid	Syngas	Biochar
Fast pyrolysis	300–1000°C	<2s	75%	13%	12%
Moderate pyrolysis	500°C	10–20s	50%	30%	20%
Slow pyrolysis	100–1000°C	5–30min	30%	35%	35%

waste (e.g., postfermentation oat, cereal husks, grapeseed, orange peel, chestnut shell, coconut, olive mill), animal by-products and waste, industrial biodegradable materials and residues (e.g., paper sludge, whiskey remains, digestates from anaerobic digestion, dense refuse-derived fuel), woody materials, municipal sewage sludge, poultry litter and manure, cattle manure, and algal biomass (Sanchez et al. 2009; Kwapiński et al. 2010; Bird et al. 2011; Ibarrola et al. 2012; Mast et al. 2013; Mohanty et al. 2013; Vassilev et al. 2013; Zhang et al. 2013). A number of factors influence the selection of feedstock materials for the production of biochar. These include the properties of feedstock (e.g., moisture content, carbon content, particle size), pyrolysis technology and process parameters applied, logistics, and targeted applications. The use of feedstock materials with high lignin content can result in the highest yields of biochar at lower temperatures (Kwapiński et al. 2010).

During the thermal conversion of selected feedstock materials, biochar is produced by the reaction of little or no oxygen with bio-oil (mixture of hydrocarbons), synthetic gas (mixture of gaseous hydrocarbons), and biochar. Temperatures applied for the production of biochar range between 100 and 1000°C (Demirbas and Arin 2002; Mohan et al. 2006; Ahmad et al. 2014). The yield of pyrolysis products is determined by the mode or process parameters applied, including temperature, heating rate, and residence time (Table 1.2). For example, conversion of biomass through slow pyrolysis can result in about 35% of biochar, 30% of bio-oil, and 35% of syngas (Lewandowski et al. 2010; Verheijen et al. 2010).

1.7 POTENTIAL APPLICATIONS OF BIOCHAR

The concept of using biochar as a tool to address environmental issues is associated with the following problems: soil degradation and food insecurity, climate change, sustainable energy production, and waste management (Lehmann and Joseph 2009). Biochar is used in a number of applications such as animal farming, soil conditioning, building, decontamination, production of compost and biogas, wastewater and drinking water treatment, textile production, and wellness (Malińska et al. 2014; Srinivasan et al. 2015). It can also be applied as a green environmental sorbent in soils for the management of water as well as in soils contaminated with organic and inorganic compounds (Beesley et al. 2010; Ahmad et al. 2014; Mohan et al. 2014). In addition, due to its large surface area and pore volume, biochar can serve as a habitat for microorganisms to stimulate soil microbial activity (Quilliam et al. 2013).

According to Xie et al. (2015) and Mohanty et al. (2018), the unique properties of biochar (e.g., high specific surface area, microporosity, and sorption capacity) can enable its application in environmental engineering and remediation, where it can be of use in the treatment of soil and groundwater and as filter media for stormwater. Due to its high specific surface area and highly carbonaceous nature, biochar is recognized as a unique adsorbent. Therefore, its application to soil, even in small amounts, can increase adsorption and consequently reduce the bioavailability of contaminants to soil-dwelling microorganisms, plants, or earthworms. In addition, biochar enhances the ability of soil to retain and slowly release plant nutrients and increases its nutrient retention capacity. According to Mohanty et al. (2018), biochar helps to maintain adequate microbiota in the soil, which can also aid in the removal of pollutants in stormwater treatment systems.

From this perspective, the aspect of biochar activation that improves its porosity appears promising. This is mainly useful in the production of energy storage devices or adsorption of organic and inorganic pollutants. Braghiroli et al. (2018, 2019) have extensively studied activated biochar resulting from the processing of sawmill residues and showed a significant reduction in pollutant concentrations in wastewater after contact with a prepared adsorbent. Some studies have also compared activated biochars obtained from forest residues and sawmill residues. Anderson et al. (2013) found that biochars from forest residues showed twofold lower porosity (total pore volume and surface area). However, compared to commercial activated carbons, forest residue-based biochars showed quite higher porosity, which indicated that they can be widely applied in the bioenergy industry. In addition, as regulations governing pollutant concentrations in discharged wastewater become more stringent, it seems that modifying the physicochemical properties of biochar will be a good option (Braghiroli and Passarini 2020).

The following properties of biochar play an important role in agricultural applications (Kwapiński et al. 2010): nutrient content, composition and availability, elemental composition, ash content and volatility, ability to absorb different organic and inorganic molecules, surface area, porosity and particle size, and bulk density.

It has been stated that plant-derived biochars can be used as soil conditioners, whereas manure-derived biochars can be used both as soil conditioners and fertilizers due to the release of nutrients (Uchimiya et al. 2010).

1.8 BIOCHAR FOR LAND RECLAMATION

Biochar introduced to soils improves its fertility, water-holding capacity, pH, and organic matter, and also prevents nutrients from leaching and binding to organic and inorganic contaminants. This suggests that biochar can be used for countering land degradation as well as for the reclamation of contaminated and degraded land. A number of studies have analyzed the effects of biochar on contaminated soils, including remediation of soils contaminated with organic and inorganic compounds (Beesley et al. 2010; Tang et al. 2013).

Beesley et al. (2010) investigated the impact of biochar amendment on mobility, bioavailability, and toxicity of inorganic and organic contaminants in a multielement-polluted soil. Their results showed that addition of biochar resulted in a 10-fold decrease in Cd in pore water and a decrease in polycyclic aromatic hydrocarbons by more than 50%, contributing to a reduction in phytotoxicity. Fellet et al. (2011) studied the effect of biochar on the amelioration of mine tailings contaminated with heavy metals for phytostabilization technology. They mixed mine tailings with 1%, 5%, and 10% of biochar obtained from prune residues from orchards and observed an increase in pH and nutrient retention in terms of cation-exchange capacity (CEC) and water-holding capacity, as well as a decrease in the bioavailability of Cd, Pb, Tl, and Zn. Based on the results, it was concluded that the addition of biochar can enhance the stability of mine tailings and allow for the formation of green cover in the phytostabilization process. Park et al. (2011) studied the potential of chicken-manure and green waste-derived biochars in the reduction of bioavailability and phytotoxicity of heavy metals and noted that these biochars were effective in the immobilization of metals in soil.

1.9 BIOCHAR FOR FOREST RECLAMATION OF DEGRADED AREAS

Biochar-based amendments are applied in areas where thousands of hectares of land are destroyed by open-cast sand mines, resulting in a significant change in landscape and degradation of the natural environment, local overexploitation of forest resources, and a decrease in the quality of water resources. Due to their adverse chemical, physical, and biological characteristics, soils in these areas pose a great challenge for restoration. We hypothesize that the application of biochar in the post-mining areas can enhance the development and quality of soil by improving its biological, physical, and chemical properties and accelerating succession and the restoration of ecosystems.

As a postmining landscape, open-cast mines are characterized by large-scale land transformation. According to Polish law, lands mined for different purposes, for example obtaining coal or minerals, must receive reclamation treatments (with lime or fertilizer, or by revegetation). Ecosystems developing in these areas represent significant carbon sinks that will increase with the development of communities and soil. Usually, organic amendment for reclamation is carried out using a mixture of local forest litter and mineral A horizons, with the addition of fertilizers and green manure.

Afforestation of the reclaimed lands is a slow process and sometimes unsuccessful. It is also associated with the limitation of water and nutrients derived from soil organic matter (post-sand-mining excavations) (Beesley et al. 2010; Pietrzykowski 2010).

So far, biochar application has not been recognized as a method of reclamation. Moreover, this concept is often disfavored by local societies including scientists, managers, and citizens. Biochar is commonly referred to as charcoal, relatively expensive, and associated with barbecue or remnants resulting from a forest fire. Studies conducted in Zambia (NGI Report) and Indonesia show that the application of biochar increases the pH and CEC of soils as well as base saturation. It was found that addition of four tons of biochar per hectare had visible effects on crop yield. Besides increasing the pH and the content of nutrients such as K, Mg, and Ca in soils, biochar improved properties including water-holding capacity, porosity, and CEC, which are important in a long-term perspective.

Biochar may also have negative effects on soil. For example, its alkaline properties may lead to unfavorable conditions for pine plantation, such as an increase in the pH of soil or contamination.

Due to the positive effects of biochar application in the afforestation of postmining areas, this method can also be recommended for improving the quality of, for example, sandy soils, which are the poorest forest sites.

One of the main factors about the potential of biochar that has to be investigated is its ability to improve the quality of postmining sites for forest reclamation and potential of carbon sequestration. According to the current knowledge and agriculture experiences, the addition of biochar alone or as a mixture with green manure to a reclaimed substrate could improve its physical, chemical, and biochemical properties, affect water quality, and also increase the content of stabilized carbon. Similarly, addition of biochar-based amendments to devastated soils will improve the tree growth response and rate of vascular plant succession.

1.10 FOREST RESIDUES AS RAW MATERIAL IN BIOCHEMICAL PROCESSES

Two methods can be used for the production of biofuel from forest residues. One is anaerobic digestion, whose end-product is methane. This biogas can be used to generate heat, electricity, and power for vehicles and machinery. The waste product resulting from anaerobic digestion is digestate, which is used as an organic fertilizer (Braghiroli and Passarini 2020). However, not all forest residues have equal efficiency to produce biogas. According to Mohsenzadeh et al. (2012), the amount of methane obtained from softwood residues is several-fold lesser compared to hardwood residues. Therefore, in regions where more wood is harvested from coniferous species and the harvested wood is soft, biogas production from forest residues may be unattractive primarily from an economic standpoint. In addition, conversion of forest residues into biogas requires more radical methods for chemical preparation (concentrated alkalis) compared to the processing of agricultural residues (Matsakas et al. 2016).

In contrast, fermentation based on enzymatic hydrolysis of wood biomass results in the production of ethanol or butanol. The formation of this type of end-product is determined by the ratio of organic matter and microorganisms (enzymes and bacteria). However, the most optimal parameters in the fermentation of forest biomass residues for producing ethanol and butanol can be achieved only after 36 to as much as 60 h (Nanda et al. 2014). Moreover, the hydrolytic enzymes used in fermentation are expensive. These negative features reduce the attractiveness of this process, despite its significant advantages, such as no requirement of biomass drying, low process temperature, and the possibility of obtaining biofuels with different properties. In addition, previous studies on the biochemical conversion of logging residues from soft poplar wood (Knoll et al. 1993) showed that the use of this type of wood is hindered by the widespread occurrence of stemwood rot, which has a higher calorific value compared to healthy wood, and is hence considered a more attractive raw material for the production of woodchip energy. At the same time, the economic attractiveness of the entire process may be reduced as rot limits the availability of healthy wood as a more desirable raw material for enzymatic digestion. Furthermore, programs aimed at increasing the proportion of deadwood in forests to enrich their biodiversity may prevent the development of biochemical processing technologies for forest residues due to the insufficient availability of raw materials in some forest regions.

1.11 FOREST RESIDUES AS A USEFUL RAW MATERIAL IN FOOD AND FEED PRODUCTION

Forest residues can also be managed by using them for breeding edible insects for food and feed. Varelas and Langton (2017) indicate that insects could be of use not only in pest control and biocontrol but also for food or feed purposes (appropriate life stages of beetles, termites, ants, and butterflies) in artificial cultures based on forest residues. They indicated that forest residues can be tested on common edible insects that are already mass-reared, such as yellowfly, house cricket, and house fly. Forest residues can provide potential substrates for the growth of various insects. Hence, the breeding of this material on appropriately prepared substrates may promote sustainable forest management, environmental protection, and global food security in the future. However, extensive research is required to analyze the future use of some insect species for food and feed purposes based on forest residues. Furthermore, studies must explore the potential risks of hazards in the insect-based food chain, which may be related to the characteristics of the insect species, breeding or processing, and the conditions of preservation and transport.

Silva et al. (2005) pointed out that forest residues can be possibly used for preparing substrates used for culturing the fruiting bodies of *Lentinula edodes* (commonly known as Shiitake mushrooms). Similarly, Philippousis (2009) reported that agroforestry residues can be potentially used for producing the fruiting bodies of *Agaricus*. Based on their results, they concluded that the cultivation of edible mushrooms on substrates derived from agroforestry residues represents an integrated system with waste disposal that contributes to achieving sustainable production, improving human health, and benefitting the environment.

1.12 SUMMARY

Forest residues (logging and wood residues) are undoubtedly a natural and renewable raw material, which can be used worldwide for various purposes. Regarding climate and environmental protection, some portion of the available wood waste can successfully replace fossil raw materials in the production of energy and fuels. In addition, utilization of the most readily available resources under controlled conditions can allow reducing air pollution caused by uncontrolled burning of logging residues in areas where the largest volumes of solid timber are harvested. On the other hand, in regions that are already degraded by mining and industrial activities, the use of forest residues as raw material can contribute to partial restoration or restoration of the lost properties of the natural environment (water, soil, air).

From a demographic point of view, forest residues can be mainly used for the cultivation of plants and fungi or for breeding of animals used to produce certain forms of food and components of certain medicines.

Finally, with regard to sustainable forest management, forest residues should be utilized without any devastating effect on the natural environment, especially in the areas where intensive penetration of woodlands is carried out for other nonproductive purposes.

1.13 LITERATURE

Ahmad, M., Rajapaksha, A.U., Lim, J.E., Zhang, M., Bolan, N., Mohan, D., Vithanage, M., Lee, S.S. and Ok., Y.S., (2014). Biochar as a sorbent for contaminant management in soil and water: A review. *Chemosphere*, 99, pp. 19–33.

Anderson, N., Jones, J., Page-Dumroese, D., McCollum, D., Baker, S., Loeffler, D. and Chung, W., (2013). A comparison of producer gas, biochar, and activated carbon from two distributed scale thermochemical conversion systems used to process forest biomass. *Energies*, 6, pp. 164–183.

Ashori, A., (2008). Wood—plastic composites as promising greencomposites for automotive industries. *Bioresource Technology*, 99, pp. 4661–4667.

Bargmann, I., Rilling, M.C., Buss, W., Kruse, A. and Kuecke, M., (2013). Hydrochar and biochar effects on germination of spring barley. *Journal of Agronomy and Crop Science*, 199, pp. 360–373.

Barrow, C.J., (2011). Biochar: Potential for countering land degradation for improving agriculture. *Applied Geography*, 34, pp. 21–28.

Beesley, L., Moreno-Jimenez, E. and Gomez-Eyles, J.L., (2010). Effects of biochar and greenwaste compost amendments on mobility, bioavailability and toxicity of inorganic and organic contaminants in a multi-element polluted soil. *Environmental Pollution*, 158(6), pp. 2282–2287.

Bird, M.I., Wurster, C.M., de Paula Silva, P.H., Bass, A.M. and de Nys, R., (2011). Algal biochar—production and properties. *Bioresource Technology*, 102(2), pp. 1886–1891.

Bis, Z., (2012). Biowęgiel—powrót do przeszłości, szansadlaprzyszłości (Biochar—a return to the past, an opportunity for the future). *CzystaEnergia*, 6 [In Polish].

Björheden, R., (2017). Development of bioenergy from forest biomass—a case study of Sweden and Finland. *Croatian Journal of Forest Engineering*, 38(2), pp. 259–268.

Braghiroli, F.L., Bouafif, H., Hamza, N., Neculita, C.M. and Koubaa, A., (2018). Production, characterization, and potential of activated biochar as adsorbent for phenolic compounds from leachates in a lumber industry site. *Environmental Science and Pollution Research*, 25, pp. 26562–26575.

Braghiroli, F.L., Bouafif, H., Neculita, C.M. and Koubaa, A., (2019). Performance of physically and chemically activated biochars in copper removal from contaminated mine effluents. *Water Air Soil Pollution*, 230, p. 178.

Braghiroli, F.L. and Passarini, L., (2020). Valorization of biomass residues from forest operations and wood manufacturing presents a wide range of sustainable and innovative possibilities. *Current Forestry Reports*, doi: 10.1007/s40725-020-00112-9.

Brick, S., (2010). *Biochar: Assessing the promise and risks to guide US policy*. New York: Natural Resources Defense Council.

Brown, R., (2009). Biochar production technology. In: J. Lehmann and S. Joseph (eds.), *Biochar for environmental management science and technology*. London: Earthscans.

Capodaglio, A.G. and Bolognesi, S., (2019). Ecofuel feedstocks and their prospects. In: *Advances in eco-fuels for a sustainable environment*. Duxford: Elsevier, pp. 15–51.

Demirbas, A. and Arin, G., (2002). An overview of biomass pyrolysis. *Energy Source*, 24, pp. 471–482.

FAO, (2020). FAOSTAT—Forestry production and trade. www.fao.org/faostat/en/#data/FO/visualize. (Accessed 2 April 2021).

Fellet, G., Marchiol, L., DelleVedove, G. and Peressott, A., (2011). Application of biochar on mine tailings: Effects and perspectives for land reclamation. *Chemosphere*, 83, pp. 1262–1267.

Glaser, B. and Birk, J.J., (2012). State of the scientific knowledge on properties and genesis of Anthropogenic Dark Earths in Central Amazonia (terra preta de Indio). *Geochimica et Cosmochimica Acta*, 82, pp. 39–51.

Han, H.-S., Jacobson, A., Bilek, E.M. and Sessions, J., (2018). Waste to wisdom: Utilizing forest residues for the production of bioenergy and biobased products. *Applied Engineering and Agriculture*, 34(1), pp. 5–10, doi: 10.13031/aea.12774.

Ibarrola, R., Shackely, S. and Hammond, J., (2012). Pyrolysis biochar systems for recovering biodegradable materials: A life cycle carbon assessment. *Waste Management*, 32, pp. 859–868.

International Energy Agency (IEA), (2017). *Key world energy statistics*. Paris, France. www.iea.org/reports. (Accessed 18 June 2021).

IRENA (International Renewable Energy Agency), (2014). Global bioenergy—supply and demand projections—a working paper for REmap 2030.88 p.

Iwakiri, S., Trianoski, R., Chies, D., Tavares, E.L., França, M.C., Lau, P.C. and Teixeira Iwakiri, V., (2017). Use of residues of forestry exploration of *Pinus taeda* for particleboard manufacture. *RevistaÁrvore*, 41(3), pp. 1–8.

Kallio, M. and Leinonen, A., (2005). Production technology of forest chips in Finland. Project Report PRO2/P2032/05. Jyväskylä.

Kim, J.K. and Pal, K., (2011). *Recent advances in the processing of wood-plastic composites*. Berlin: Springer.

Kizha, A. R. and Han, H.-S., (2016). Processing and sorting forest residues: Cost, productivity and managerial impacts. *Biomass Bioenergy*, 93, pp. 97–106.

Knoll, C.S., Wong, B.M. and Roy, D.N., (1993). The chemistry of decayed aspen wood and perspectives on its utilization. *Wood Science Technology*, 27, pp. 439–448.

Kwapiński, W., Byrne, C.M.P., Kryachko, E., Wolfram, P., Adley, C., Leahy, J.J., Novotny, E.H. and Hayes, M.H.B., (2010). Biochar from biomass and waste. *Waste Biomass Valorization*, 1, pp. 177–189.

Lehmann, J. and Joseph, S. (eds.), (2009). *Biochar for environmental management: Science and technology*. London: Earthscan.

Leinonen, A., (2004). Harvesting technology of forest residues for fuel in the USA and Finland. VTT Tiedotteita—Research Notes 2229.132 p. + app. 10 p., Espoo.

Lewandowski, W.M., Radziemska, E., Ryms, M. and Ostrowski, P., (2010). Nowoczesne metodytermochemicznekonwersjibiomasy w paliwagazowe, ciekłeistałe (Modern thermochemical methods of biomass conversion to gaseous, liquid and solid fuels). *Proceedings of ECOpole*, 4(2), pp. 453–457.

Malińska, K. and Dach, J., (2014). Możliwości wykorzystania biowęgla w procesie kompostowania (Possibilities of using biochar in the composting process). *Inżynieria Ekologiczna*, 36, pp. 28–39.

Mast, R.E., Kumar, S., Rout, T.K., Sarkar, P., George, J. and Ram, L.C., (2013). Biochar from water hyacinth (*Eichornia crassipes*) and its impact on soil biological activity. *Catena*, 111, pp. 64–71.

Matsakas, L., Rova, U. and Christakopoulos, P., (2016). Strategies for enhanced biogas generation through anaerobic digestion of forest material—an overview. *BioResources*, 11(2), pp. 5482–5499.

Migneault, S., Koubaa, A. and Perré, P., (2014). Effect of fiber origin, proportion, and chemical composition on the mechanical and physical properties of wood-plastic composites. *Journal of Wood Chemistry and Technology*, 34, pp. 241–261.

Mohan, D., Pittman, C.U. and Steele, P.H., (2006). Pyrolysis of wood/biomass for bio-oil: A critical review. *Energy Fuels*, 20, pp. 848–889.

Mohan, D., Sarswat, A., Ok, Y.S. and Pittman, Jr. C.U., (2014). Organic and inorganic contaminants removal from water with biochar, a renewable, low cost and sustainable adsorbent—A critical review. *Bioresource Technology*, 160, pp. 191–202.

Mohanty, P., Nanda, S., Pant, K.K., Naik, S., Kozinski, J.A. and Dalai, A.K., (2013). Evaluation of physiochemical development of biochars obtained from pyrolysis of wheat straw, timothy grass and pinewood: Effects of heating rate. *Journal of Analytical and Applied Pyrolysis*, 104, pp. 485–493.

Mohanty, S.K., Valenca, R., Berger, A.W., Yu, I.K.M., Xiong, X., Saunders, T.M. and Tsang, D.C.W., (2018). Plenty of room for carbon on the ground: Potential applications of biochar for stormwater treatment. *Science of The Total Environment*, 625, pp. 1644–1658, doi: 10.1016/j.scitotenv.2018.01.037.

Mohsenzadeh, A., Jeihanipour, A., Karimi, K. and Taherzadeh, M.J., (2012). Alkali pretreatment of softwood spruce and hardwood birch by NaOH/thiourea, NaOH/urea, NaOH/urea/thiourea, and NaOH/PEG to improve ethanol and biogas production. *Journal of Chemical Technology & Biotechnology*, 87, pp. 1209–1214.

Moriana, R., Vilaplana, F. and Ek, M., (2015). Forest residues as renewable resources for bio-based polymeric materials and bioenergy: Chemical composition, structure and thermal properties. *Cellulose*, 22, pp. 3409–3423, doi: 10.1007/s10570-015-0738-4.

Nanda, S., Dalai, A.K. and Kozinski, J.A., (2014). Butanol and ethanol production from lignocellulosic feedstock: Biomass pretreatment and bioconversion. *Energy Science & Engineering*, 2, pp. 138–148.

Nguyen, Q.N., Cloutier, A., Achim, A. and Stevanovic, T., (2015). Effect of process parameters and raw material characteristics on physical and mechanical properties of wood pellets made from sugar maple particles. *Biomass Bioenergy*, 80, pp. 338–349.

Nguyen, Q.N., Cloutier, A., Achim, A. and Stevanovic, T., (2016). Fuel properties of sugar maple and yellow birch wood in relation with tree vigor. *BioResources*, 11, pp. 3275–3288.

Park, J.H., Choppala, G.K., Bolan, N.S., Chung, J.W. and Chuasavathi, T., (2011). Biochar reduces the bioavailability and phytotoxicity of heavy metals. *Plant Soil*, 348, pp. 439–451.

Pattiya, A., (2018). Fast pyrolysis. In: *Direct thermochemical liquefaction for energy applications*. Duxford: Elsevier, pp. 3–28.

Pellet Fuels Institute, (2020). www.pelletheat.org/legislationregulations. (Accessed 18 June 2021).

Philippousis, A.N., (2009). Production of Mushrooms Using Agro-Industrial Residues as Substrates. In: P. Singh nee' Nigam and A. Pandey (eds.), *Biotechnology for agro-industrial residues utilisation*. Springer Science+Business Media B.V., doi.org/10.1007/978-1-4020-9942-79.

Pietrzykowski, M., (2010). Scots pine (*Pinus sylvestris* L.) ecosystem macronutrients budget on reclaimed mine sites—stand trees supply and stability. *Natural Science*, 2(6), pp. 590–599, doi.org/10.4236/ns.2010.26074.

Pinheiro Pires, A.P., Arauzo, J., Fonts, I., Domine, M.E., Fernández Arroyo, A., Garcia-Perez, M.E., et al. (2019). Challenges and opportunities for bio-oil refining: A review. *Energy Fuels*, 33, pp. 4683–4720.

Quilliam, R.S., Helen, C., Glanville, H.C., Wade, S.C. and Jones, D.L., (2013). Life in the 'charosphere'—Does biochar in agricultural soil provide a significant habitat for microorganisms? *Soil Biology and Biochemistry*, 65, pp. 287–293, doi.org/10.1016/j.soilbio.2013.06.004.

Rhen, C., Ohman, M., Gref, R. and Wästerlund, I., (2007). Effect of raw material composition in woody biomass pellets on combustion characteristics. *Biomass Bioenergy*, 31, pp. 66–72.

Richardson, J., Björheden, R., Hakkila, P., Lowe, A.T. and Smith, C.T. (eds.), (2002). *Bioenergy from sustainable forestry. Guiding principles and practice*. Dordrecht: Kluwer Academic Publishers.

Ross, R.J., (2010). *Wood handbook: Wood as an engineering material*. Madison: FPL-GTR-190.

Routa, J., Asikainen, A., Björheden, R., Laitila, J. and Röser, D., (2013). Forest energy procurement: State of the art in Finland and Sweden. *WIREs Energy Environ*, 2, pp. 602–613, doi: 10.1002/wene.24.

Saddem, M., Koubaa, A., Bouafif, H., Migneault, S. and Riedl, B., (2019). Effect of fiber and polymer variability on the rheological properties of wood polymer composites during processing. *Polymer Composites*, 40, pp. 609–616.

Sanchez, M.E., Lindao, E., Margaleff, D., Martinez, O. and Moran, A., (2009). Pyrolysis of agricultural residues from rape and sunflower: Production and characterization of bio-fuels and biochar soil management. *Journal of Analytical and Applied Pyrolysis*, 85, pp. 142–144.

Shackley, S., Carter, S., Knowles, T., Middelink, E., Haefele, S., Sohi, S., Cross, A. and Haszeldine, S., (2012). Sustainable gasification-biochar systems? A case-study of environmental and health and safety issues. *Energy Policy*, 42, pp. 49–58.

Shmulsky, R. and Jones, P.D., (2011). *Forest products and wood science: An introduction*. Oxford: Wiley-Blackwell.

Silva, E.M., Machuca, A. and Milagres, A.M.F., (2005). Effect of cereal brans on *Lentinula edodes* growth and enzyme activities during cultivation on forestry waste. *Letters in Applied Microbiology*, 40, pp. 283–288.

Singh, R., Singh, B., Gupta, M. and Tarannum, H., (2020). Composite building materials from natural fibers/agro-forest residues. *Indian Journal of Engineering & Materials Sciences*, 27, pp. 137–149.

Speight, J.G., (2019). Unconventional gas. In: K. Hammon (ed.), *Natural Gas*. Cambridge: Elsevier, pp. 59–98.

Srinivasan, P., Sarmah, A.K., Smernik, R., Das, O., Farid, M. and Gao, W., (2015). A feasibility study of agricultural and sewage biomass as biochar, bioenergy and biocomposite feedstock: Production, characterization and potential applications. *Science of The Total Environment*, 512–513, pp. 495–505, doi.org/10.1016/j.scitotenv.2015.01.068.

Tang, J., Zhy, W., Kookana, R. and Katayama, A., (2013). Characteristics of biochar and its application in remediation of contaminated soil. *Journal of Bioscience and Bioengineering*, 116(6), pp. 653–659.

Uchimiya, M., Lima, I.M., Klasson, K.T. and Wartelle, L.H., (2010). Contaminant immobilization and nutrient release by biochar soil amendment: Roles of natural organic matter. *Chemosphere*, 80, pp. 935–940.

Varelas, V. and Langton, M., (2017). Forest biomass waste as a potential innovative source for rearing edible insects for food and feed—A review. *Innovative Food Science and Emerging Technologies*, 41, pp. 193–205, doi: 10.1016/j.ifset.2017.03.007.

Vassilev, V., Martos, E., Mendes, G., Martos, V. and Vassileva, M., (2013). Biochar of animal origin: A sustainable solution to the global problem of high-grade rock phosphate scarcity? *Journal of the Science of Food and Agriculture*, 93(8), pp. 1799–1804.

Verheijen, F.G.A., Jeffery, S., Bastos, A.C., van der Velde, M. and Diafas, I., (2010). *Biochar application to soils—A critical scientific review of effects on soil properties.* Process and Functions. EUR 24099 EN. Luxembourg: European Commission.

Verkerk, P.J., Fitzgerald, J.B., Datta, P., Dees, M., Hengeveld, G.M., Lindner, M. and Zudin, S., (2019). Spatial distribution of the potential forest biomass availability in Europe. *Forest Ecosystems*, 6(5), doi: 10.1186/s40663-019-0163-5.

Xie, T., Reddy, K.R., Wang, C., Yargicoglu, E. and Spokas, K., (2015). Characteristics and applications of biochar for environmental remediation: A review. *Critical Reviews in Environmental Science and Technology*, 45(9), pp. 939–969, doi: 10.1080/10643389.2014.924180.

Xiu, S. and Shahbazi, A., (2012). Bio-oil production and upgrading research: A review. *Renewable & Sustainable Energy Reviews*, 16, pp. 4406–4414.

Zhang, W., Mao, S., Chen, H., Haung, L. and Qiu, R., (2013). Pb(II) and Cr(II) sorption by biochars pyrolyzed from the municipal wastewater sludge under different heating conditions. *Bioresource Technology*, 147, pp. 545–552.

Zion Market Research, (2017). Wood plastic composites market (polyethylene, polypropylene, polyvinyl chloride and others) for building & construction, automotive, electrical and other applications: Global market perspective, comprehensive analysis and forecast, 2016–2022. www. zionmarketresearch.com/report/wood-plastic-composites-market. (Accessed 16 June 2021).

2 Commercial Production of Antioxidants from Fresh Agricultural Waste

Bornita Bose and Harshata Pal

CONTENTS

2.1 INTRODUCTION

Statistics predicted a total world population of 7.9 billion as of April 2021, out of which more than 800 million people do not get adequate food, and an average of

DOI: 10.1201/9781003245773-2

1.3 million people survive on less than $1 per day. This extensively increased population has made it significantly important to expand the food production on a large scale. Over years of research, it has been found that there are over 350,000 plant species existing in the world; however, only about 80,000 of them are suitable for human consumption (Füleky, no date). This edible plant population, often produced naturally through agriculture, is processed industrially to transform it into consumable products. This agro-industrial processing produces around 25–30 percent of non-edible portions from the plant by-products. The disposal of these non-edible portions has led to a huge economic loss, because of which the scientists formulated a potential alternative to produce effective bioactive compounds, namely, antioxidants from the processed plant by-products (Ajila et al., 2010).

The fresh agricultural wastes are often derived from various streams of biowastes that includes agro-industrial products—these essentially cover a diverse range of raw materials that have the potential to serve as value added products; food wastes—a wide variety of products, such as fruits, vegetables, fish, milk, meat and wine, are categorized under food wastes (Baiano, 2014); the fruit and vegetable wastes account for approximately 40–50 percent of the total agricultural wastes per year; agroforestry—the wastes from agroforestry include the crop plant stems such as rice, wheat, sugarcane, etc. (Sládková et al., 2016). Hence, in an aim to reduce the environmental impacts, these wastes are recycled to produce antioxidants.

Bioactive compounds can be defined as essential constituents that are naturally present in food and agricultural products in small quantities. These compounds most commonly include secondary metabolites like alkaloids, antibiotics, plant growth factors, mycotoxins, phenolic compounds, food pigments, etc. Among these, phenolic compounds and flavonoids are considered to be antioxidants that are naturally present in higher orders of plants. The unutilized parts of plant tissue products are high in antioxidants, including flavonoids, carotenoids, vitamins, polyphenols and anthocyanins (Kris-Etherton et al., 2002). Along with this, the crude extracts obtained from processing of agricultural plant products contain high concentrations of phenolic compounds with the high potential of functioning as preservatives and in the industries of functional foods and nutraceuticals. Other than this, processing of fruits and vegetables produces few by-products like peels, seeds, stones and kernels are rich sources of antioxidants. These have been obtained from certain economically valuable fruits like grapes (Lafka, Sinanoglou and Lazos, 2007), pomegranate (Ingh and Urthy, 2002), citrus fruits (Chemistry, 2008), banana, mangosteen, rambutan and star fruit (Shui and Leong, 2006). Among all the fruits, the most abundant fruit crop is the citrus family, which is processed not only to produce juice but also mandarin segments and jam in the canning industry (Kumar et al., 2017). A list of the most common crops that generate agricultural biowastes are mentioned in Table 2.1.

The antioxidants obtained from various agricultural by-products or biowastes help in enriching foods with nutrition, and prevent occurrence of any disease caused from oxidative stress (Joshi, Kumar and Kumar, 2012). Free radicals such as hydroxyl radical, superoxide anion and peroxide anion causes denaturation of proteins, DNA oxidation, fluidity of membranes, lipid peroxidation and alteration in functioning of platelets, and hence increases the probability of several chronic health issues, such as inflammation, cancer, atherosclerosis and other problems. However, consumption

TABLE 2.1
Agricultural Biowastes Produced from Various Crop Species

Crop species	Biowastes produced
Cotton	Stalks
Tomato	Peels and skin
Apple	Pomace, fruit dust
Olive	Leaves and pomace
Coconut	Shell, husk, fronds
Mango	Seeds
Corn	Stover, cob, leaves, stalks
Coffee	Husk, hull, ground
Spinach	Leftover wastes
Rice	Stalks, hull/husk, bran, straw
Broccoli	Leaves
Pomegranate	Peels
Potato	Peels
Passion fruit	Rinds
Nuts	Hulls
Grape	Pomace
Macadamia	Skin
Sugarcane	Bagasse
Artichoke	External and internal bracts
Agricultural crops	Mixed waste including crop waste
Peanuts	Shells
Hazelnut and chestnut	Shells
Fruits and vegetables	Peel, seeds, stones, rind, kernel, stalk
Mixed type	Wastes derived from agricultural crops and others that include non-organic wastes

Adapted from source: (Azeez, Narayana and Oberoi, 2017)

of foods containing antioxidants recycled from agricultural biowastes has shown to have restricted the formation of these harmful free radicals inside the human body (Fridovich, 1978). They have proved to have potent functionality in the pharmaceuticals and functional food production industries as well.

Instead of disposing these agricultural wastes, they are exposed to pre-treatment methods, various physical and chemical extraction methods and thereafter, they are exclusively utilized in different commercially valuable industries. However, a few important points that are to be kept in mind while commercially producing antioxidants from agricultural biowastes via pre-treatment and extraction techniques are as follows:

- The methods should be cost-effective,
- They should be economically inexpensive,

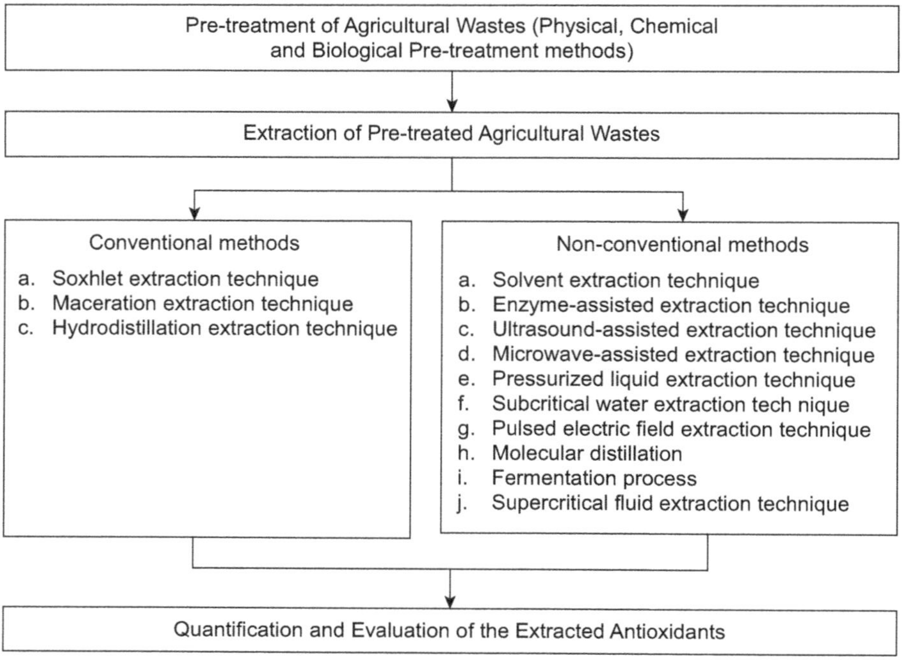

FIGURE 2.1 Graphical representation of the pre-treatment and extraction of fresh agricultural wastes, quantification and evaluation of the extracted antioxidants (self-made).

- The materials required should be easily available and less expensive,
- Ease of transport,
- Minimization of energy input,
- Highly efficient

The complete process of commercial production of antioxidants is mainly based on three general stages: pre-treatment, extraction, purification and quantification, which has been further illustrated in Figure 2.1.

2.2 PRE-TREATMENT METHODS

As a first step, the agricultural biowastes are exposed to different methods of pre-treatment. The pre-treatment methods are absolutely necessary because of the fact that the cellular structure of agricultural biowastes consist of hemicellulose, cellulose and lignin. Generally, due to the presence of hemicellulose and cellulose in the cell wall, they can easily undergo lignification. Therefore, as an initiative to separate hemicellulose and cellulose from each other, various physical, chemical and biological or enzymatic pre-treatment methods are employed. The physical and chemical methods of treatment are usually used to break down the cellular materials present in the agro-waste residues. On the other hand, the biological method involves several enzymes, such as lignin peroxidase, phytase, manganese peroxidase and phytase that are produced by micro-organisms. These enzymes perform bleaching, delignification, etc. It is very important for agro-wastes to undergo effective pre-treatment and so to ensure these, a few key parameters are to be considered:

- Even though we are trying to separate hemicellulose and cellulose to prevent lignification, it should be made sure that the fraction of hemicellulose is preserved to produce maximum yield of fermentable sugars,
- It should also be made sure that the loss of carbohydrate is limited to ensure the minimization of inhibitor formation because of degradation products,
- Decreasing the input of energy,
- Cost-effective and economically efficient process (Azeez, Narayana and Oberoi, 2017).

Following are the physical, chemical and biological methods of pre-treatment of the agricultural biowastes.

2.2.1 Physical Methods of Pre-Treatment

The physical method of pre-treatment involves the methods that do not include the application of any chemicals. Hydro-thermolysis (hydrothermal treatments), steam explosion, comminution (mechanical size reduction) and microwave are some of the most regularly used pre-treatment strategies. These methods vary depending on the type of cells that are to be treated for the production of antioxidants.

Generally, the cellulose fibers are excellent absorbers of water, which leads to swelling around the amorphous parts of the cellulose fibers along with the formation of crystalline parts that counterbalances the amorphous regions. Hence, on the basis of this property, the pre-treatment of cellulose fibers is best to be carried out by means of a method that would result in swelling. This desirable swelling, ideal for effective hydrolysis, can be obtained with the help of physical pre-treatment techniques such as milling, ultrasonic and stream treatments. This exposes the number of glycosidic bonds needed for substantial biological or chemical methods of pre-treatment (Azeez, Narayana and Oberoi, 2017). On the other hand, for lignocellulosic biomass, the most suitable pre-treatment methods include extrusion processing (Zheng and Rehmann, 2014), (Yoo et al., 2011) and microwaves treatment. Microwaves can be defined as radio waves with frequency ranging from 0.3GHz to 300GHz and size of range 1 m to 1 mm. These have the potential to interact with the organic substances by getting absorbed by fats, sugars and water. Upon getting absorbed, microwaves generate high levels of energy and cause localized heating that results in destruction of lignocellulose structure, thus making it easier to access the hemicellulose and cellulose walls for further biological or enzymatic hydrolysis (Sarkar et al., 2012). For low lignin-containing cells, particularly banana peels, orange peels, etc., it is majorly dependent on hydrothermal treatments like hot water compression or liquid hot water treatment (Singh et al., 2011; Palacios et al., 2021).

2.2.2 Chemical Methods of Pre-Treatment

Chemical pre-treatments are characterized by the presence of chemicals such as alkalis, acids, ammonia and lime or calcium hydroxide. The most frequently applied chemical pre-treatments include alkali and acid hydrolysis. These chemical-based pre-treatments have been particularly employed to retrieve the sugar molecules from hemicellulose and cellulose polymers present in lignocellulosic biomass in addition to cellulosic delignification. The following are some of the chemical pre-treatment methods that are often used:

- Alkali hydrolysis: These methods engage alkaline compounds such as calcium, aluminium and sodium hydroxide. Application of these has specifically helped in bringing about alterations in the internal structure of lignocellulosic biomass that, most importantly, include swelling of the cellulosic wall accompanied by partial decrystallization, solvation of the hemicellulose and cellulose and degradation of lignin barrier (Sills and Gossett, 2011). Often, sodium hydroxide treatment has been beneficial for crops bearing a maximum of 26 percent of lignin, such as softwood, hardwood, switch-glass, wheat straw, etc. (Zhao et al., 2008). Contrarily, the use of other alkali pre-treatments have been successful on crop species like wheat, rice straw, corn stover (Zhu, Wan and Li, 2010), bagasse (Asgher et al., 2013), cotton stalk (Kaur et al., 2012) and also switch-grass (Hu, Wang and Wen, 2008).
- Acid hydrolysis: Generally, the cellulosic biomass is treated by means of acids, particularly sulphuric acid and hydrochloric acid. This method is performed at very low temperatures with high concentrations of acid, ranging from 30–70 percent. However, high acid concentrations increase the probability of extremely corrosive reactions. In an attempt to reduce the chances of corrosive reactions, expensive non-corrosive (carbon-brick or ceramic materials) or non-metallic equipments are specially crafted, which make it economically risky. Moreover, this also demands highly-skilled labour, advanced operating systems along with environmental costs. All of these factors have provided points of disadvantage for acid hydrolysis pre-treatment when compared to dilute acid hydrolysis pre-treatment (Wyman, 1999). Not only this, but incubation period and temperature are also two important deciding factors in alterations occurring in the internal structures of cellulose fibers. Acid pre-treatment has been tested on a diverse range of agricultural wastes including hardwoods and grass (herbaceous plants) and has, in fact, resulted in effective solubilization of the hemicellulose wall (Azeez, Narayana and Oberoi, 2017).

- Ammonia pre-treatment: This pre-treatment method involves treatment of aqueous ammonia at high temperatures. This results in adequate reduction of lignin content, followed by some loss of hemicellulose content, while the cellulose material gets decrystrallized. The ammonia pre-treatment is broadly categorized into three groups of treatment which are as follows:
    a. Soaking in Aqueous Ammonia (SAA) treatment: In SAA treatment, the cellular structure is exposed to ammonia at a low temperature. This enables minimum interaction between lignin and hemicellulose/cellulose fibers, which results in increase in pore size and surface area, thereby causing efficient removal of lignin from the cell wall. The remining hemicellulose and cellulose fibers are then treated with commercial enzymes that hydrolyzes them to form economically valuable fermented sugars (Kim, Taylor and Hicks, 2008).
    b. Ammonia Recycle Percolation (ARP) treatment: While performing ARP treatment, a flow-through column reactor is used wherein the cellular biomass is exposed to aqueous ammonia (Hyun et al., 2003).
    c. Ammonia Fiber Explosion (AFE) treatment: AFE treatment is, so far, considered as the most potential ammonia treatment method where liquid anhydrous ammonia is administered on the cellular biomass at a high pressure, ranging from 250–300 psi and a high temperature of 60–100 °C. Even though the temperature is held constant, the pressure is rapidly discharged. The synergistic effect produced by high temperature and ammonia has been a very effective technique in causing lignocellulose swelling, further disrupting the cellular architecture of lignocellulose or lignin that finally leads to hydrolysis of hemicellulose and cellulose decrystallization (Teymouri et al., 2018).
- Pre-treatment utilizing oxidation reagents: Sometimes, delignification or swelling and disruption of lignocellulose fibers or lignin is carried out by treating them with oxidizing agents such as ozone, hydrogen peroxide, oxygen which results in conversion of lignin to acids (Nakamura, Daidai and Kobayashi, 2004). However, the main problem with this technique is that the acids produced from here is capable of causing damage and degradation of a major fraction of the hemicellulose content of the cell wall and thus, making it unavailable during the fermentation process. Hence, it is very important to remove these acids, or else they function as inhibitors in the fermentation process, thereby hampering the entire process.
- Pre-treatment using organic solvents: This pre-treatment method, also known as organosolv pre-treatment, includes the application of a mixture of organic solvents such as methanol, ethylene glycol, ethanol and acetone that exists in combination with water (Ichwan and Son, 2011). This process is conducted in the presence of catalysts such as organic or inorganic acids at a temperature range between 100-250 °C. This leads to breaking down of the internal bonds present in lignin along with the bonds present between hemicellulose and lignin, followed by glycosidic bond hydrolysis existing

in hemicellulose and also in cellulose but to a minimal extent. As a result of all these steps, the cellular biomass rich is cellulose fibers is prepared for further treatment by enzymatic or biological hydrolysis (Azeez, Narayana and Oberoi, 2017).

2.2.3 Biological Methods of Pre-Treatment

The biological methods of pre-treatment outweigh the aforementioned physical and chemical pre-treatment strategies on the basis of the parameters that biological pre-treatment methods do not require exclusive, highly qualified infrastructure and are eco-friendly, less complicated, economically viable and reduces the odds of health hazards. All of these parameters are increasing the scopes of research on biological pre-treatment methods. The most commonly employed methods involve ligninolytic microorganisms that degrade woods like soft-rotor brown fungi and white fungi along with bacteria. Fungi are often the main sources of cellulase production, along with some other bacteria that have the potential to produce commercial enzymes. These commercial enzymes are capable of performing complete hydrolysis of crystallized cellulose under in vitro conditions. These methods basically work by modifying the cellular structure or chemical composition of the lignocellulosic material. The cellulose fibers extracted from the physical and chemical methods of pre-treatment are then treated by cellulases, produced by various microorganisms, that mainly degrade the highly structured polymers of cellulose into smaller sugar molecules. These smaller sugar molecules then become capable of passing through the cell wall of microbes, which in turn gets disposed out in the medium. The successful degradation of cellulose using enzymes is only possible with the contribution of a minimum of three kinds of enzymes, namely, β-glucosidase, endoglucanases (Cx) and exoglucanase (C1) (Bhavsar et al., 2015). The biological conversion of cellulose fibers into economically and commercially useful products usually involves multiple steps, including pre-treatment (physical, chemical and biological), enzymatic hydrolysis followed by fermentation processes. This series of steps help in making the cellulose material more prone to digestion by enzymes, which provides better results.

A few disadvantages of using biological methods of pre-treatment include the rate of process, which is comparatively slower; the growth conditions should be specially controlled and the area required for treatment should be large enough. Moreover, sometimes the application of ligninolytic microorganisms not only degrade lignin but also effects cellulosic and hemicellulosic fractions of the biomass, which is not desirable. Due to some of these reasons, it is not an industrially favourable technique yet; however, working on these shortcomings can turn biological pre-treatment into a very potential method (Eggeman and Elander, 2019; Ichwan and Son, 2011).

Besides all these methods, some other pre-treatment techniques that are often employed include alcohol precipitation, which increases the ease of separating smaller molecules such as phenolic compounds rich in antioxidants from macromolecules, and others such as micro-filtration, foam mat, electro-osmotic dewatering, etc. Mostly, pre-treatments involving heat are not favoured as they, instead of speeding up the process, render it ineffective. Since the pre-treatment strategies act as a deciding parameter in producing effective results for antioxidant production, the methods engaged in performing pre-treatments should be adopted very carefully.

2.3 TECHNIQUES OF EXTRACTION

Extraction techniques are primarily employed to extract bioactive materials, particularly antioxidants from a diverse range of plant species. By definition, extraction is a technique for separation of the soluble solutes, i.e., the bioactive substances from the by-products of plants by using certain specific solvents via standard procedures. The rich sources of oxidants are phenolic compounds that include several compounds, out of which the essentials are phenolic acids, flavonoids and tannins. Phenolic acids can be classified as those bioactive phenolic compounds that are found in plants in addition to food products, these can be subdivided into two groups: hydroxycinnamic acid and hydroxybenzoic acid. Simultaneously, flavonoids can be described as the largest class of plant phenolic compounds which consist of an approximate quantity of eight thousand compounds, including flavones, flavanols, isoflavones, flavones and anthocyanins. Additionally, the extent of efficacy depends on certain important parameters: pre-treatment, temperature, agitation rate, solvent, etc. For that matter, it has been evident over years that the recovery of hemicellulose and pectin is most likely when it is treated with complex solvent in combination with ethanol followed by treatment with alkali (Koubala et al., 2008). In addition to this, carotenoids are seen to be liposoluble much more in non-polar or polar aprotic solvents (Strati and Oreopoulou, 2011), whereas solubilizing phenols in polar protic solvents has been easier than any other solvents (Tsakona, Galanakis and Gekas, 2012). There are also several other environment-friendly solvents that are emerging as an alternative to the conventional solvents, like surfactants, terpenes or alcohols, deep and naturally occurring eutectic solvent, ionic liquids, etc., that have been used for decades now.

The extraction techniques relevant in antioxidant recovery from agricultural by-products are predominantly solid-liquid techniques which engage organic solvents in systems of heat-flux. These extraction techniques can be broadly catalogued into two categories: Conventional or classical methods and Non-conventional methods. Conventional or classical extraction techniques are those techniques that are applied in terms of small-scale production of antioxidants from various agricultural materials. The potency of these techniques is highly dependent on the type of solvents used in the procedure. The type of solvent can be selected based on the factor of molecular affinity between the given solute and the selected solvent and the extent of the solvent's toxicity and its property of feasibility and safety in accordance with the environment. All these factors in combination help to create effective and promising results. The techniques included under conventional extraction process are (a) Soxhlet extraction technique, (b) Maceration technique, and (c) Hydrodistillation technique. In comparison, the introduction of non-conventional techniques can be strongly related back to the disadvantages faced while using conventional extraction techniques. The few drawbacks, turned to challenges now, are degradation of thermolabile materials, highly purified and costly solvent requirement, prolonged period of extraction and selectivity, huge quantities of solvent evaporation, high amounts of solvents required and low antioxidant yields (Luque de Castro and García-Ayuso,

1998). As an objective to overcome all these limitations, non-conventional methods, including enzyme-assisted extraction (EAE), microwave-assisted extraction (MAE), ultrasound-assisted extraction (UAE), pressurized low-polarity water extraction, pressurized liquid extraction (PLE), pulsed electric field (PEF) extraction, molecular distillation and supercritical fluid extraction (SFE), are popularly practiced. Furthermore, they meet the standard regulations set by U.S. Environmental Protection Agency (2015), which aims at:

- synthesis of less hazardous and harmful chemicals,
- design for renewable feedstock use, catalysis, reduction of derivatives and energy efficiency,
- designing safer solvent auxiliaries and chemicals,
- design to analyze time for prevention of pollution, atom economy, and to prevent degradation which thus results in inherently safer and protective chemistry for any kind of accident prevention.

Due to this fact, among all the techniques mentioned here, some are also labelled as 'green techniques'(Selvamuthukumaran and Shi, 2017).

2.4 CONVENTIONAL OR CLASSICAL TECHNIQUES OF EXTRACTION

2.4.1 Soxhlet Extraction Technique (SE)

The Soxhlet technique of extraction, initially planned to essentially extract lipids, was first and foremost proposed by a German chemist named Franz Ritter von Soxhlet in 1879. Its modern application extends to the recovery of valuable antioxidants for various agricultural by-products. This method typically involves the containment of dry sample in a thimble, which is followed by the localization of the thimble in a distillation flask that already contains the specific solvent to be used. It is then exposed to heating and condensation to a level of overflow. Afterward, the aspiration of the solution present in the thimble takes place by means of a siphon that results in unloading of the solution again back into the distillation flask. This solution bears the extracted solutes back into the bulk liquid from which only the solvent moves to the soil bed while the solute is left behind in the distillation flask. This process is repeated several times for exhaustive extraction till it is successfully completed (Selvamuthukumaran and Shi, 2017; Azeez, Narayana and Oberoi, 2017). Numerous phenolic compounds have been extracted via Soxhlet technique from agro-wastes, such as spent ground coffee, grape pomace and hazelnut wastes (Manna, Bugnone and Banchero, 2015).

2.4.2 Maceration Technique

This technique has been used from ancient times for the homemade preparation of tonics, thus making it a promising and inexpensive method of obtaining bioactive compounds like antioxidants and essential oils from various agricultural wastes.

Maceration is mostly employed in terms of small-scale production, where it usually is composed of plant materials that are grounded, which in turn helps increase the surface area due to smaller sizes. This increase in surface area provides the opportunity of uniform mixing of the solute with a suitable solvent, mostly menstruum, in a closed vessel. Following the processing, the solvent is drained off, leaving behind the marc, meaning the solid residue produced from this extraction technique. The marc is then again filtered to extract the leftover large quantities of clogged solutions. The recovered pressed-out liquid is further filtered to separate the impurities for processing henceforth (Selvamuthukumaran and Shi, 2017; Azeez, Narayana and Oberoi, 2017).

2.4.3 Hydrodistillation Technique

Hydrodistillation, or HD, can be defined as the traditional extraction technique employed to replenish essential oils and antioxidants from fresh agricultural wastes by utilizing water as the solvent. The three physicochemical processes involved in HD are (a) hydrodiffusion, (b) hydrolysis and (c) decomposition by heat. This process has to be performed before the fresh agricultural plant materials get dehydrated, and also does not specifically involve the usage of organic solvents. There are different types of hydrodistillation; namely, steam and water distillation, water distillation and direct steam distillation (Vankar, 2004). This technique begins with the packing of plant agricultural materials in a motionless compartment, followed by the addition of an adequate amount of water. The components are then brought to a boil. As an alternative, direct steam is also administered in the plant sample. Simultaneously, steam and hot water act as the two main effective factors for the release of bioactive compounds from the agricultural products. On the other hand, indirect cooling though water executes condensation of oil and water vapour mixture. This condensed mixture passes from a condenser to another component, a separator, wherein the essential oils and antioxidants get separated automatically and immediately from water (Silva et al., 2005). However, one of the drawbacks of this technique is that several thermolabile compounds cannot be used here because of the high temperature of extraction, thus a portion of the volatile compounds are lost (Selvamuthukumaran and Shi, 2017).

2.5 NON-CONVENTIONAL TECHNIQUES OF EXTRACTION

2.5.1 Solvent Extraction (SE)

Solvent extraction (SE) is a technique that is predominantly associated with the usage of different types of solvents, such as ethanol, methanol, ether, acetonitrile, hexane, chloroform and benzene in different proportions in combination with water. This solvent-water combination is executed to extract both non-polar and polar organic compounds selectively, depending on the need and interest of the process (Starmans and Nijhuis, 1996). These compounds may include fatty acids, phenols, alkaloids, oils, organochlorine pesticides, aromatic hydrocarbons and many others (Plaza et al., 2010). The agro-wastes that have already undergone pre-treatments are treated with different solvents at varying temperatures. The temperatures and extraction time vary depending on the type of solvent employed; for example, extraction of antioxidants mediated by methanol is most effective at a temperature of 60°C. Afterward, the solvent-treated agro-wastes are centrifuged, which results in the formation of a supernatant and a pellet. The supernatant is removed and the pellet containing the extract along with some solid residues is filtered, from which the solid residues are removed, leaving behind only the essential antioxidant extracts. These antioxidants are later processed to incorporate them in functional foods, food supplements or additives. However, this process demands enormous quantities of solvents over longer periods of time for it to be successful. Additionally, the solvents used are not completely safe for either humans or the environment, and therefore they have to be completely washed off after extraction before the final product is released for further processing. Hence, these drawbacks limit the application of the SE technique, and in order to increase its efficiency, it is used along with other extraction techniques, such as ultrasound-assisted extraction, supercritical fluid extraction or microwave extraction techniques (Gahler, 2012). Several antioxidant products, mainly phenolics, have been extracted from potato peels, pomace and seeds of guava, leaves and pomace of olive (Khalifa et al., 2016), grape pomace, apple pomace (Rana et al., 2015), hazelnut and chestnut shells, and flavonoids and phenols from pomegranate peels through the solvent extraction technique. Sequential solvent extraction techniques, which involve the use of different solvents at different steps to extract bioactive compounds, have been employed to recover antioxidants like seeds and peels of acerola.

2.5.2 Enzyme-Assisted Extraction (EAE)

The enzyme-mediated pre-treatment has been thus far considered a very promising and novel method of extracting high yields of antioxidants from fresh agricultural waste. Most often, the enzyme-assisted extraction technique is facilitated by the application of various enzymes such as α-amylase, cellulase, β-gluconase, β-glucosidase, xylanase and pectinase. These certain specific enzymes enhance the production process of antioxidants by acting on the cell wall to degrade it, which further helps in depolymerizing the lipid bodies and polysaccharides present in the cell wall. The process of depolymerization results in the hydrolyzation of ester-linked phenolic acids, which releases the linked complex antioxidant compounds. This extraction method was initially designed to enhance the composition of antioxidants in black carrot juice along with obtaining vegetable oils, but the application of enzyme-assisted extraction has been expanded in recent years to produce economically valuable antioxidants, such as gallic acid from fresh agricultural wastes, which can be further utilized for functional food preparation, especially propyl gallate and pyrogallol. These extracted antioxidants can also function as intermediates for the production of trimethoprim, antibacterial drug synthesis, and in the pharmaceutical industry.

The EAE technique is carried out through either of the two pathways: (a) enzyme-assisted cold pressing (EACP); and (b) enzyme-assisted aqueous extraction (EAAE). Both techniques have been widely applied for the extraction of a wide range of antioxidants and vegetable oils. The EAAE technique is mostly implemented because of its efficacy in degrading the cell wall, which happens to be one of the crucial steps in any extraction technique. Several commercially important phenolic antioxidants have been extracted from fresh wastes of a diverse range of agricultural plants, such as grape pomace, five kinds of citrus fruits including grapefruit, Yen Ben lemon, orange, Meyer lemon and mandarin, and raspberry and watermelon rind (*Citrullus lanatus*), by means of various enzymes including celluzyme, a combination of cellulolytic and pectinolytic enzymes in the ratio of 2:1 pectinex, celluclast and noveform. While the EACP technique does not involve the availability of polysaccharide-protein colloid in comparison to the EAAE technique, enzymes are particularly administered to carry out hydrolyzation of the seed cell walls in this system. For a particular plant nutrient with potential antioxidant properties, known as lycopene, the extraction process was carried out in a two-step process. A total of four solvents were tried out to understand the efficacy of the extraction process for lycopene. The two-step process of extraction is accompanied by the EAE technique implemented on extracting lycopene thereafter. Firstly, the samples of fresh agricultural wastes are collected and then exposed to enzyme treatment, wherein the enzymes included are pectinase and cellulase. Secondly, the method of lycopene extraction is achieved by applying solvents (Selvamuthukumaran and Shi, 2017; Azeez, Narayana and Oberoi, 2017).

2.5.3 Ultrasound-Assisted Extraction (UAE)

The UAE extraction technique can be vividly divided into two major stages: (a) performing diffusion across the cell wall, and (b) disintegrating the cell wall, followed by washing of the cell components. To begin with, ultrasound can be defined as a special kind of sound wave that is beyond the capacity of hearing by humans, with a frequency ranging from 20 kHz to 100 MHz. This wave has a special characteristic of creating a phenomenon known as acoustic cavitation, whereby the waves are able to move through a medium by producing expansion and compression. The UAE technique, an efficient alternative to conventional SE technique, is specifically based on the principle of acoustic cavitation. This phenomenon, exhibiting adequate amounts of ultrasound intensity, in turn results in production, followed by growth and collapsing of microbubbles or cavities in a given liquid or solvent. During the process, the kinetic energy of motion gets converted to an enormous amount of heat energy that further heats the constituents, i.e. cells inside bubbles produced. The generated bubbles most often possess a pressure of 1000 atmosphere, temperature of 5000K and cooling and heating rate that levels off above 1010K/s. The ultrasonic waves work by producing effects of expansion and compression cycles when they pass through the liquid. So, once the formation of bubbles is done, they absorb the enormous amount of energy generated and undergo growth in the expansion cycles and then contract during the compression cycles. These expansion and compression cycles can in turn induce a collapse of bubbles, resulting in shock waves exhibiting enormous levels of temperature and pressure. The collapse of bubbles allows the cells inside the bubbles to get smacked to the solid matrix and disintegrated, due to which the required components present inside the cells get released outside (Selvamuthukumaran and Shi, 2017; Azeez, Narayana and Oberoi, 2017).

The UAE technique is only valid for those media which involve liquids or a mixture of solid constituents in liquids, as only these have the property of cavitation.

Pressure, temperature, sonication time, turbulence and frequency play the roles of crucial parameters in this technique that helps in accelerating the entire extraction process, along with particle size, presence of moisture level in the sample taken, degree of milling and solvent used, which universally enhances the extraction efficacy. The ultrasound-assisted extraction has been widely employed for the extraction of several products, such as polysaccharide-protein complexes, essential oils, proteins, sugars, carotenoids and most recently antioxidants, such as phenolic acids. The extraction of phenolic acids has not been as effective as expected under controlled and optimum conditions, but research is still ongoing. The major boons of UAE technique are facilitated mixing, reduction in energy, time, thermal gradients, equipment size, extraction temperature, processing steps and production cost, enhanced energy transfer and production and effective use of solvents. The UAE technique has been accounted for the recovery of carotenoids from tomato seeds and skin, and pomegranate peels (Sood and Gupta, 2015), phenols from artichoke waste (Llorach et al., 2002), potato peels, and flavonoids and pro-anthocyanins along with phenols from macadamia skin (Dailey and Vuong, 2015).

2.5.4 Microwave-Assisted Extraction (MAE)

By definition, microwaves, with a frequency range of 300 MHz to 300 GHz, are electromagnetic fields that consist of two oscillating fields which are localized in such a way that they are perpendicular to each other, like magnetic fields and electric fields. The microwave-assisted extraction technique is based on the principle of direct application of microwaves on substances that are polar in nature. This technique is an improved and advanced technology compared to the conventional SE methods. As the first step, the moisture present in the cells is heated to convert it into vapours. These vapours accumulate and further create high pressure against the cell wall. As the pressure increases along the cell wall, there are alterations in the physical characteristics and architecture of the cell wall and the cell organelles inside. This improves the feature of porosity in the biological matrix. The result of all these steps is better penetration of the solvent used for the extraction procedure. The stages followed in the MAE technique are the most effective for the extraction of desirable products with antioxidant properties, such as terpenoids, alkaloids and saponins, phenolic compounds and carotenoids. It has helped in providing high yields of antioxidants from fresh agricultural wastes. Moreover, it has been found that the presence of solvents like methanol or ethanol during the extraction of antioxidants like phenolic compounds has been more effective than water. The solvents are most commonly associated with a high dissipation factor, which plays a crucial role in antioxidant extraction by MAE technique. Water has a high dielectric constant, but a low dissipation factor compared to methanol or ethanol. Even though the high dielectric constant also plays an important role in the extraction procedure, this makes water less compatible for being able to heat up the moisture inside the biological matrix taken and to bring about pressure change that further initiates the leaching of the desirable components bearing antioxidant properties. On the other hand, solvents having a high dielectric constant, such as polar solvents and water, are better capable of absorbing high microwave energy emitted compared to other non-polar solvents. Therefore, it is often recommended to use solvents with a combination of high dissipation factor and high dielectric constant. This combination can be attained by mixing water with solvents such as methanol or ethanol. Other than dissipation factor and dielectric constant, some other parameters that affect the efficacy of the extraction process are moisture content, frequency, size of the particles present in the sample taken, solvent concentration, type of solvent selected, solubility of the solvent used, solid to liquid ratio, pressure and temperature maintained during the recovery process, power, exposure time and number of times the extraction cycles are repeated. Some of the advantages of the MAE method over traditional methods include higher extraction efficacy, less risk to the environment, reduced environmental pollution and a shorter period of extraction processing. The commercially valuable products recovered containing higher properties of antioxidants are often achieved from the peel extracts of certain fresh agricultural biowastes, including tomatoes (Strati and Oreopoulou, 2011), citrus mandarin (Singh et al., 2011), onions, skins of peanut, grapes and fruit dust of wild apple. Some antioxidants like anthocyanins are extracted from grape skin, and silymarin is extracted from milk thistle seeds. Other antioxidants such as tannins and pro-anthocyanins are extracted from mango seeds (Ajila et al., 2010), and glucosinolates are recovered from broccoli leaves by the process of MAE (Selvamuthukumaran and Shi, 2017; Azeez, Narayana and Oberoi, 2017).

2.5.5 Pressurized Liquid Extraction (PLE)

Pressurized liquid extraction technique is an advanced kind of combination of pressurized and accelerated conventional SE techniques. Hence, it is also referred to as accelerated fluid extraction, pressurized fluid extraction, subcritical water extraction (SWE), HP solvent extraction and enhanced solvent extraction. This technique is carried out at high pressure between 1450 and 2175 psi and high temperature ranging from 50 to 200°C by utilizing organic solvents that are liquid in nature. This ensures the extraction rate will increase rapidly. In this technique, the polarity of the solvent is a vital factor and is dependent on temperature. With an increase in temperature, there is a lowering in the dielectric constant of the particular solvent used in the extraction process, which in turn decreases the solvent polarity. By maintaining the extraction temperature, the polarity of the solvent can be balanced to that of the bioactive compounds to be extracted. The high pressure provided here is capable of forcing the liquid, i.e. the solvent used, into the sample biological matrix. This makes the rate of extraction process faster in the presence of a minimal amount of solvents and produces a high yield of antioxidants. One major advantage of PLE compared to SE is that it works in combination with temperature and HP, which facilitates the processing rate of extraction. Regardless of all the advantages available, PLE is unfavourable for the treatment of thermolabile compounds due to the high temperatures to which the biowaste materials are exposed. The PLE technique has been practiced to extract bioactive compounds from marine sponges, flavonoids from spinach and phenolics from passionfruit rinds (Silva et al., 2005) by using a mixture of water and ethanol, polyphenols from skin and pomace of grapes and other grape by-products, several antioxidants from industrially produced apple pomace and tricin, a type of flavonoid from the leaves of black bamboo (Selvamuthukumaran and Shi, 2017; Azeez, Narayana and Oberoi, 2017).

2.5.6 Pressurized Low-Polarity Water Extraction or Subcritical Water Extraction (SWE)

The pressurized low-polarity extraction technique, also called SWE, has recently become an environmentally-friendly green technology for the extraction and fractionation of antioxidants and other bioactive materials from fresh plant wastes in high yields. This extraction process is solely based on the principle of utilizing hot water as an extractant under high pressure. The high pressure is capable of maintaining water in the liquid state. The instrumentation of this technique is composed of a high-pressure pump in addition to a water reservoir, an oven, a valve or restrictor. The water reservoir contains the water which is then passed through the high-pressure pump, which exerts the high pressure on water to maintain its liquid state. This exerted pressure is maintained by the valve or restrictor. The oven is where the actual process of extraction takes place and the antioxidant components are separated from the wastes. After separation of the desirable products, they get stored in a vial that is located at the edge of the extraction instrumentation. Occasionally, the instrumentation also consists of a cooling system that carries out the rapid cooling of the products recovered. Normally, an approximate pressure of 5 MPa or 50 bar is exerted on water at a temperature range of 100–250°C so that water exists in its liquid state and does not get vapourized. The temperature might vary till 374°C, which is the upper limit of critical temperature, and pressure can range from 10–60 bar, depending on the need. The pressure is slowly increased until it reaches the desirable range where it does not get vapourized anymore. The pressure change does not influence the physical and chemical properties of the solvent, i.e. water, much; however, the temperature shift from 25°C to 250°C results in changes in surface tension, dielectric constant and viscosity. The temperature can be shifted according to the various class of antioxidant products needing to be extracted. With a shift in temperature, permittivity is affected inversely, along with surface tension and viscosity of the polar substances, and the diffusion rate is directly affected. Water generally has a dielectric constant of 80°C at room temperature; when the temperature is changed to 250°C the constant drops down to 30. At this temperature, water acts like that of other organic solvents like methanol and ethanol. Due to this property, SWE can be implemented to extract non-polar compounds like phytochemicals, which are usually water-insoluble at higher temperatures, whereas extraction from polar compounds is carried out at lower temperatures. The SWE technique has been applied to recover antioxidants from several vegetable wastes and other biowastes; for example, performing SWE at a temperature of 160°C has resulted in high yields of total content of phenolic compounds along with a high capacity of antioxidants per gram of canola seed meal. Using this method, it is also possible to extract phenolics from immature fruits along with citrus fruits (Chemistry, 2008). Recovery of antioxidants like pro-anthocyanins and catechins from wine-based products is done by following the SWE technique at three temperatures of 50°C, 100°C and 150°C consecutively. Like this, numerous other antioxidants exhibiting unique polymerization degrees have been extracted from fresh agricultural wastes by performing SWE extraction at various temperatures. Some of the advantages of the SWE technique over other conventional techniques are that it allows the utilization of a minimal volume of solvents and enhances faster and higher yields of antioxidants (Selvamuthukumaran and Shi, 2017; Azeez, Narayana and Oberoi, 2017).

2.5.7 Pulsed Electric Field (PEF) Extraction

The pulsed electric field works on the principle of application of the electric field on the living cells, which separates the desired antioxidants from the other biowastes. The PEF method is helpful in performing several processes including drying, diffusion, pressing and extraction. In this process, the electric field is applied on the living cells and on the basis of the dipole nature of molecules, the charge separates them. As the dipole nature of the molecules present in the living cells varies, when the electric potential passes along the cell membrane, the different molecules are separated based on their charge. The transmembrane electric potential is gradually increased until it exceeds the critical value of an average of one voltage. At this point, repulsion occurs between the charge-induced molecules which results in the formation of pores in the weaker parts of the cells. The pore formation leads to a sudden increase in permeability. The PEF treatment of plant biowastes is achieved by arranging a simple circuit composed of exponential decay pulses. Here, a treatment chamber consisting of two electrodes is present, in which the biowastes are placed in either a batch or continuous approach. The efficacy of PEF extraction techniques depends on certain parameters like pulse number, field strength, extraction temperature, specific energy and properties of the materials to be treated. The PEF technique works by degrading the structural foundation of plant membranes, which facilitates mass transfer in the extraction process. This helps reduce the time and enhance the effectiveness of the process. Moreover, exposing the plant materials to a moderate electric field with potential ranging from 500 and 1000 V/cm for 10–4 or 10–2 seconds along with a slight increase in temperature can damage and degrade the cell membranes, but minimizes the depletion of the thermolabile materials (Selvamuthukumaran and Shi, 2017; Azeez, Narayana and Oberoi, 2017).

2.5.8 Molecular Distillation

Molecular distillation works on the principle of separating the desirable antioxidants from the mixed constituents in the sample taken by using partial evaporation, which is employed to extract antioxidants from fresh agricultural wastes. Normally, distillation is defined as a process used to separate and purify components from a mixture by means of heating and cooling, which produces vapour. Vapour has relatively higher vapour pressure than any other solvent, which makes it a volatile compound. When a comparison can be drawn between the transfer distance and free mean path (meaning the average path that a molecule is able to travel without colliding with any other molecule in vapour state), then that type of distillation is known as molecular distillation. There are two distinct types of evaporators included under molecular distillation: namely, short-path evaporators (SE) and thin-film evaporators (TFE). In both types of evaporators, there is a rotor wiper system in which the sample is exposed to agitation, along with vacuum pumps that generate high vacuum. The operating pressure in both the evaporators differs on the basis of extraction requirements. The operating pressure of TFE can reduce up to 1 to 100 mbar, while the operating pressure of SE can reduce to 0.0001 mbar. In TFE, the distance between the condenser and vacuum is not filled by any other unit. On the other hand, the centre of the evaporator unit is interfered with by a condenser in SE, which exceedingly reduces the distance between the condensation and boiling surfaces and minimizes the pressure drop. There are definite parameters which are important factors in determining the output of the extraction process. These are mainly flow rate, wiper speed, vacuum and temperature. The process of molecular distillation quite likely occurs at very low temperatures, which further reduces the issues related to thermal decomposition. A high amount of vacuum helps eliminate the oxidation reaction which might have existed in the presence of air. It has also been found that faster speed has an effect on the viscosity and thickness of the film, whereas a high rate of flow decreases the molecules' residence time to be vapourised (Selvamuthukumaran and Shi, 2017).

2.5.9 Fermentation Process for Extraction

Extraction of antioxidants using the fermentation process is a newly emerging technique that provides a good alternative to the conventional technique of extraction. The fermentation process can be broadly categorized into two types: (a) Solid-State Fermentation (SSF), and (b) Submerged Fermentation (SmF). In SSF, there is minimal to no involvement of water, whereby the microbial growth as well as the formation of desirable products take place on solid media, but to compensate for the absence of water, the substrate consists of adequate moisture so the microorganisms have the optimum conditions to grow and metabolize. The substrate that is commonly used is derived from agro-industrial and agricultural residues, which makes the SSF process cost-effective. Comparatively, SmF is marked by the presence of a type of liquid where the microorganisms are grown in a liquid media enriched with the essential nutrients. In a liquid culture, the desired products only begin to form when there is an exhaustion in the quantity of one of the essential nutrients like nitrogen, carbon or phosphate, provided to the media. Through either of the fermentation processes, the resultant antioxidant products are generated only after complete microbial growth in the media used (solid or liquid) are in the form of secondary metabolites. For some time, SmF has been applied to extract numerous bioactive compounds; however, recently it has been observed that SSF is industrially replacing SmF due to various advantages that it offers over SmF. Firstly, SSF requires minimal to zero volume of water, which impacts the economy of the process positively. Due to this, the process of SSF demands a smaller size of fermenter, lower costs of sterilization and reduction in downstream processing and stirring. In total, SSF is shown to have higher productivity, efficacy and yield and better quality of products than SmF. Despite the many advantages of SSF, there is one major issue with this process involving the scaling-up, which is largely due to culture homogeneity and heat transfer reasons. Therefore, it can be concluded that the fermentation process has been effective in producing high-quality extracted antioxidant products by excluding any type of toxicity related to the extractants, particularly the solvents that are organic in nature.

Numerous antioxidant-bearing products have been extracted from fresh agricultural wastes through the fermentation process, such as various phenolic compounds that include hydrolysable tannins (pedunculagin, catechin, ellagic and gallic acid esters of glucose, punicalin, epicatechin, punicalagin), anthocyanins that are originated from cyanidin, delphinidin and pelargonidin, and other lignans such as matairesinol, secoisolariciresinol, isolariciresinol, pinoresinol, medioresinol and syringaresinol, pertaining to anti-mutagenic, anticancer, antioxidant and anti-inflammatory activities, obtained from pomegranate wastes. Some phenolic compounds and flavonoids have been extracted from onion paste via a solid-liquid extraction method. Pomegranate husks, functioning as the substrate, are treated by SSF in addition to *Aspergillus niger GH1*, wherein ellagitannin acyl hydrolase biologically converts ellagitannin of pomegranate husks into tons of ellagic acid. This extraction process of ellagic acid from pomegranate husks is considered an economically viable technique with low expenses and abundant pomegranate husk sources. The ellagitannin acyl hydrolase functions as an enzyme during SSF and is responsible for mobilizing the phenolic compounds along with other antioxidants from the agro-wastes. Additionally, other enzymes such as tannin acyl hydrolase, α-amylase, β-glucosidase and laccase, are responsible for carrying out the same function. Other sources of ellagic acid are cranberry pomace, which is produced as a by-product after processing of cranberry juice, and Lentinus edodes, treated by its esterase enzyme in association to SSF. Abundant amounts of ferulic acid can be obtained from green coconut husks, which are available sufficiently as a result of agro-industrial processing. The coconut husks are treated with ligninolytic enzymes under SSF to isolate ferulic acid from the husk cell wall, followed by biological conversion of ferulic acid to vanillin via *Phanerochaetechrysosporium*-a basidiomycete. Biological conversion of tannin as the substrate to produce gallic acid is performed by SSF in the presence of *Rhizopus oryzae*. Tannin can be derived from the solid residue that is produced after oil recovery from Teri pod available in India. Sometimes, lignocellulosic enzymes produced by fungi are also used for the extraction of antioxidants. These lignocellulosic enzyme-producing fungi consist of two systems: (a) a ligninolytic system that is capable of degrading lignin, which exposes the phenyl rings, thus releasing the phenolic contents; and (b) a hydrolytic system that has the feature of degrading the polysaccharides. Most of the agricultural wastes, including fruit and vegetable wastes like stover, bagasse, husks, straw, cobs and cereals, are lignocellulosic in nature, i.e. their cell walls are made up of lignin, hemicellulose and cellulose. The lignin portion of the cell wall is a harbour of phenolic compounds such as vanillic acid, ferulic acid, syringic acid, p-hydroxybenzoic acid and p-coumaric acid, so there is the application of ligninolytic enzymes along with some filamentous fungi such as *Trametes hirsute*, white-rot fungi *Phanerochaetechrysosporium*, *Bjerkanderaadusta* and *Trametes versicolor* in association with SSF that together produces a synergistic effect to degrade lignin.

2.5.10 Supercritical Fluid Extraction (SCFE)

Supercritical fluid extraction (SCFE) is an environmentally-friendly technology that is dependent on the properties exhibited by fluids, including dielectric constant, density, viscosity and diffusivity, and some other altering conditions like temperature and pressure, which are adjusted to meet the supercritical fluid state (SF). The supercritical fluid state (SF) can be defined as the state where gas and liquid behave identically to each other. Considering the properties and conditions, the density of a liquid is similar to that of SF and the viscosity of a gas is similar to that of SF; hence, a fluid is selected which has a combination of liquid and gas states. According to the requirements, there are several compounds that can be considered for use as SFs, like nitrogen, ethylene, fluorocarbons, methane, xenon, carbon dioxide, etc. Among these, carbon dioxide is considered to be the most suitable and popular SF because it enables the process to be carried out at a low cost and maximum safety. It also reduces any sort of alteration in the functional properties of the antioxidants that are being extracted. Occasionally, when only carbon dioxide is used in this technique, known as subcritical Soxhlet extraction (SDCS), green technology has shown to have the potential extract high quantities of bioactive products, like total tocotrienol that includes α-, β-, γ-, δ-tocopherol, γ-oryzanol, total tocopherol, and α-, γ-, δ-tocotrienol from rice bran oil, approximately ten times more than hexane. Additionally, SFs have certain characteristic features, like their density, that allow them to have better transport pathways compared to liquid solvents, and their density is also adaptable by varying pressure and temperature. Carbon dioxide at a supercritical state is highly beneficial as it is non-toxic, stable, non-explosive, GRAS, cheap and solubilizes lipophilic materials easily so that they can be discarded from the resulting products. Moreover, it exists in the form of gas at room pressure and room temperature, which makes it an effective alternative to organic solvents. Carbon dioxide also comes with some advantages, like lower polarity that restricts it from being able to extract highly polar substances from biological matrixes. In relation to this problem, few solvents such as isopropanol, methanol, acetonitrile, hexane, dichloromethane or ethanol are added to supercritical carbon dioxide in small amounts. These solvents, termed as co-solvents, enhance the process selectivity and solubility. Out of all these co-solvents, ethanol is the preferable solvent for mixture with carbon dioxide in SCFE as it has the property of miscibility and is comparatively less toxic. The combination of ethanol and carbon dioxide has been applied to extract leaf extracts of oregano possessing high levels of antioxidant properties, and this method has proven to be highly beneficial and effective. The technique of supercritical fluid extraction is generally performed by placing the raw biowastes in an extractor vessel that is highly controlled by pressure and temperature. The raw biowastes, consisting of the desired constituents and the fluid, pass to the separators. The separators are made up of a tap at the bottom part, where the desired products move out after separation, while the fluid regenerated is discarded into the surroundings. The SCFE method is a potential alternative for traditional methods of extraction and is thus widely employed for the recovery of essential and commercially valuable antioxidants and other bioactive products, sometimes even in the form of dry powders as microparticles. The main advantages of implementing this technology in extracting commercial antioxidants are that it causes lower pollution and harm to the environment, increased rate of extraction, high selectivity and utilization of organic solvents that are non-toxic and stable. The SCFE technique has been successfully administered on spent ground coffee, grape pomace and hazelnut waste (Manna, Bugnone and Banchero, 2015) for the extraction of phenolic compounds. A set of agricultural wastes that are being used to extract antioxidants by supercritical fluid extraction (SCFE) has been listed in Table 2.2.

TABLE 2.2
Showing the Extraction of Antioxidants from Agricultural Wastes Under Proper Pressure and Temperature Conditions in the Presence of Required Co-Solvents

Agricultural wastes	Pressure (Bar)	Temperature (°C)	Co-solvents	Quantity of co-solvents taken (in percentage %)	Target antioxidants
Tomato paste waste	300	880	Ethanol	5	Carotenoids
Red grape pomace	100–250	45	Methanol	5	Pro-anthocyanins
Grape seeds	130	30–40	Ethanol	30	Anthocyanins
Spearmint leaves	100–300	40–50/50–60	Ethanol	10	Flavonoids
Tomato skin	350	75	Ethanol	10	Carotenoids
Green tea leaves	310	60	Ethanol	10	Catechins

Adapted from source: (Azeez, Narayana and Oberoi, 2017)

Depending on the type of antioxidant to be extracted, newer extraction technologies are yet to be developed for obtaining the best results of high yield in a shorter extraction period, at lower expense, that is eco-friendly, requires lower solvent concentrations, and maintains a good balance between extraction pressure and temperature and modern reactor designs.

2.6 QUANTIFICATION AND EVALUATION OF THE EXTRACTED ANTIOXIDANTS

The methods of quantification can be carried out by performing either chromatographic techniques, such as high-pressure liquid chromatography (HPLC), thin layer chromatography (TLC) and mass spectroscopy (MS), or non-chromatographic techniques like Fourier-transform infrared spectroscopy (FTIR) and NMR spectroscopy. The principle of chromatographic techniques is based on the separation of the molecules present in the mixture being applied to the solid itself or onto the surface and the fluid stationary phase, i.e. the stable phase, when it moves with the help of a mobile phase (Coskun, 2016); the spectroscopy technique is based on the principle that an electromagnetic radiation beam is passed through the sample and its response is recorded on the basis of wavelength of the radiation. Other than these two techniques, there is another method where a combination of chromatography and spectroscopy, termed as hyphenated techniques like liquid chromatography-mass spectroscopy, is followed. The hyphenated techniques exploit the benefits of both techniques, where spectroscopy selectively identifies and quantifies the antioxidants and chromatography evaluates the portions of chemical components present in the mixture. Some methods are favoured for the evaluation of the bioactivities shown by the antioxidant products for all of their antioxidant activities (such as capability of ferric reducing antioxidant power (FRAP), cupric reducing antioxidant capacity (CUPRAC), 2,2′-azino-bis 3-ethylbenzothiazoline-6-sulphonic acid (ABTS), Oxygen radical absorbance capacity (ORAC), and 2,2-diphenyl-1-picrylhydrazyl (DPPH)), lipid oxidation inhibition (tThiobarbituric acid reactive substances (TBARs) and value of peroxide), antiproliferative activity, antimicrobial activity and many other factors (Socaci, 2017).

2.7 CONCLUSION

From this chapter, it can be concluded that agricultural wastes including peels, shells, seeds, husks, leaves and pomace are a rich source of various essential antioxidants like phenolic compounds, flavonoids, carotenoids, vitamin C, vitamin E, fibres and many other types of antioxidants which, after extraction, are used for the preparation of important food supplements, additives and functional foods. Carotenoids, a characteristic group of pigments with antioxidant properties, are widely employed as food colourants and as nutritional precursors for vitamin A that improves the shelf life of the food products. The recovered carotenoids are incorporated in food products such as refined vegetable oils and macaroni. Polyphenols or phenolic compounds that are extracted from various agro-wastes such as grape pomace, olive oil and peel of carob seeds, are sometimes integrated in patties made of lab meat as natural antioxidants. These antioxidant compounds are also applied in pharmaceutical, nutraceutical, cosmetics and food industries. Additionally, they are also used as natural sweeteners in the form of 'sugar syrup'. The main aim of recycling fresh agro-wastes to produce commercially important antioxidants is to make an effective use of the wastes produced by utilizing some modern, green and environmentally-friendly methods. Hence, the agricultural wastes should not be treated as non-valuable wastes anymore; instead, they should be considered as a potential and valuable source of nutritional compounds that can be exploited for their benefits as shown in Figure 2.2.

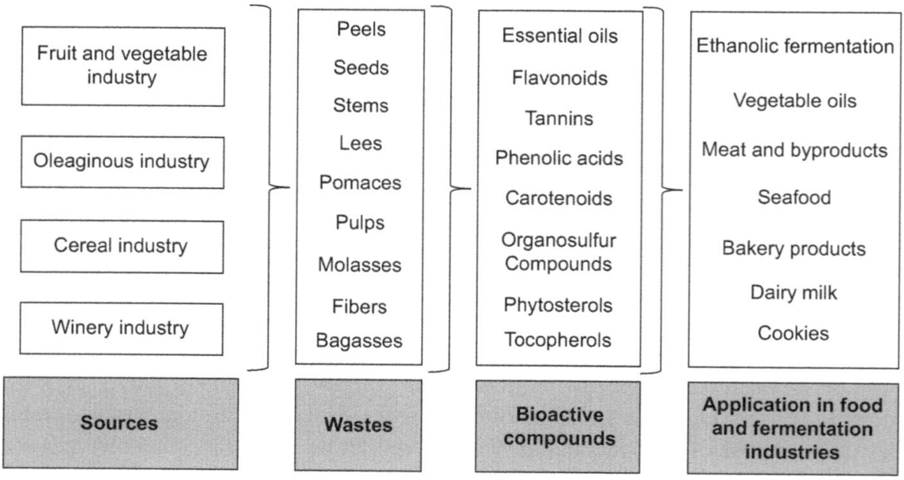

FIGURE 2.2 Application of natural antioxidants obtained from fresh agricultural wastes in industries like fermentation and food.

Source: (Shirahigue and Ceccato-Antonini, 2020) www.scielo.br/img/revistas/cr/v50n4//1678-4596-cr-50-04-e20190857-gf2.jpg

2.8 FUTURE PERSPECTIVES

Considering the present conditions, the key limitation faced while commercially producing the antioxidants from fresh agricultural waste is the scaling-up process. Even though the scaling-up process limits the face of potentiality depicted by this emerging technology, with proper management skills and improved pre-treatment and extraction strategies, the limitations can be overcome with ease. With increasing environmental pollution, the major objective of every researcher is to develop technologies that are environmentally-friendly, cost-effective, have increased efficacy, minimal loss of bioactive compounds while processing, decreased consumption of energy, increased durability and selectivity, and therefore advanced, sustainable technologies that have the capacity of meeting these standards can build the future of antioxidant extraction from agro-wastes.

2.9 REFERENCES

Ajila, C. M. *et al.* (2010) 'Mango peel powder: A potential source of antioxidant and dietary fiber in macaroni preparations', *Innovative Food Science & Emerging Technologies*, 11(1), pp. 219–224. https://doi.org/10.1016/j.ifset.2009.10.004.

Asgher, M. *et al.* (2013) 'Alkali and enzymatic delignification of sugarcane bagasse to expose cellulose polymers for saccharification and bio-ethanol production', *Industrial Crops & Products*, 44, pp. 488–495. doi: 10.1016/j.indcrop.2012.10.005.

Azeez, S., Narayana, C. K. and Oberoi, H. S. (2017) 'Extraction and utilisation of bioactive compounds from agricultural waste', In: Q. V. Vuong (ed.), *Utilisation of bioactive compounds from agricultural and food waste*, pp. 127–158. Boca Raton, FL: CRC Press.

Baiano, A. (2014) 'Recovery of biomolecules from food wastes—a review', *Molecules*, 19(9), pp. 14821–14842. doi: 10.3390/molecules190914821.

Bhavsar, N. H. *et al.* (2015) 'Original research article production, optimization and characterization of fungal cellulase for enzymatic saccharification of lignocellosic agro-waste', *International Journal of Current Microbiology and Applied Sciences*, 4(3), pp. 30–46.

Chemistry, F. (2008) 'Minerals, phenolic compounds, and antioxidant capacity of citrus peel extract by hot water', 73(1). doi: 10.1111/j.1750-3841.2007.00546.x.

Coskun, O. (2016) 'Separation tecniques: Chromatography', *Northern Clinics of Istanbul*, 3(2), pp. 156–160. doi: 10.14744/nci.2016.32757.

Dailey, A. and Vuong, Q. V. (2015) 'Effect of extraction solvents on recovery of bioactive compounds and antioxidant properties from macadamia (Macadamia tetraphylla) skin waste ', *Cogent Food & Agriculture*, 1(1), p. 1115646. doi: 10.1080/23311932.2015.1115646.

Eggeman, T. and Elander, R. T. (2019) 'Process and economic analysis of pretreatment technologies', 96(2005), pp. 2019–2025. doi: 10.1016/j.biortech.2005.01.017.

Fridovich, I. (1978) 'The biology of oxygen radicals: The superoxide radical is an agent of oxygen toxicity; superoxide dismutases provide an important defense', *Science*, 201(4359), pp. 875–880. https://doi.org/ 10.1126/science.210504

Füleky, G. (no date) 'M SC PL O E—C EO M SC PL O E—C EO', I.

Gahler, M. (2012) 'More relevant research for the EU arctic policy', 82, pp. 37–43. doi: 10.1007/978-3-642-24203-8_5.

Hu, Z., Wang, Y. and Wen, Z. (2008) 'Alkali (NaOH) pretreatment of switchgrass by radio frequency-based dielectric heating', pp. 71–81. doi: 10.1007/s12010-007-8083-1.

Hyun, T. *et al.* (2003) 'Pretreatment of corn stover by aqueous ammonia', 90, pp. 39–47. doi: 10.1016/S0960-8524(03)00097-X.

Ichwan, M. and Son, T. W. (2011) 'Study on organosolv pulping methods of oil palm biomass', International Seminar on Chemistry, November, pp. 364–370.

Ingh, R. P. S. and Urthy, K. N. C. H. M. (2002) 'Studies on the antioxidant activity of pomegranate (Punica granatum) peel and seed extracts using in vitro models', pp. 81–86. doi: 10.1021/jf010865b.

Joshi, V. K., Kumar, A. and Kumar, V. (2012) 'Antimicrobial, antioxidant and phyto-chemicals from fruit and vegetable wastes: A review', *International Journal of Food and Fermentation Technology*, 2(2), p. 123.

Kaur, U. *et al.* (2012) 'Ethanol production from alkali- and ozone-treated cotton stalks using thermotolerant Pichia kudriavzevii HOP-1', *Industrial Crops & Products*, 37(1), pp. 219–226. doi: 10.1016/j.indcrop.2011.12.007.

Khalifa, I. *et al.* (2016) 'Optimizing bioactive substances extraction procedures from guava, olive and potato processing wastes and evaluating their antioxidant capacity', *Journal of Food Chemistry and Nanotechnology*, 2(1), pp. 170–177. doi: 10.17756/jfcn.2016-027.

Kim, T. H., Taylor, F. and Hicks, K. B. (2008) 'Bioethanol production from barley hull using SAA (soaking in aqueous ammonia) pretreatment q', 99, pp. 5694–5702. doi: 10.1016/j.biortech.2007.10.055.

Koubala, B. B. *et al.* (2008) 'Physicochemical properties of pectins from ambarella peels (Spondias cytherea) obtained using different extraction conditions', *Food Chemistry*, 106(3), pp. 1202–1207. doi: 10.1016/j.foodchem.2007.07.065.

Kris-Etherton, P. M. *et al.* (2002) 'Bioactive compounds in foods: Their role in the prevention of cardiovascular disease and cancer', *The American Journal of Medicine*, 113(9), pp. 71–88. https://doi.org/10.1016/S0002-9343(01)00995-0.

Kumar, K. *et al.* (2017) 'Food waste: A potential bioresource for extraction of nutraceuticals and bioactive compounds', *Bioresources and Bioprocessing*. doi: 10.1186/s40643-017-0148-6.

Lafka, T., Sinanoglou, V. and Lazos, E. S. (2007) 'Food chemistry on the extraction and antioxidant activity of phenolic compounds from winery wastes', 104, pp. 1206–1214. doi: 10.1016/j.foodchem.2007.01.068.

Llorach, R. *et al.* (2002) 'Artichoke (Cynara scolymus L.) Byproducts as a potential source of health-promoting antioxidant phenolics', *Journal of Agricultural and Food Chemistry*, 50(12), pp. 3458–3464. doi: 10.1021/jf0200570.

Luque de Castro, M. D. and García-Ayuso, L. E. (1998) 'Soxhlet extraction of solid materials: An outdated technique with a promising innovative future', *Analytica Chimica Acta*, 369(1–2), pp. 1–10. doi: 10.1016/S0003-2670(98)00233-5.

Manna, L., Bugnone, C. A. and Banchero, M. (2015) 'Valorization of hazelnut, coffee and grape wastes through supercritical fluid extraction of triglycerides and polyphenols', *Journal of Supercritical Fluids*, 104, pp. 204–211. doi: 10.1016/j.supflu.2015.06.012.

Nakamura, Y., Daidai, M. and Kobayashi, F. (2004) 'Ozonolysis mechanism of lignin model compounds and microbial treatment of organic acids produced', *Water Science and Technology*, 50(3), pp.167–172. https://doi.org/10.2166/wst.2004.0188.

Palacios, A. S. *et al.* (2021) 'Ethanol production from banana peels at high pretreated substrate loading: Comparison of two operational strategies', *Biomass Conversion and Biorefinery*, 11(5), pp.1587–1596. https://doi.org/10.1007/s13399-019-00562-7.

Plaza, M. *et al.* (2010) 'Screening for bioactive compounds from algae', *Journal of Pharmaceutical and Biomedical Analysis*, 51(2), pp. 450–455. doi: 10.1016/j.jpba.2009.03.016.

Rana, S. *et al.* (2015) 'Functional properties, phenolic constituents and antioxidant potential of industrial apple pomace for utilization as active food ingredient', *Food Science and Human Wellness*, 4(4), pp. 180–187. doi: 10.1016/j.fshw.2015.10.001.

Sarkar, N. *et al.* (2012) 'Bioethanol production from agricultural wastes : An overview', *Renewable Energy*, 37(1), pp. 19–27. doi: 10.1016/j.renene.2011.06.045.

Selvamuthukumaran, M. and Shi, J. (2017) 'Recent advances in extraction of antioxidants from plant by-products processing industries', pp. 61–81. doi: 10.1093/fqs/fyx004.

Shirahigue, L. D. and Ceccato-Antonini, S. R. (2020) 'Agro-industrial wastes as sources of bioactive compounds for food and fermentation industries', *Ciencia Rural*, 50(4). doi: 10.1590/0103-8478cr20190857.

Shui, G. and Leong, L. P. (2006) 'Food chemistry residue from star fruit as valuable source for functional food ingredients and antioxidant nutraceuticals', 97, pp. 277–284. doi: 10.1016/j.foodchem.2005.03.048.

Sills, D. L. and Gossett, J. M. (2011) 'Bioresource Technology Assessment of commercial hemicellulases for saccharification of alkaline pretreated perennial biomass', *Bioresource Technology*, 102(2), pp. 1389–1398. doi: 10.1016/j.biortech.2010.09.035.

Silva, L. V. *et al.* (2005) 'Comparison of hydrodistillation methods for the deodorization of turmeric', *Food Research International*, 38(8–9), pp. 1087–1096. doi: 10.1016/j.foodres.2005.02.025.

Singh, H. *et al.* (2011) 'Bioresource technology enhanced ethanol production from Kinnow mandarin (Citrus reticulata) waste via a statistically optimized simultaneous saccharification and fermentation process', *Bioresource Technology*, 102(2), pp. 1593–1601. doi: 10.1016/j.biortech.2010.08.111.

Sládková, A. *et al.* (2016) 'Yield of polyphenolic substances extracted from spruce (Picea abies) bark by microwave-assisted extraction', *BioResources*, 11(4), pp. 9912–9921. https://doi.org/10.15376/biores.11.4.9912-9921

Socaci, S. (2017) 'Antioxidant compounds recovered from food wastes world ' s largest science, technology & medicine open access book publisher', (August). doi: 10.5772/intechopen.69124.

Sood, A. and Gupta, M. (2015) 'Extraction process optimization for bioactive compounds in pomegranate peel', *Food Bioscience*, 12, pp. 100–106. doi: 10.1016/j.fbio.2015.09.004.

Starmans, D. A. J. and Nijhuis, H. H. (1996) 'Extraction of secondary metabolites from plant material: A review', *Trends in Food Science and Technology*, 7(6), pp. 191–197. doi: 10.1016/0924-2244(96)10020-0.

Strati, I. F. and Oreopoulou, V. (2011) 'Effect of extraction parameters on the carotenoid recovery from tomato waste', *International Journal of Food Science and Technology*, 46(1), pp. 23–29. doi: 10.1111/j.1365-2621.2010.02496.x.

Teymouri, F. *et al.* (2018) 'Optimization of the ammonia fiber explosion (AFEX) treatment parameters for enzymatic hydrolysis of corn stover', 96(February 2005), pp. 2014–2018. doi: 10.1016/j.biortech.2005.01.016.

Tsakona, S., Galanakis, C. M. and Gekas, V. (2012) 'Hydro-ethanolic mixtures for the recovery of phenols from mediterranean plant materials', *Food and Bioprocess Technology*, 5(4), pp. 1384–1393. doi: 10.1007/s11947-010-0419-0.

U.S. EPA. (2015) Report on the 2015 U.S. Environmental Protection Agency (EPA) International Decontamination Research and Development Conference. U.S. Environmental Protection Agency, Washington, DC, EPA/600/R-15/283.

Vankar, P. S. (2004) 'Essential oils and fragrances from natural sources', *Resonance*, 9(4), pp. 30–41. doi: 10.1007/bf02834854.

Wyman, C. E. (1999) 'Technical progress, opportunities, and commercial challenges', *Annual Review of Energy and the Environment*, 24(1), pp. 189–226.

Yoo, J. *et al.* (2011) 'Bioresource technology thermo-mechanical extrusion pretreatment for conversion of soybean hulls to fermentable sugars', *Bioresource Technology*, 102(16), pp. 7583–7590. doi: 10.1016/j.biortech.2011.04.092.

Zhao, Y. *et al.* (2008) 'Enhanced enzymatic hydrolysis of spruce by alkaline pretreatment at low temperature', 99(6), pp. 1320–1328. doi: 10.1002/bit.21712.

Zheng, J. and Rehmann, L. (2014) 'Extrusion pretreatment of lignocellulosic biomass: A review', pp. 18967–18984. doi: 10.3390/ijms151018967.

Zhu, J., Wan, C. and Li, Y. (2010) 'Bioresource technology enhanced solid-state anaerobic digestion of corn stover by alkaline pretreatment', *Bioresource Technology*, 101(19), pp. 7523–7528. doi: 10.1016/j.biortech.2010.04.060.

3 Utilization of Agricultural Residues for Energy Production

Felicia O. Afolabi, Paul Musonge, Edward Kwaku Armah, Opeyemi A. Oyewo, Sam Ramaila and Lydia Mavuru

CONTENTS

3.1 INTRODUCTION

Fossil fuels form a large fraction of the world's basic energy supply, consisting of two-thirds of the total energy generated. Nuclear power is composed of 5%; however, renewable energy has 3%, consisting of wind, solar, geothermal, tidal and hydropower. About 10% is attributed to biomass and waste. Considering the reports

DOI: 10.1201/9781003245773-3

on world resources and reserves, biomass is less utilized compared to fossil fuels, gas and coal. Renewable energy resources became very important for the movement and power generation to reduce the climatic effect and economic, political and environmental challenges related to the combustion of fossil fuel (Bajwa et al., 2018). Bioenergy refers to the use of materials of biological origin, such as plant and manure from livestock, used to generate alternative fuels, also known as biofuels, and the generation of electric power. These biofuels provide a means of sustainable transportation and electricity. The sales of the local resources that produce the biofuels are immensely beneficial to the communities, and the environment also benefits from the low carbon emission compared to fossil fuels.

The generation of bioenergy is one of the policies used by many developed countries to reduce global warming caused by greenhouse gases (GHGs). The USA, the European Union, China and others are proposing regulations to gradually switch to an alternative energy source and promote the implementation of renewable sources (Gabrielle et al., 2014; Jiang et al., 2019). Facing this reality, research focus has shifted to generating energy using various accessible biomass such as agricultural wastes (Bharathiraja et al., 2018; Bajwa et al., 2018; Aqsha et al., 2017; Kumar et al., 2018). A vast amount of these wastes are produced every year (Negri et al., 2020). Therefore, the presence of these residues in the environment without adequate discarding measures may lead to environmental pollution issues and adverse impacts on human and animal health. In addition, these wastes have an impact on climatic conditions, which consequently elevate the greenhouse gases generated.

Agricultural residues have varying compositions of high proteins, sugar and minerals which can serve as animal feed. These agricultural residues may be converted directly into energy through direct combustion and/or other advanced techniques such as pyrolysis to obtain products in liquid, solid or gaseous forms. Agricultural residues are cellulosic biomass such as crop residues which include wheat straw, corn stover and rice straw. These resources are obtained from crop production; their collection and utilization for energy generation establish sustainable practices that do not lead to the food crisis and land competition. Many countries have made use of agricultural residues as another energy resource (Negri et al., 2020; Kumar et al., 2018; Jiang et al., 2019). The selection of appropriate agricultural residues and manure for the generation of energy depends on the kind and resources available in a particular area. Practically, the application of agricultural wastes in the production of energy improves both local and regional economies.

The agricultural residue or waste is lignocellulosic biomass comprising leaves, stovers, straws, husk, pods, seeds, roots, cobs and seed pods (Zabed et al., 2016). The use of agricultural residues for energy production is an immediate research focus throughout the world. The application of agricultural residues as a source of energy generation depends on the quantity of the agricultural residues, their geographical locations, evaluation of various techniques for the conversion of agricultural residue into energy and potential power generation. Agricultural residues can be transformed into different essential products for the generation of energy via processes including thermochemical and biochemical conversion processes. Thermochemical conversion processes comprise liquefaction, pyrolysis, gasification and combustion. Thermochemical conversion is a method mostly used to generate heat and electricity, which is referred to as combustion. Biochemical conversion technology converts agricultural residues into biofuel through biological pre-treatments. This process produces biofuels and biochemicals which include ethanol, acetone, biogas, butanol, hydrogen and various organic acids (Yue Li et al., 2019).

Considering the other advantages that come with the utilization of agricultural residues for energy, some European countries have installed huge plants for heat and energy generation utilizing agricultural residues. For example, France has huge agricultural land among the European countries that focuses on producing biodiesel and ethanol (Tursi, 2019).

3.2 AGRICULTURAL RESIDUES

Agricultural wastes or residues refer to materials that are highly rich in carbon content produced in the process of harvesting and processing farm crops. Agricultural residues are obtained as unwanted products from food crops, which include rice, maize, barley, chickpeas, sunflower wheat and many others. Agricultural residues can be sub-divided into primary or secondary residues. Primary residues are obtained in the process of harvesting crops on the field, also known as field residues, while secondary residues are produced from the processing of crops, referred to as process residues (Sadh et al., 2018). The primary or field residues occasionally serve as fertilizers or manure for farmland, for controlling erosion and feed for livestock. In addition, a large portion of these residues is burnt on the farm when another farm year is about to commence. The field residues include leaves, stalks, seed pods and stems.

Process residues are produced when the crop has undergone processing to form an alternative resource. These residues are molasses, seeds, bagasse, roots, pulp, peels, straw, husks, stubble, etc. Process residues provide a high source of energy. Agricultural residues are recognized based on their availability, abundance and characteristics compared to other solid fuels such as charcoal, wood and char. One of the major advantages of using agricultural residues is because the production of foods is not affected and can be utilized economically as an efficient resource for the generation of energy.

3.2.1 Crop Residues

Crop residues, referred to as remnants, are produced during crop processing operations such as corn, wheat and rice, differently from the grains. These are unwanted residues obtained from agricultural produce on the field such as leaves, straws, seed pods and stovers. Agricultural residues generated by crop residues are mostly available and low-cost waste, which are readily transformed into various essential products. These leftovers are important resources for the generation of energy. These resources are generally used as animal bedding, burnt or left on the farmland to decay. Recently, scientific discovery has proven that these residues can be used for electricity generation and provide both economic and environmental advantages.

Residues offer many advantages, including erosion control, manure for farmland, soil carbon depletion mitigation and energy generation. The number of residues generated depends on many conditions relative to the farmland; however, the removal of residues needs to be sustainably considered, as too much removal can expose the land to severe erosion and too many residues can hinder the soil from being used in the next farming season. It could also lead to excessive exposure to the farmland, thereby worsening the environment. The excessive withdrawal of residues for energy generation and implementation can adversely affect other agricultural practices. To reduce such effects, different strategies can be adopted by the farmers such as the use of no-till farming and covering of crops to minimize water pollution and soil erosion. This will improve agricultural production capacity and also increase the availability of residues for energy generation.

Crop residues can be utilized in four main ways, as shown in Figure 3.1. Some developing countries still practice straw burning while an amount of crop residues is stored at landfills.

Agricultural crop residues can be transformed into energy and biofuels can be obtained through processes like thermochemical, methanation and fermentation (Tursi, 2019). The most common process is thermochemical. Crop residues contain high mineral content; hence the combustion process is challenging. To enhance the efficacy of crop residues, many techniques are utilized such as pyrolysis, gasification and liquefaction and combustion.

3.2.2 Waste from Livestock

Large-scale production of livestock generates a huge amount of manure, which can be converted into energy production. Besides, smaller-scale production of livestock enables farmers to produce biogas from manure with the help of an anaerobic digester, which produces environmental and economic returns. Biogas can be a source of power and heat, which can further be treated and commercialized as an alternative natural gas. The application of anaerobic digesters for the generation of biogas from livestock manure is highly advantageous and helps to control environmental pollution. It improves water quality, reduces unpleasant greenhouse gases produced from the manure and helps the farmers in nutrients' fixation to the soil. Mostly, the livestock manure is utilized very closely to the place it is being produced and could also be combined with the production of the crop.

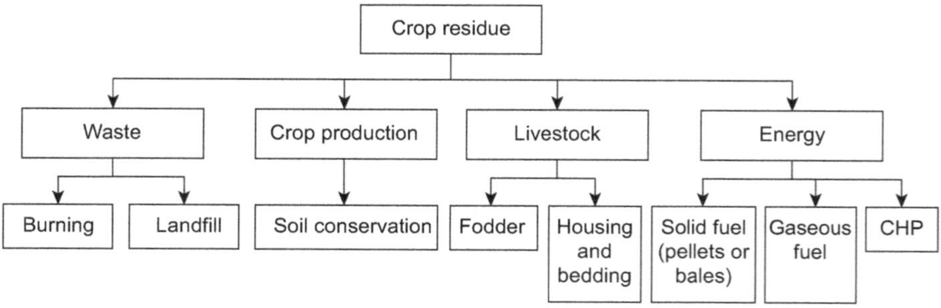

FIGURE 3.1 Crop residue utilization.

Furthermore, the use of a particular residue for energy generation depends on factors such as technical restrictions, scattered abundance, ecosystem functionality, and other needs such as fertilizer, animal feeds, domestic cooking and heating relative to the application of the resource.

3.2.3 Physical and Chemical Properties

The properties of agricultural residues' both physical and chemical features are necessary to evaluate their potential as feedstock for transformation into energy. Agricultural residues majorly include proximate, lignocellulose composition, ultimate, and biochemical analysis. The compositions such as fixed carbon content, moisture, volatile solids and ash content are referred to as proximate analyses. For the thermochemical process, the raw material is required to have low moisture and ash content. On the contrary, high levels of volatile solids and fixed carbon represent increased energy level and organic matter of the main material, which becomes highly appropriate for the biochemical conversion process. The biofuel potency of the raw material is obtained as the ultimate analysis (Singh et al., 2017). Hence, the ratio of C/N obtained from the ultimate analysis explains the applicability of the material for biogas as well as biohydrogen production. Table 3.1 shows the compositions of some agricultural residues. Aside from the elementary analysis, analysis of the composition is required for the design of the biofuel production process. Composition analysis majorly comprises the lignocellulosic content which includes lignin, hemicellulose and cellulose, while biochemical analysis of the raw material includes lipids, proteins and carbohydrates, etc.

From Table 3.2, it is obvious that most agricultural residues contain about 86% lignocellulosic content comprising lignin, hemicellulose and cellulose. A high amount of hemicellulose and cellulose content results in the generation of bio-alcohols. The crop residue comprises 30–50% cellulose, 20–38% hemicellulose and lignin composition range from 5–24% as a factor of various sources of origin. The livestock wastes such as cattle manure contain 50–60% carbohydrate, which is an efficient main stock for the production of biogas.

3.3 BIOENERGY CONVERSIONS

Biomass can be transformed into a range of products for use in the production of power and chemicals. Several factors determine the choice of biomass conversion. Quality and amount of biomass raw material, accessibility, end-product choices, economical factor, and environmental challenges are some of the factors that have been considered in the past. The following are some of the bioenergy conversion methods discussed:

3.3.1 Thermochemical Method

Processes such as gasification, combustion and pyrolysis are considered thermochemical biomass conversions which are illustrated in Figure 3.2. The combustion process for heat and electricity generation is the most widely used for the conversion

TABLE 3.1
Ultimate and Proximate Analysis of Agricultural Residues

	Ultimate Analysis (wt.)					Proximate Analysis (wt.)					
Raw material	C	H	O	N	S	Moisture	Fixed Carbon	Ash	Total Volatile Solid (VS)	Heating values (MJ/Kg)	Reference
Crop Residue											
Rice straw	33.8–42.0	5.0–7.0	33.0–41.0	0.2–1.0	0.1–0.2	4.0–6.0	15.1	8.1–15.8	71.5–92.9	14.7–15.7	(Kumar et al., 2018) (Bhuiyan et al., 2018)
Wheat straw	42.0–46.7	5.1–6.3	34.1–51.4	0.4–0.5	0.1–0.3	4.4–8.4	17.3	7.3–12.8	74.4–92.7	17–18.9	(Bridgeman et al., 2008) (Kumar et al., 2018)
Corn stover	35.2–45.6	5.4–6.3	43.4–45.7	0.3–0.8	0.1–0.3	5.3–7.4	16.9	4.2–6.3	86.5–96.8	16.2–16.5	(Kumar et al., 2018) (Vassilev et al., 2010)
Barley straw	49.4	6.2	44.0	0.7	0.13	-	19.79	5.3–9.8	76.2	16.42	(Serrano et al., 2011) (Bhuiyan et al., 2018)
Oat straw	49.0	6.6	45.0	0.5	0.1	5.38	19.53	6.0	74.48	-	(Mani et al., 2011) (Vassilev et al., 2010) (Bhuiyan et al., 2018)
Livestock Waste											
Cattle manure	29.0	323	2.46	0.82	19.30	38.6	-	45.1	50–72	-	(Rokni et al., 2018)

TABLE 3.2
Lignocellulosic and Biochemical Analysis of Agricultural Wastes

	Lignocellulosic Composition (wt.)			Biochemical Composition (wt.)			
Feedstocks	**Cellulose**	**Hemicellulose**	**Lignin**	**Carbohydrates**	**Protein**	**Lipids**	**References**
Crop Residue							
Rice straw	30.0–52.2	19.9–31.8	7.1–12.9	-	-	5.9	(Vassilev et al., 2012) (Banik and Nandi, 2004)
Wheat straw	32.9–44.5	37.8–33.2	8.5–22.3	-	3.48	5.34	(Vassilev et al., 2012) (Negri et al., 2020)
Corn stover	31.3–49.4	21.1–26.2	3.1–8.8	7.9	3.6–8.7	0.7–1.3	(Li Xiuji et al., 2015)(Vassilev et al., 2012)
Barley straw	29.0–48.6	35.8–29.7	6.7–21.7	-	3.62	1.91	(Aqsha et al., 2017)(Vassilev et al., 2012)
Oat straw	37.6–44.8	23.3–33.4	12.9–21.8	-	5.34	1.65	(Mani et al., 2011)(Aqsha et al., 2017)
Livestock Waste							
Cattle manure	33.0	24.5	43.0	62.46	15.09	7.0	(Vassilev et al., 2012)
Pig manure	-	-	-	52.08	23.9	14.3	(Pattanaik et al., 2019)

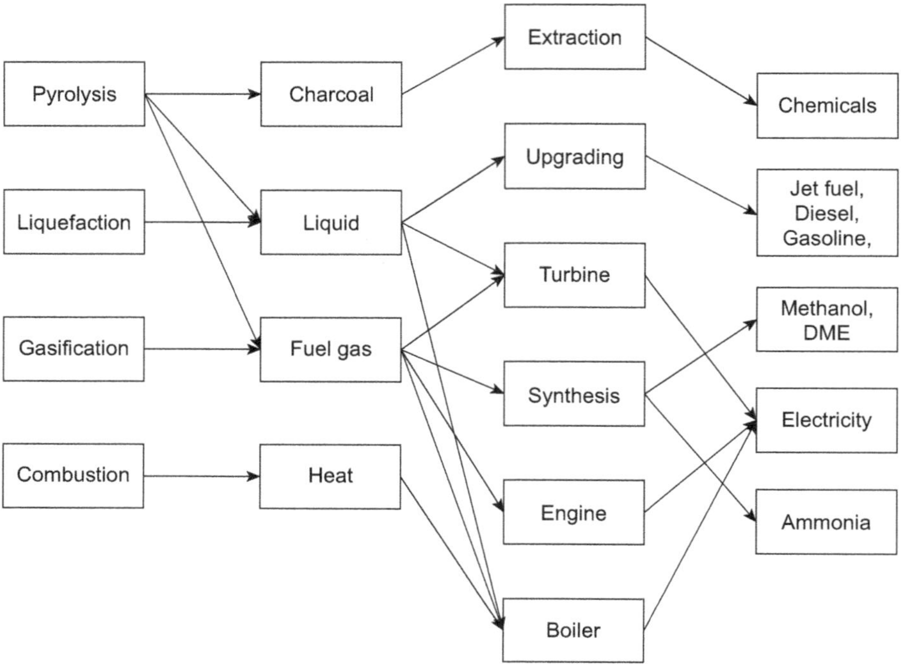

FIGURE 3.2 Thermochemical conversion processes and end products.

of biomasses in industries. Most biomass thermochemical conversions are done with or without catalysts, though catalyst usage has different effects on the end-products (Kore et al., 2013).

3.3.1.1 Gasification

France and England had separately discovered the mechanism of biomass gasification as far back as the 1700s. It took another 60 years for this technique to gain widespread acceptance. Until natural gas was discovered 30 years later from oil fields, the gasification process thrived. Also, since 1970, liquid fuels supplanted natural gas for cooking and illumination because of oil discovery. The gasification of solid biomass fuels using air, CO_2 or steam, which are referred to as gasifying agents, produces a combination of gases such as CO_2, CH_4, H_2 and CO (Sikarwar et al., 2016). A typical biomass conversion process is depicted in Figure 3.3. At a ranging temperature of 800 to 1300°C, the operation is then carried out. Due to the versatility of the gasification technique and the numerous uses of the generated syngas, biomass gasification may now be combined with several industrial processes and power generation systems (Sansaniwal et al., 2017). However, knowledge with regards to the factors influencing the feedstock selection and potential is required for the gasification process evaluation. As a result, low-volatility feedstocks are

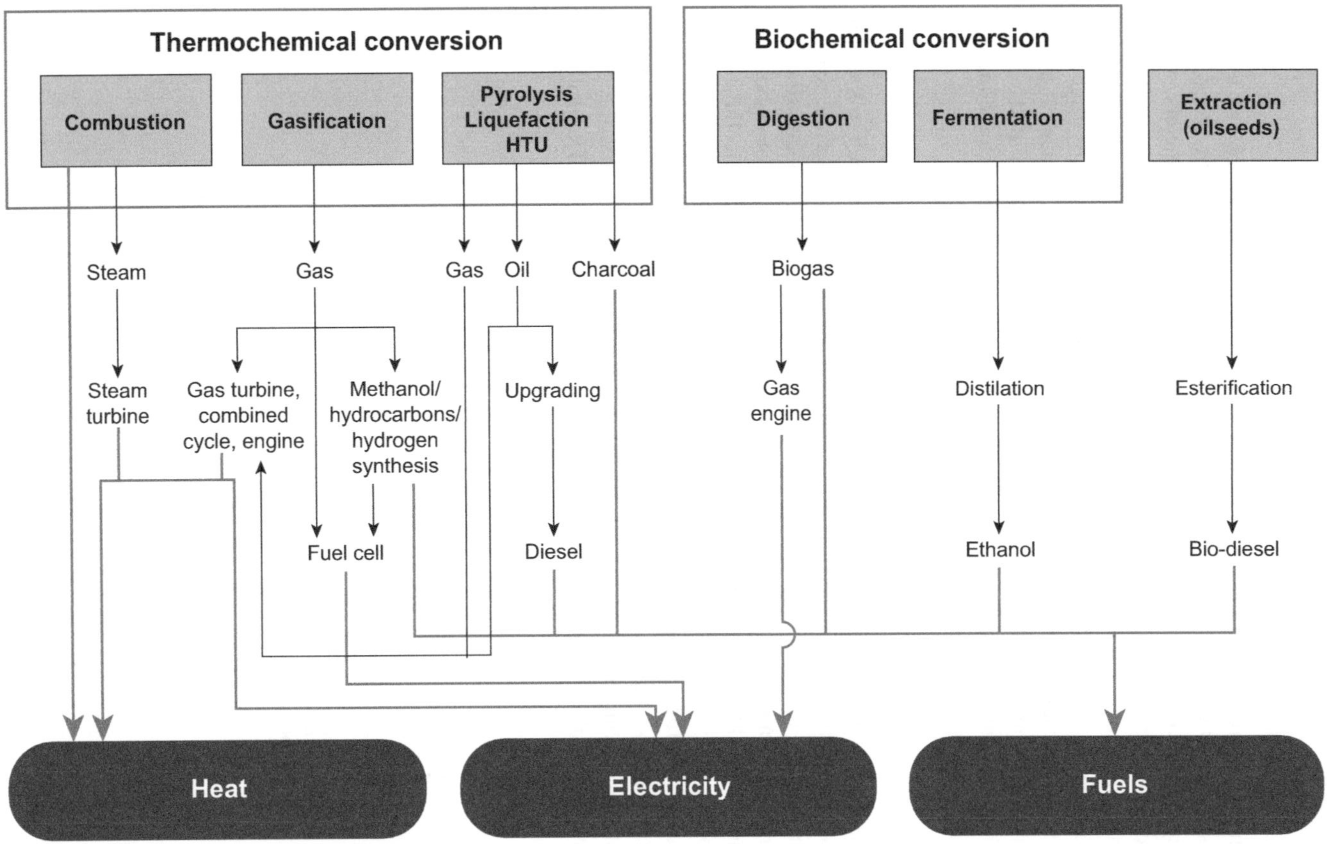

FIGURE 3.3 Some biomass conversion pathways.

preferable for partially oxidized gasification, while feedstocks with high volatility are better for indirect gasification.

3.3.1.1.1 Hydrothermal Gasification

The term "hydrothermal gasification" uses water with high pressures and temperatures to break down biomass (Kruse, 2008). Many reactions taking place in the biomass produce products which are influenced primarily by parameters such as temperature, treatment time and strain. Figure 3.4 depicts the water phase diagram, in which ice turns to liquid water at atmospheric pressure of 0.101325 MPa and a temperature of 273.15 K. The boiling point of water is consequently affected by heat due to increased pressure as the boiling point increases and reduces at a lower pressure. As compared to vapour or steam volume, the volume of liquid water is unaffected by increased pressure. As a result, as pressure rises, the volume increase associated with the phase shift shrinks.

3.3.1.2 Pyrolysis

This is the thermal depolymerization of organic matter without oxygen or the presence of nitrogen. When heated in an inert atmosphere, it is an exothermic reaction that makes use of heat ranging from 209 to 435 kJ/kg, and it has been utilized to obtain vapours and residues with high content of carbon from a range of wood-based and agricultural biomass (Demirbas and Arin, 2002). Cellulose, hemicellulose, and lignin polymer make up the vapours. Bio-oil, a free-flowing organic liquid, can be made from these vapours. Biochar is made from carbon leftovers, on the other hand. The lignocellulosic component of the biomass feed determines the distribution of polymeric compounds in bio-oil. Many scientists have investigated the pyrolysis properties of cellulose, hemicellulose, and lignin individually. Hemicellulose

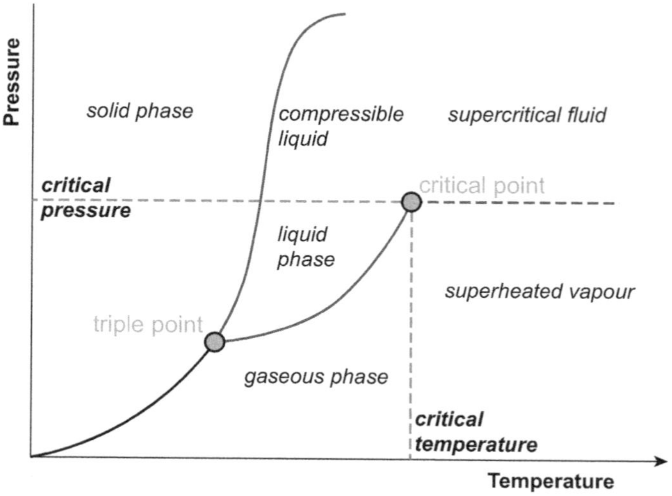

FIGURE 3.4 Phase diagram of water.

decomposes between 220°C and 315°C, cellulose between 314°C and 400°C, and lignin between 160°C and 900°C, resulting in a solid residue with a maximum proportion of roughly 40% (Karimi and Taherzadeh, 2016). From an energy standpoint, cellulose pyrolysis is an endothermic reaction, while hemicellulose and lignin reactions are exothermic.

3.3.1.2.1 Fast Pyrolysis

Because of the increased demand for biomass-based fuel oil, this sort of pyrolysis, a cutting-edge technique, has recently gained traction. Quick pyrolysis is a continuous procedure aimed at preventing non-condensable chemicals from forming from the pyrolytic fractions. In the process, the factors that result in high oil production were closely examined, with high heat transfer rates being the most important. To reach this characteristic, the biomass feed could be finely powdered. The finely powdered biomass feed is suddenly heated to temperatures of 450°C to 600°C for less than 2 seconds (Karimi and Taherzadeh, 2016; Demirbas and Arin, 2002).

3.3.1.2.2 Intermediate Pyrolysis

Intermediate pyrolysis takes place at temperatures between 300 and 500°C, in contrast to rapid pyrolysis. The liquid products derived throughout the operation have reduced viscosity and contain less tar. The reactions that take place during intermediate pyrolysis, on the other hand, are better controlled, which allows for a wider range of operating parameter adjustments of the process optimization. Big sizes of biomass feed that are often coarse, sliced, shredded, or crushed are suitable, despite the process yielding a low product of up to 55% for liquids (Yang et al., 2013).

3.3.1.2.3 Slow Pyrolysis

The heating of a biomass origin in the absence of air with no condensation of the pyrolyzed products is referred to as slow pyrolysis. Though the majority of the current literature on the process focuses on its application to make solid fuels like charcoal and biochar, it can also be used to make liquid fuels and biogas (Yang et al., 2013). Lower temperatures between 0.1°C to 2°C have been documented in previous research (Karimi and Taherzadeh, 2016). Slow pyrolysis is one of the oldest technologies for biomass transformation which produces charcoal or biochar as an end product. The steam or vapours obtained throughout the operation were not generally condensed; however, they may be employed to heat the process either directly or indirectly. The moisture content levels of 15% to 20% have been seen, which influences the characteristics of the solid fuels obtained.

3.3.2 Biochemical Methods

Biochemical techniques of biomass conversion aid in the transformation of biomass using biological pre-treatments. Using a range of microbes and enzymes, these biological pre-treatments were developed to convert biomass into a variety of end products and intermediates. The process produces biogas, hydrogen, and a variety of acids, among other fuels and chemicals (Armah et al., 2019). This strategy, on the other hand, attempted to generate items that could be used to replace both petroleum-based and grain-based products. Biomass biochemical conversion is uncontaminated, pure, and effective in contrast with other biochemical conversion technological approaches.

3.3.2.1 Anaerobic Digestion

Anaerobic digestion (AD) is an environmentally friendly and economical technology for processing lignocellulosic and other wastes to recover energy in the form of biofuels. This method not only decreases garbage production but also turns it into biofuel (Armah et al., 2019). Energy-dense methane is released when lignocellulosic biomass is digested (CH_4). Within the same variety, CH_4 yield per unit area is typically used to determine an individual feedstock's energy output, which varies greatly among species and maturity, position, as well as inputs (such as fertilizer, water, and so on) (Søndergaard et al., 2015). AD is a process of producing biogas from biomass that has been biologically treated. It's done in two stages at temperatures ranging from 30 to 35°C or 50 to 55°C. Acid-forming bacteria break down complex organics in biomass into smaller molecules, namely acetic and propionic acids, as well as volatiles, in the first process. The second stage entails converting such acids into carbon dioxide CO_2 and methane CH_4, often known as biofuel, utilizing methane-producing microorganisms. In most cases, both steps of biogas processing are completed in a single tank. After the total AD process is completed, approximately 60% of the biogas generated is CH_4, 35% is CO_2, and the remaining 5% is a combination of other gases which are H_2, NH_3, CO and H_2S. (Tsapekos et al., 2015; Karimi and Taherzadeh, 2016). Another metabolic mechanism that has been found to convert basic sugars to smaller molecular weight compounds like alcohols and acids is referred to as fermentation. The production of enzymes generated from microbes is usually helpful. During fermentation, biochemical processes could be used to convert biomass to alcohols.

3.4 BIOENERGY & BIOFUEL POTENTIALS/APPLICATIONS

To meet the United Nations Development Programme's Millennium Development Goals, over two billion people will need modern energy services (United Nations Department of Economic and Social Affairs, 2010). The absence of affordable and reliable power and liquid transportation fuels has a severe influence on developing countries, especially countries with high population density with limited access to energy. Solid fuels are used by over 2 billion people, mostly for cooking and heating the house. Emissions and health hazards have emerged from the increase in combusting biomass. Long-term health expenditures are incurred as a result of the burning of solid fuels in inadequately ventilated buildings, with many individuals bearing the brunt of the costs (Tsapekos et al., 2015). Bioenergy was pursued by developed countries to address the threat of environmental contamination caused by CO_2 emissions, as well as to potentially cut CO_2 emissions and provide domestic energy (Armah et al., 2020). Tsapekos et al. (2015) investigated many energy crops capable of producing lignocellulosic biomass. The exploration of specific energy crops in developing countries has the potential to dispense with food crops, leading to undesired conflict of food energy. Food security and provision of energy from these crops are enhanced when expended farmlands are explored for growing crops after deforestation, thus leading to CO_2 emissions due to unsustainable land use (Sansaniwal et al., 2017). As a result, the dual cropping phase offers numerous opportunities, such as increasing agricultural output while ensuring food production by producing bioenergy from agricultural waste. Despite the numerous benefits of utilizing agricultural residues as a waste product, Molino et al. (2016) feel that removal of specific kinds of agricultural waste from the farm could lead to severe environmental challenges. In many regions of the world, the use of modern biomass has skyrocketed in recent decades. Many governments have set aggressive biomass usage goals in light of Kyoto GHG reduction commitments, but rising oil prices have prompted renewed interest in bioenergy. Biomass provides for less than ten percent of overall energy sources in wealthy countries, whereas it accounts for 20 to 30% in poor ones. In some countries, biomass meets 50 to 90% of total energy needs. Biomass combustion accounts for over 90% of existing biomass-derived secondary energy carriers. Domestic combustion (heating, cooking), trash incineration, industrial utilization of process wastes, and state-of-the-art furnace and boiler designs for efficient power generation all play a role in certain situations and markets. Despite starting from a far smaller basis, worldwide ethanol output has more than doubled since 2000, while global biodiesel production has nearly tripled.

3.5 ENVIRONMENTAL IMPACT ASSESSMENT OF BIOMASS UTILIZATION FOR ENERGY PRODUCTION

The application of biomass for energy production has become fundamental to the improvement of civilization. Biomass is stated as one prospective source of alternative and/or renewable energy and the transformation of plant-based material into a suitable form of energy (Sheltawy et al., 2016). A quick review of the biomass major conversion processes to energy with particular regard to the production routes is presented in this chapter. However, improving the utilization of biomass for energy sources or electric power production is a factor associated with the accessibility to supply and the level of fuel cycle sustainability and management (McKendry, 2002). The environmental impact associated with all the elements present in each biomass must be assessed to screen the proposed technology before use. The influence on the environment is associated with the biomass production cycle and conversion technologies.

Apart from the issues related to biomass production, the influence of the combustion systems is well defined with stringent environmental regulations already in place to determine the effectiveness and eco-friendliness of the existing electricity production system (Grammelis et al., 2006). However, the environmental impact associated with the formation of biomass aspect remains unresolved. As a result, few questions on biomass have been raised in terms of environmental risk regarding biomass reuse (Beagle and Belmont, 2019). It has been stated that biomass used is mostly already harvested material or consumed over time; thus, it is composed of the recycled residues (Bajwa et al., 2018). This practice often affects the fuel composition of biomass resources on a weight basis, and also reduces the standard of energy, mostly concerning smaller components, including ash and trace metals. Major deviations in the waste/residue pattern are causing high expenditure in particulate and gaseous emissions control (Leal Filho et al., 2016).

In addition, ashes obtained from pure wood residue are sent to the forestlands in Scandinavia as a soil amendment, for example; however, urban wood wastes are being disposed of in landfills as a result of the heavy metals contained (Iakovou et al., 2010). A huge improvement in the quantity of bioenergy utilized is predetermined for two decades from now, which leads to increased need for an allocated fuel supply chain (DFSS)(Magelli et al., 2009). The proposed crops grown in DFSS are presently in the developmental stage and can either be classified as short-rotation woody crops (SRWC), such as poplar or eucalyptus, or herbaceous energy crops (HEC), such as switchgrass. Neither SRWC nor HEC needs exhaustive water or fertilizer usage, and the production techniques are developed to reduce operations on the land to minimize the threat of erosion. This requires plantings on a huge proportion of the land presently dedicated to conserve land and regulate uneconomical foodstuff production (Beagle and Belmont, 2019).

These environmental impacts could be averted, as reported by many researchers (Karka et al., 2019). Karka and his co-workers reported that life cycle assessment (LCA) is one of the suitable methodologies for the determination and assessment of environmental impacts of processes because it makes use of a large amount of often secluded process data at preliminary design stages. This proposed method reported a break away from conventional LCA work, which supports the decision at the preliminary stages by presuming the least use of data accessible to the most dominant LCA impacts (Karka et al., 2019), which provides useful feedback to process design before the commencement of waste conversion process.

3.6 PROSPECTS ON WASTE CONVERSION TO ENERGY

Many techniques and approaches could be applied to enhance and improve waste management and transformation choices at various levels in developing societies. However, preliminary and emergency steps for the choice of techniques should be made available in the future, before waste conversion. The strategic environmental assessment, environmental impact assessment, life cycle assessment, cost analysis, life cycle costing, risk analysis, material accounting, substance control analysis, energy analysis and environmental management systems should be critically analyzed and put in place (Mubeen and Buekens, 2019).

The characteristics used for these analyses are the kinds of impacts, which include the focus under the study/method to be used whether it is procedural or analytical (Sheltawy et al., 2016). The different approaches could be investigated as systems analysis techniques. For example, the European Commission suggested a thematic strategy on waste where life cycle assessment and life cycle thinking take notable positions. Hence, evidence has shown that life cycle analyses provide policy-relevant and consistent results (Han et al., 2019). Nevertheless, studies will oftentimes give rise to reproval since they are simplified reality and include vagueness. Therefore, improved refractories and systems for various *Waste* to *Energy* applications, especially life cycle analysis implementation, are proposed for future application. Assumptions could be probed and become hard to widely apply from case studies to policies, but if choices are made, the less might gain advantages above the perfect basis.

3.7 CONCLUSION

Energy can be generated from biomass and used as a solid, liquid, or gaseous fuel for many purposes, including heating, electricity, cooking, and combined heat and power. These resources are prevalent globally, which could be exploited for energy production as well as environmental protection. Currently, there is a lot of interest in using lignocellulosic biomass as a raw material for bioenergy generation to develop more sustainable energy production methods. As seen in this chapter, the majority of biomass conversion technologies are aimed toward the development of sophisticated ways of creating energy fuels to alleviate the worldwide energy shortfall. Also, biomass is a renewable energy source in recent times. This study agreed that thermochemical and biochemical processes for turning biomass into energy were developed many years ago; however, progress crept after fossil fuels were discovered.

3.8 REFERENCES

Aqsha, A., Tijani, M. M., Moghtaderi, B. & Mahinpey, N. 2017. Catalytic pyrolysis of straw biomasses (wheat, flax, oat and barley) and the comparison of their product yields. *Journal of Analytical and Applied Pyrolysis*, 125, 201–208.

Armah, E. K., Chetty, M. & Deenadayalu, N. 2019. Biomethane potential of agricultural biomass with industrial wastewater for biogas production. *Chemical Engineering Transactions*, 76, 1411–1416.

Armah, E. K., Chetty, M. & Deenadayalu, N. 2020. Biogas production from sugarcane bagasse with South African industrial wastewater and novel kinetic study using response surface methodology. *Scientific African*, 10, 1–9.

Bajwa, D. S., Peterson, T., Sharma, N., Shojaeiarani, J. & Bajwa, S. G. 2018. A review of densified solid biomass for energy production. *Renewable and Sustainable Energy Reviews*, 96, 296–305.

Banik, S. & Nandi, R. 2004. Effect of supplementation of rice straw with biogas residual slurry manure on the yield, protein and mineral contents of oyster mushroom. *Industrial Crops and Products*, 20, 311–319.

Beagle, E. & Belmont, E. 2019. Comparative life cycle assessment of biomass utilization for electricity generation in the European Union and the United States. *Energy Policy*, 128, 267–275.

Bharathiraja, B., Sudharsana, T., Jayamuthunagai, J., Praveenkumar, R., Chozhavendhan, S. & Iyyappan, J. 2018. Retracted: Biogas production—A review on composition, fuel properties, feed stock and principles of anaerobic digestion. *Renewable and Sustainable Energy Reviews*, 90, 570–582.

Bhuiyan, A. A., Blicblau, A. S., Islam, A. K. M. S. & Naser, J. 2018. A review on thermochemical characteristics of coal/biomass co-firing in industrial furnace. *Journal of the Energy Institute*, 91, 1–18.

Bridgeman, T. G., Jones, M. J. & I. Shield, P. T. W. 2008. Torrefaction of reed canary grass, wheat straw and willow to enhance solid qualities and combstion properties. *Science Direct*, 844–856.

Demirbas, A. & Arin, G. 2002. An overview of biomass pyrolysis. *Energy Sources*, 24, 471–482.

Gabrielle, B., Bamière, L., Caldes, N., De Cara, S., Decocq, G., Ferchaud, F., Loyce, C., Pelzer, E., Perez, Y., Wohlfahrt, J. & Richard, G. 2014. Paving the way for sustainable bioenergy in Europe: Technological options and research avenues for large-scale biomass feedstock supply. *Renewable and Sustainable Energy Reviews*, 33, 11–25.

Grammelis, P., Skodras, G. & Kakaras, E. 2006. An economic and environmental assessment of biomass utilization in lignite-fired power plants of Greece. *International Journal of Energy Research*, 30, 763–775.

Han, D., Yang, X., Li, R. & Wu, Y. 2019. Environmental impact comparison of typical and resource-efficient biomass fast pyrolysis systems based on LCA and Aspen Plus simulation. *Journal of Cleaner Production*, 231, 254–267.

Iakovou, E., Karagiannidis, A., Vlachos, D., Toka, A. & Malamakis, A. 2010. Waste biomass-to-energy supply chain management: A critical synthesis. *Waste Management*, 30, 1860–1870.

Jiang, Y., Havrysh, V., Klymchuk, O., Nitsenko, V., Balezentis, T. & Streimikiene, D. 2019. Utilization of crop residue for power generation: The case of Ukraine. *Sustainability*, 11.

Karimi, K. & Taherzadeh, M. J. 2016. A critical review of analytical methods in pretreatment of lignocelluloses: Composition, imaging, and crystallinity. *Bioresource Technology*, 200, 1008–1018.

Karka, P., Papadokonstantakis, S. & Kokossis, A. 2019. Environmental impact assessment of biomass process chains at early design stages using decision trees. *The International Journal of Life Cycle Assessment*, 24, 1675–1700.

Kataki, R., Chutia, R. S., Mishra, M., Bordoloi, N., Saikia, R. & Bhaskar, T. 2015. Feedstock suitability for thermochemical processes. In: *Recent Advances in Thermo-Chemical Conversion of Biomass*. Elsevier.

Kore, S., Assefa, A., Matthias, M. & Spliethoff, H. 2013. Steam gasification of coffee husk in bubbling fluidized bed gasifier. Proceedings of the Fourth Internation Conference on Bioenvironment, Biodiversity and Renewable energies; ISBN. Citeseer, 2013. Citeseer, 978–981.

Kruse, A. 2008. Supercritical water gasification. *Biofuels, Bioproducts and Biorefining: Innovation for a Sustainable Economy*, 2, 415–437.

Kumar, S., Paritosh, K., Pareek, N., Chawade, A. & Vivekanand, V. 2018. De-construction of major Indian cereal crop residues through chemical pretreatment for improved biogas production: An overview. *Renewable and Sustainable Energy Reviews*, 90, 160–170.

Leal Filho, W., Brandli, L., Moora, H., Kruopienė, J. & Stenmarck, Å. 2016. Benchmarking approaches and methods in the field of urban waste management. *Journal of Cleaner Production*, 112, 4377–4386.

Li Xiuji, Dang Feng, Zhang Yatian, Zou Dexun & Hairong, Y. 2015. Anaerobic digestion performance and mechanism of ammoniationn pretreatment of corn stover. *Bioresources*, 10, 5777–5790.

Magelli, F., Boucher, K., Bi, H. T., Melin, S. & Bonoli, A. 2009. An environmental impact assessment of exported wood pellets from Canada to Europe. *Biomass and Bioenergy*, 33, 434–441.

Mani, T., Murugan, P. & Mahinpey, N. 2011. Pyrolysis of oat straw and the comparison of the product yield to wheat and flax straw pyrolysis. *Energy & Fuels*, 25, 2803–2807.

Mckendry, P. 2002. Energy production from biomass (part 2): Conversion technologies. *Bioresource Technology*, 83, 47–54.

Molino, A., Chianese, S. & Musmarra, D. 2016. Biomass gasification technology: The state of the art overview. *Journal of Energy Chemistry*, 25, 10–25.

Mubeen, I. & Buekens, A. 2019. Chapter 14 — Energy from Waste: Future prospects toward sustainable development. In: Kumar, S., Kumar, R. & Pandey, A. (eds.) *Current Developments in Biotechnology and Bioengineering*. Elsevier.

Negri, C., Ricci, M., Zilio, M., D'imporzano, G., Qiao, W., Dong, R. & Adani, F. 2020. Anaerobic digestion of food waste for bio-energy production in China and Southeast Asia: A review. *Renewable and Sustainable Energy Reviews*, 133.

Pattanaik, L., Pattnaik, F., Saxena, D. K. & Naik, S. N. 2019. Biofuels from agricultural wastes. *Second and Third Generation of Feedstocks*, 103–142.

Rokni, E., Ren, X., Panahi, A. & Levendis, Y. A. 2018. Emissions of SO2, NOx, CO2, and HCl from Co-firing of coals with raw and torrefied biomass fuels. *Fuel*, 211, 363–374.

Sadh, P. K., Duhan, S. & Duhan, J. S. 2018. Agro-industrial wastes and their utilization using solid state fermentation: A review. *Bioresources and Bioprocessing*, 5.

Sansaniwal, S., Pal, K., Rosen, M. & Tyagi, S. 2017. Recent advances in the development of biomass gasification technology: A comprehensive review. *Renewable and Sustainable Energy Reviews,* 72, 363–384.

Serrano, C., Monedero, E., Lapuerta, M. & Portero, H. 2011. Effect of moisture content, particle size and pine addition on quality parameters of barley straw pellets. *Fuel Processing Technology*, 92, 699–706.

Sheltawy, S., Al-Sakkari, E. & Fouad, M. 2016. Waste to energy trends and prospects: A review. In: Ghosh, S. (eds.) *Waste Management and Resource Efficiency*. Springer. https://doi.org/10.1007/978-981-10-7290-1_56

Singh, Y. D., Mahanta, P. & Bora, U. 2017. Comprehensive characterization of lignocellulosic biomass through proximate, ultimate and compositional analysis for bioenergy production. *Renewable Energy*, 103, 490–500.

Søndergaard, M. M., Fotidis, I. A., Kovalovszki, A. & Angelidaki, I. 2015. Anaerobic co-digestion of agricultural by products with manure for enhanced biogas production. *Energy & Fuels*, 29, 8088–8094.

Tsapekos, P., Kougias, P. G. & Angelidaki, I. 2015. Anaerobic mono- and co-digestion of mechanically pretreated meadow grass for biogas production. *Energy & Fuels*, 29, 4005–4010.

Tursi, A. 2019. A review on biomass: Importance, chemistry, classification, and conversion. *Biofuel Research Journal*, 6, 962–979.

United Nations Department of Economic and Social Affairs. 2010. *100% Renewable Electricity Supply by 2050* [Online]. Available: https://sustainabledevelopment.un.org/index.php?page=view&type=99&nr=24&menu=1449 [Accessed 2/11/2017].

Vassilev, S. V., Baxter, D., Andersen, L. K. & Vassileva, C. G. 2010. An overview of the chemical composition of biomass. *Fuel*, 89, 913–933.

Vassilev, S. V., Baxter, D., Andersen, L. K., Vassileva, C. G. & Morgan, T. J. 2012. An overview of the organic and inorganic phase composition of biomass. *Fuel*, 94, 1–33.

Yang, Y., Brammer, J., Ouadi, M., Samanya, J., Hornung, A., Xu, H. & Li, Y. 2013. Characterisation of waste derived intermediate pyrolysis oils for use as diesel engine fuels. *Fuel*, 103, 247–257.

Yue Li, Y. C. & Wu, J. 2019. Enhacement of methane production in anaerobic digestion process: A review. *Applied Energy*, 240, 120–137.

Zabed, H., Sahu, J. N., Boyce, A. N. & Faruq, G. 2016. Fuel ethanol production from lignocellulosic biomass: An overview on feedstocks and technological approaches. *Renewable and Sustainable Energy Reviews*, 66, 751–774.

4 Conversion of Agricultural and Forest Wastes into Water Purifying Materials

Donald Tyoker Kukwa, Maggie Chetty,
Ifeanyi Anekwe and Jeremiah Adedeji

CONTENTS

DOI: 10.1201/9781003245773-4

4.1 INTRODUCTION

The world population over one decade, between 2010 and 2020, saw a rise from 6.957 billion to 7.795 billion, representing a whopping 12.04% increase (Gu *et al.* 2021). This represents a yearly average population increase of 1.13 ± 0.06% and imposes stress on the utility services during the same period. Table 4.1 shows the world population growth between 2010 and 2020 inclusive. The human population on the six blocks of the world in 2020 is shown in Table 4.2, with Asia taking the lead. The potable water sector is most hit by this population explosion, as the technology of producing treated water and distributing the same through reticulation is getting more expensive by the day.

As a result, water and wastewater treatment materials have been in short supply, prompting a variety of studies on the materials' nature, efficacy, availability, and long-term viability. Agricultural and forest residues and wastes are renewable,

TABLE 4.1
World Population Growth Between 2010 and 2020 with Water Demand (United Nations Department of Economic and Social Affairs 2020)

Year	World Population (Billion)	Yearly change (%)	Population density (P/km²)	Urban population (%)
2010	6.957	1.24	47	51.7
2015	7.379	1.19	50	54.0
2016	7.464	1.14	50	54.4
2017	7.548	1.12	51	54.9
2018	7.631	1.10	51	55.3
2019	7.713	1.08	52	55.7
2020	7.795	1.05	52	56.2

TABLE 4.2
Population Density, 2020 (United Nations Department of Economic and Social Affairs 2020)

Region	Population (Billion)	Yearly change (%)	Land Area (%)	Population density (/km²)	World abundance (%)
Asia	4.641	0.86	31033131	150	59.5
Africa	1.341	2.49	29648481	45	17.2
Europe	0.748	0.06	22145786	34	9.6
Latin America/ Caribbean	0.654	0.9	20139452	32	8.4
North America	0.369	0.62	18651660	20	4.7
Oceania	0.043	1.31	8486460	5	0.5

sustainable materials that have been used in commercial and industrial applications, such as water and wastewater treatment.

Crop residues and wastes generated during farm management and harvesting include leaves, seeds, stems, straw stalks, stubble, and others, whereas crop processing wastes include peels, roots, bagasse, husks, peanut shells, coconut shells, palm kernel shells, and others. Plantations of perennial crops such as rubber, coconut, palm tree, cashew, and banana generate trash through tree cutting and replanting.

Both poultry and livestock wastes are considered animal wastes. These include waste from both animal management and slaughterhouses.

Animal wastes include droppings, dung, eggshells, sawdust, and other materials, whereas slaughterhouse wastes include feathers, blood, rejected tissues, bone, skin, fur, and other materials.

Logging and wood processing waste, such as sawmilling and plywood and particleboard production, contribute to forestry waste.

It's worth noting that 99.9% of these residues and wastes are carbon-based compounds that can be processed to meet the end-use requirements.

Groundwater serves as a source of drinking water for at least half of the world's population and accounts for 43% of all irrigation water (Siebert *et al.* 2010). Globally, 2.5 billion people rely exclusively on groundwater to meet their daily basic water demands (UNESCO 2015). This implies that half the world's population relies exclusively on surface water, which requires more rigorous treatment.

It is projected that global water demand will reach 4,350 billion cubic meters in terms of withdrawal by 2040. A fundamental consideration for the sizing of any water system, or its parts, is an estimate of the amount of water expected to be used by the consumers of the system. This chapter is intended to provide the basic water treatment materials sourced from agriculture and forestry resources.

4.2 SOURCES OF AGRICULTURAL WASTE RESOURCES

Agricultural waste can be attributed to the production of residues from agricultural and forestry production and processing activities. Agricultural waste products were once thought to be waste that needed to be disposed of, but it is now widely recognized that they are valuable natural resources rather than waste. Agricultural residues are a significant component of lignocellulose penetrating the soil and can be used to improve soil quality and performance. These wastes (plant residue) degradation can have both benefits and drawbacks on crop yield, and the objective is to maximize the positive impact of plant residue degradation while minimizing the negative effect. Management of crop residue helps improve various physical, chemical, and biological soil processes. It protects the soil against wind and rain erosion, maintains soil moisture, and increases infiltration and aeration in the soil structure. Proper management of crop residues contributes to the addition of soil organic matter and provides microbes with food. Soil, water, sunlight, and air are regarded as the main factors for constant food production and the sustainability of the ecosystem. Agricultural waste can be grouped into crop waste, poultry waste, livestock waste, etc.

4.2.1 Crop Waste

Crop waste is unwanted residue generated due to agricultural activities such as fertilizer application, harvesting, and crop processing. Some crop wastes can serve as animal feed and building materials in developing countries, as well as in industrial applications such as paper production (Powell and Unger 1997). Crop waste can be grouped into primary and secondary, depending on the means of production, and can be used as a soil amendment in small amounts.

4.2.1.1 Primary Crop Waste

Primary crop waste is residue that remains after a crop has been harvested in an agricultural field or after pastures have been grazed. These wastes consist of stalks and stubble (stems), leaves, corn stover, and seed pods, among other residues. The leftovers may either be ploughed into the field or burnt first. No-till, strip-till, or reduced-till cultivation practices, on the other hand, aim to increase crop residue cover while irrigation quality and erosion mitigation can also be improved from good field residue management. To determine residue distribution, simple line transect measurements can be employed.

4.2.1.2 Secondary Crop Waste

Secondary crop wastes are residues that remain after a crop has been transformed or processed into a useful resource. Husks, nuts, bagasse, molasses, straws, hulls, and roots are among the residues. They can also be used as animal feed, soil amendments, fertilizers, and industrial materials. These crop residues are sometimes regarded as waste products due to their economic value.

4.2.1.3 Overview of Some Crop Wastes

i. Rice Husks: Also known as rice hulls, these are the coating on the seed or grain of rice, formed from solid components (lignin and silica) which protects the grain during the growth period and are removed during the milling process. These are among the most popular agricultural wastes, with about one-fifth of the global net rice production. Given recent environmental contamination, attempts have been made to use burn husks under regulated temperatures as a substitute cementing and wastewater purification material (Rafiee *et al.* 2012). Net rice production in the US accounts for one-fifth of global production.
ii. Coconut shell (CNS) is an agricultural waste that can be found in large amounts in tropical countries around the world. The waste from this resource can be used in the treatment of industrial wastewater in an attempt to minimize global warming and protect the environment. This material has been investigated for the removal of various heavy metals, and it was discovered that CNS can be applied as a cost-effective adsorbent for organic and wastewater treatment.
iii. Corn, also known as maize, is a cereal grass that is commonly cultivated for food and livestock fodder. Corn is one of the world's most

important grain crops, alongside wheat and rice. The waste (corncob and corn stover—stalks, leaves, husks, and tassels) produced by these crops contributes to global warming since the entire waste is burned to ashes. Corncobs have been investigated as a biosorbent material for pollutant removal (Mohammed *et al.* 2014).

iv. Palm Oil Fuel Ash (POFA) is a waste produced when oil palm fibres, shells, and empty fruit bunches are burned in a furnace at 900–1000°C to generate energy for the crude palm oil extraction process. Approximately 5% of the material is collected as ash waste and disposed of in an open space, posing a significant health risk (Mohammed *et al.* 2014). Several studies have looked into the potentials of palm oil fuel ash for wastewater treatment as a response to these issues (Hamada *et al.* 2018).

v. Bagasse: Sugar is derived from a bamboo-like plant known as sugarcane, and the remains after sucrose are obtained is known as bagasse. It is the dry pulpy fibrous residue left over after processing sugarcane or sorghum stalks to obtain juice. It is a byproduct of the sugar industry that was previously dumped in industrial landfills or burned. It can be quickly broken down and foamed into any container form or size. Bagasse has a clear application as an absorbent in the production of biodegradable packaging, particularly in the food industry, where low-cost, lightweight packaging is critical. It has a range of advantages over traditional polystyrene or ethanol-based packaging materials, in addition to being entirely biodegradable.

4.2.2 Poultry Waste

The poultry market is among the world's biggest and most rapidly expanding agricultural sectors. The United States and other countries are the leading producers of poultry meat and eggs. On a global scale, poultry meat and egg production are increasing at a rate of about 5% per year. Large quantities of solid waste are produced by the poultry industry. Poultry litter, also known as broiler litter, is a combination of faeces, spilt feed, feathers, manure, sawdust, wood shavings, and other materials used as bedding in poultry activities. There are a variety of nutrients in this combination that can be used for other purposes (Thyagarajan *et al.* 2013). Poultry wastes pollute the atmosphere by emitting unpleasant odours and encouraging the breeding of flies and rodents. The proper use of this waste or by-products boosts monetary performance while minimizing adverse effects. Inappropriate management and improper handling of poultry slaughterhouse waste products can result in the occurrence of poultry farm diseases. This leads to direct mortality losses and lower productivity. As a result, early waste disposal using a coordinated system is an effective poultry waste management technique for ensuring safe and productive poultry operations. There are numerous waste management alternatives and applications for poultry residue, including land utilization of waste as an organic fertilizer, livestock feed, biofuel production, and industrial applications.

4.2.3 Livestock Waste

Livestock industries produce meat, milk, and eggs, as well as vast amounts of waste that, if not properly handled, can be detrimental to the environment. An increase in demand for animal products is driven by a growing population, industrialization, and the rise in per capita incomes. These rises in demand and consumption of animal products are expected to continue in the coming years, which means an increase in the production of livestock waste. Waste from the livestock industries is directly related to animal raising. To maximize the value of the ever-increasing demand for livestock products, livestock handlers have resorted to effective collection and processing of large quantities of waste which can either be in liquid (wastewater or blood) or solid form (faeces, etc.). Moreover, from the perspectives of cost, environmental protection, and biosafety, the disposal of these wastes remains a difficult task. The question remains on how these animal wastes can be handled without jeopardizing food production, natural resources, or human health.

4.2.3.1 Breeding and Maintenance Waste

In certain production systems, animals are reared mainly for human consumption or other purposes such as labour sources, raw materials, or fertilizer. During the production, processing, transportation, and marketing of animals, by-products are produced, which can become waste if not properly handled. Leftover meal, wastewater, hatchery wastes, and animal manure (animal faeces and urine) are some of the wastes produced from animal production and management activities. Secondary sources of animal manure include beddings, vomit, wash water, precipitation, spilt meal and water. Before the advent of organic fertilizers, animal manure was the primary source of soil fertility improvement. Despite the importance of organic fertilizers in agricultural development, manure continues to be a vital fertilizer source, particularly in places where organic fertilizers are easily obtainable.

4.2.3.2 Slaughterhouse Waste

The word "slaughterhouse waste" applies to all waste generated in an abattoir as a result of butchering, as well as waste generated during slaughterhouse operations. Slaughterhouse wastes include inedible animal parts, blood, and other animal by-products from the processing of meat. Animal tissues that are not edible (organs, integuments, ligaments, tendons, blood vessels, feathers, bone) can account for up to 45% or more of the slaughtered animal waste. The residual portion, on the other hand, isn't just suitable for disposal, as pet food industries have been known to buy vast quantities of abattoir waste. These may be utilized as a standalone product or as part of an animal feed supplement.

However, a significant amount of slaughterhouse waste is generated globally, and its disposal poses a major challenge for the meat industry. Due to the large volumes of animal waste, as well as legal constraints and increasing treatment costs, inappropriate and hazardous disposal of these wastes is inevitable. These activities can then result in severe environmental issues. Slaughterhouse waste, on the other hand, can be used as an energy source, potentially reducing the need for fossil fuels, which are presently used to meet the Earth's energy demand.

4.3 SOURCES OF FOREST WASTE RESOURCES

Forestry has waste resources emanating from logging and processing activities. These waste resources are, however, categorized into primary and secondary resources.

4.3.1 Primary Forest Wastes

Leaves, branches, and barks that fall off trees, residues such as limbs, tops, cone and culled trees from logging of trees, dead trees, diseased and poorly formed trees, as well as non-merchantable trees all form the primary forest waste resources (Sahin 2020). Generally, most forest wastes contain cellulose, hemicellulose, and lignin.

Cellulosics are abundant natural materials that develop into functional foods and novel biomaterials and are transporters of bioactive compounds because they have a wide range of physicochemical properties, as well as edibility and biocompatibility (Tavker *et al.* 2019). Because of its low density, chirality, hydrophilicity, biodegradability, nontoxicity, and low cost, cellulose is a biodegradable and plentiful organic polymer whose composites have a lot of application interest. It has a long straight-chain polymer that is densely packed with intermolecular and intramolecular hydrogen bonds. Due to its high mechanical strength, it is employed as a reinforcing material in the manufacturing of various composites (Krishnan 2013; Peresin *et al.* 2010).

4.3.2 Secondary Forest Wastes

Wastes from the processing of wood are the tree bark and sawdust.

4.4 AGRICULTURAL AND FOREST WASTE-BASED WATER TREATMENT MATERIALS

Agricultural and forest waste includes things used as adsorbents, coagulants, and photocatalysts, like cassava peel, citrus peel, banana peel, rice husk, chicken feather, fruit pit shell, pomegranate peel, tea and herbal tea waste, potato peels, almond shell, soybean hulls, sugarcane bagasse, papaya wood, tomato leaf powder, pinecone, tobacco stem biomass, coffee waste, parsley stalks, cucumber peel, Eucalyptus *sheathiana* bark biomass, *Malaleucadiosmifolia* leaf, eggshell, sawdust of oil palm tree, sawdust, water hyacinth, Giombo persimmon seed, bamboo waste, cedar bark, mango kernel, Ulmas tree leaves, breadnut peel, phoenix tree leaves, corncob, cotton waste, wheat straw, pineapple leaves, guava leaves, coconut coir, tamarind wood, hazelnut husks, date stone, coconut bunch waste, grapefruit peel, peach gum, cherry sawdust, cedar bark, chitosan, wood apple shell, pumpkin husk, pistachio shell, neem black, and poplar leaf (Abdolali *et al.* 2017; Afroze and Sen 2018; Fan *et al.* 2016; Gupta *et al.* 2009; Hosseinabady *et al.* 2018; Marshall and Champagne 1995; Ziarati *et al.* 2020).

4.4.1 Agricultural and Forest Waste-Based Adsorbents

Some agricultural and forest wastes possess abilities which are the main characteristic found in common adsorbents used in the industries for the treatment of water and wastewater. Some agricultural wastes have relatively good surface area and pore network distribution. There are a variety of preparations that are used before using these wastes as biosorbents, depending on the type of agricultural or forest waste. Treatments include shredding, crushing into powder, chemical treatment, or heating at various temperatures for opening the pore or increasing the surface area of the specified waste. The treatment is for modification of physical, chemical, or biological characteristics of the waste, such as improvement of the surface area, pore volume, enhanced catalytic oxidation ability, increased functional affinity for specific metal, and improved uptake of organics. Hydrochloric acid (HCl), Sulphuric acid (H_2SO_4), Phosphoric acid (H_3PO_4), Nitric acid (HNO_3), Glutamic acid, Zinc chloride ($ZnCl_2$), O_3, CO_2, SO_2, NH_3, H_2O_2, Sodium oxide (NaOH), Ammonium sulphite $(NH_4)_2SO_3$, Calcium oxide (CaO), Calcium chloride ($CaCl_2$), Methanol, Ethanol, Epichlorohydrin, triethylamine, Cationic surfactant, and magnetic iron oxides are some of the modifying agents used over the years. Physical methods such as irradiation, steaming, hydrothermal combustion, pyrolysis, milling, and extrusion have also been utilized to improve the properties of these wastes for adsorption. Biological agents include fungi (white or brown rot), bacteria (actinomycetes), and enzymes (laccase, manganese peroxide, lignin peroxidase).

Agricultural and forest wastes have been used in the adsorption of the following metals: chromium, zinc, cobalt, lead, nickel, copper, iron, mercury, cadmium, and zirconium. They are also used in dyes such as methylene blue, toluidine blue, Congo red, crystal violet, basic green 4, acid blue 92, red 14, reactive black 5, methyl violet, acid yellow 132, acid green 25, light green, malachite green, bismark brown, remazol brilliant blue, and direct blue. The adsorption removal capacity for dyes ranges between 26.6-1639.9 mg/g, and for heavy metals between 1.6-302 mg/g, based on various wastes used and the modification process employed. The operating parameters for the adsorption process using agricultural waste adsorbents range from 1-12 for pH, 9°C-60°C for temperature, 0.01-20 g/L for adsorbent dosage.

4.4.2 Agricultural and Forest Waste-Based Membranes

Ioannou *et al.* (2015) investigated adsorbent materials generated from agricultural wastes because of their low cost, abundance, and environmental friendliness. Particles of banana peel, tea waste, and shaddock peel were used as fillers in polyether sulfone (PES) to create porous mixed matrix membranes (MMMs) (Lin *et al.* 2014). The produced MMMs were used in the batch or flow adsorption of methylene blue (MB) and methyl violet 2B dyes. Desorption was also used to rejuvenate the membranes. The saturated dye adsorption capabilities for MMMs in a batch method were found to be 294–340 mg/g for methylene blue and 308–370 mg/g for methyl violet. The desorption rate reached 95%, resulting in membrane rejuvenation. After three adsorption/desorption cycles, high dye removal and recovery efficiency could be maintained.

4.4.3 Agricultural and Forest Waste-Based Photocatalysts

A photocatalyst is a substance that absorbs light to raise its energy level, then transfers that energy to a responding substance to cause a chemical reaction (Oshida 2013). The use of agricultural and forest waste for the photocatalysis process involves the extraction of the cellulose component or liquid extract. In the extraction of the cellulose component, the waste is usually washed, crushed, or milled, then boiled to remove water-soluble polysaccharides, and treated with acid or alkali for the hydrolysis process (Wulandari *et al.* 2016). Once the cellulose component is extracted, it may be further bleached to get a colourless solution (Tavker *et al.* 2019). Solution of the extract will be filtered or centrifuged to remove excess lignin present; the filtered solution will be further mixed as support for photocatalyst material such as titanium dioxide (TiO_2), copper indium sulphide (CuInS), silver phosphate (Ag_2PO_3), molybdenum sulphide, and palladium (Ajmal *et al.* 2019; Tavker *et al.* 2019, 2020).

4.4.4 Agricultural and Forest Waste-Based Coagulant

Coagulants in water treatment aid in the reduction of colour, turbidity, organic compounds, clay particles, algae, and microbiological content in the water, thereby lowering the risk of waterborne disease (Vara 2012). The introduction of the agricultural and forest waste-based coagulant for treatment of water and wastewater helps solve the health effect concerns raised about chemical and synthetic polymer which are commonly used (Choy *et al.* 2014; Mohd-Asharuddin *et al.* 2017; Ramavandi 2014). The following are agricultural and forest wastes that have been utilized as coagulants for treatment of water or wastewater: rice husk, cassava peel, activated poultry feathers, moringa seed, and bones

Cassava peel has been reported to have the binding capability for metal when in solution because it contains carboxyl, hydroxyl, and amino groups.

4.5 APPLICATION OF AGRICULTURAL AND FOREST WASTE-BASED MATERIALS IN WATER TREATMENT

The application of organic solid waste from the agriculture sector as possible biosorbents for the removal of contaminants (dye and heavy metal) from polluted water is receiving a lot of publicity. The method of extracting heavy metals using biosorbents is called biosorption. Biosorption uses physicochemical (e.g., chelation, adsorption, precipitation) processes to attach and eliminate pollutants from water. It is regarded as a significant metal and dye remediation system based on the following benefits: (i) no substrate is needed, (ii) simple process, and (iii) cost-effective approach (iv) biosorbent reuse in different cycles, (v) biosorbent environmentally friendly nature, and (vi) biosorbent regeneration for relevant industrial application. Furthermore, the use of these agricultural organic wastes will help minimize the volume of solid waste, resulting in a more effective waste recycling and management system (Vieira and Volesky 2000; Fomina and Gadd 2014). The main constituents of agro-waste materials are lignin and cellulose, but they can also contain other polar functional groups like alcohol, aldehydes, phenolic, and ether groups, which contribute to their contaminant removal capabilities (Hossain *et al.* 2012).

The adsorption process occurs in various phases, involving the movement of adsorbates to the outside of the absorbent's surface. The adsorbates are then transferred around adsorbent pores, and some of the adsorbates are eventually adsorbed on the internal surface of the adsorbent through a single or multi-layer adsorption (Vieira and Volesky 2000; Bharathi and Ramesh 2013; Rehman *et al.* 2019). During contaminant removal from polluted water, the physical shape of adsorbent material is generally essential. The physical forms of adsorbents made from agricultural sources can be classified into powder or granule (carbon) form, fibre form, film or membrane form, and hydrogel form.

4.5.1 Carbon Filters

Agricultural by-products may provide a low-cost, sustainable source of activated carbon filters. These residue materials also pose a problem in waste management because they have little to no economic value. Hence, these cost-effective by-products must be valorized. As a result, agricultural waste conversion into activated carbons will add significant beneficial value, minimize waste disposal costs, and, most significantly, provide a low-cost equivalent to currently available industrial activated carbons. Different kinds of carbon have been produced from agricultural residue; bagasse, coir pith, banana pit, date pits, sago waste, silk cotton hull, corncob, cob cot, bagasse, fruit stones, nutshells, pinewood, sawdust, coconut tree, coir pith, and soy hull have been applied for wastewater remediation (Singh *et al.* 2003; Mohan *et al.* 2006). Some agricultural wastes used as adsorbents were shown in Table 4.3.

TABLE 4.3
Agricultural Waste as Absorbents for Contaminant Removal from Wastewater

Adsorbates (water contaminant)	Agricultural Waste Type	Contaminant Removal Efficiency (%)	References
Oil	Wool, Bark, Rice Husk, Banana peel, Kapok, Oil Palm leaves, Sugarcane Bagasse	75–99	(Brandão *et al.* 2010; Teli and Valia 2013; Wang *et al.* 2013)
Heavy metals	Sugarcane bagasse, Rice Husk, Soyabean hulls, Orange peels	34–95	(Wong *et al.* 2003; Gönen and Serin 2012; Kong *et al.* 2014)
Dye	Maize cob and Husk, Oil palm biomass, coconut coir, kapok fibre, Jute	66–99	(Igwe and Abia 2007; Ahmad *et al.* 2011; Etim *et al.* 2016)
Ionic (NO_3^-, NO_2^-, PO_4^-, NH_4^+)	Wheat straw, Almond shell, Oakwood, Coconut coir pith	25–99	(Anirudhan and Unnithan 2007; Rezaee *et al.* 2008; Wang *et al.* 2015)

Animal bone charcoal, often known as bone char, is a porous, black granular material made by charring animal bones. Its structure varies according to its forming; however, it is mainly composed of 57–80% tricalcium phosphate, 6–10% calcium carbonate and 7–10% carbon. Its primary functions are filtration and decolourization. Bone char contains tricalcium phosphate, which can be used to remove fluoride and metal ions from water, making it beneficial for water treatment. Although bone chars have a lower surface area than activated carbons, they have significant adsorptive capabilities for some metals, particularly those in group 12 (copper, zinc, and cadmium). Additional highly toxic metal ions can also be removed, such as those for arsenic and plum. The use of nanofiltration in Tanzania shows the practical example of using bone char in water purification (Deydier *et al.* 2003; Chen *et al.* 2008)

Palm kernel shell charcoal has a higher fixed coal content, which is approximately 90%. Charcoal is commonly used to heat homes throughout the winter because of its environmental benefits and high caloric value. It is directly used as a fuel in the steel melting industry. It is used to make high-quality activated carbon and can be degraded into ethanol and other industrial goods. It is also used for water purification, air cleansing, and industrial decolourization. Palm kernel shell activated carbon, like coconut shell charcoal, is a new great adsorbent with superior adsorption and purifying abilities (Cakbentra 2017).

Activated carbon (AC) is usually applied for the removal of contaminants from gas and liquid phases, including heavy metals and organic pollutants. Presently, ACs are made from agricultural wastes including nut shells, fruit stones, palm, coconut, and other lignocellulose biomass. Since there is a need for a low-cost, readily sourced feedstock for the production of biochar and AC, using agricultural residue may not only be beneficial as an adsorbent for cleaning polluted water, but it also helps in reducing the pollution generated by this waste material. Biochar is charcoal produced from agricultural waste biomass through pyrolysis (burning in the absence of oxygen) and can be applied as a soil supplement as well as an adsorbent of contaminants from soil and water. Biochar, like many other activated carbon filters, can remove a wide range of organic and inorganic pollutants from water (Singh *et al.* 2003; Mohan *et al.* 2006). The physicochemical properties of biochar and AC are affected by the properties of the feedstock and the pyrolysis parameters, with temperature being a key factor.

Ricordel *et al.* (2001) reported that a carbon filter prepared from peanut husk was a viable adsorbent for heavy metal (Pb, Cd, Zn, and Ni) removal from a solution, which validates its application as a potential alternative for wastewater treatment. The absorbent from an agro-waste showed higher adsorption capacity than the industrial-based activated carbon filters (Ricordel *et al.* 2001). Also, Juang *et al.* (2002) reported that the adsorption ability of bagasse-based activated carbon filters was very high for dye-to-carbon adsorption attributed to the significant surface area. A good carbon adsorbent should have a porous texture as well as a large surface area. Aside from the physical composition, the chemical properties of the surface have a great impact on the adsorption potential of a given carbon filter, which adds to the uncertainty of the adsorption process; it hinges on many factors, including electrostatic and non-electrostatic interactions (Moon and Lee 2005; Park *et al.* 2007; Blecken *et al.* 2009).

4.5.2 Membrane Filters

Agricultural waste adsorbents can be applied as a membrane. Membrane mechanisms like microfiltration, nanofiltration, ultrafiltration, and reverse osmosis are particularly successful in contaminant removal from wastewater (Cicek 2003; Ahmad *et al.* 2006; Das and Gebru 2017; Abd Rahman *et al.* 2018). Owen *et al.* (1995) reported that the membrane filtration approach may be a preferable option to the conventional method of wastewater treatment because it is more cost-effective and efficient. Additionally, the membrane method is capable of extracting solid components as well as other inorganic pollutants. The use of chitosan augmented membrane filtration for the removal of metal ions (Cu and Zn) from wastewater recorded appreciable removal efficiency according to Juang and Shiau (2000).

Nanofiltration can be applied to eliminate the smallest portion or charged solute pollutant by minimizing the membrane pores and altering the membrane surface with an anionic or cationic layer. As a result, adsorption processes (physical, chemical, and ionic adsorption) can be used (Barakat 2011). A hydrogel, made from polysaccharide-based materials, is a modernized type of adsorbent. To make a water-insoluble crosslinked network, hydrogels are often crosslinked with other functional groups or binding agents. The nano-cellulose substance could be used as a filler, binder, or reinforcement agent (Carpenter *et al.* 2015)

4.5.3 Fibre Filters

Agro-based fibres including kenaf, roselle, and tobacco have been shown to absorb heavy metal ions from wastewater filtration systems. In comparison to agro-based fibre, most wood-based fibres have a lower potential for filtering heavy metal. The characteristics of agro-waste fibres vary greatly depending on fibre morphology, fibre cell dimensions, microfibril angle, and chemical properties (Rowell *et al.* 2000; Osorio *et al.* 2010). In agricultural waste-based fibre composites, the fibre dimensions are critical since they indicate the potential durability of material (Rowell *et al.* 2000). Agricultural waste fibre properties show a wide variety of fibre length, breadth, and thickness. Plant fibre is a naturally occurring composite made up of different chemical compounds, mostly cellulose, hemicellulose, and lignin, with a small amount of pectin. Plant fibres are also known as cellulosic or lignocellulosic fibres because of their chemical constituents (Siqueira *et al.* 2010). The amount of these chemicals in the cell wall layer is influenced by the fibre's origin, cultivation environment, and extraction process. These factors chemically and physically affect the surface characteristics of the fibre. Scientists aim to improve the ability of sorption of lignocellulosic fibre materials with chemical treatment; however, bark and aquatic fibres have been more effective (Bledzki and Gassan 1999; Rowell *et al.* 2000).

Jute fibre is a type of plant fibre that may be spun into strong, coarse threads. Jute fibres are recognized for being soft, lengthy, and lustrous. The principal producers of this fibre are considered to be plants of the genus Corchorus. Jute fibres are commonly utilized in the making of gunny fabric, hessian cloth, and burlap fabric. Jute fibres are 100% biodegradable and recyclable, which makes them environmentally beneficial. Jute fibres provide high thermal and acoustic insulating characteristics, with mild moisture resorption and no skin irritations (Yeh 1988).

Banana fibre, also known as Musa fibre, is one of the strongest natural fibres on the planet. The natural fibre is made of banana tree stem and is incredibly durable. This material is biologically degradable. Banana fibre is similar to natural bamboo fibre, but it is reported to have higher spinnability, fineness, and tensile strength. Banana fibres can be used as a natural absorbent, a bioremediation agent for bacteria in natural water purifiers, and in the production of mushrooms. They are also used in the production of handicrafts, high-quality paper cards, tea bags, string thread, high-quality fabric material, currency notes, and good rope for tying purposes (Hendriksz 2017).

Some modifications or treatments, typically by physical (physical adsorption, ionic fluid modifications, and plasma treatment) and chemical means, are necessary to improve the treatment of agro-waste fibres as adsorbents (Esterification, Oxidation of cellulose, Alkaline treatment, Silylation, and Grafting). Plasma treatment of agro-waste fibre has been proposed as a feasible alternative for water treatment involving electrical gas discharges and free radicals. In theory, this procedure was supposed to cause physical changes to the surface of the fibre. Furthermore, it may aid in cleansing, an etching effect, grafting, and functionalization effect (Mukhopadhyay and Fangueiro 2009; Yang *et al.* 2015). Essentially, surface cleaning and etching may promote physical adhesion on the fibre surface, while grafting or crosslinking may reinforce the surface. As a result, plasma fibre modifications are dependable, reproducible, and precise when tracked with a plasma diagnostic system. Also, plasma treatments can aid in the modification of a variety of surface properties as required by their usage. The downside of this approach is that only the side of the adsorbent that practically faces the plasma jet will be modified (Mukhopadhyay and Fangueiro 2009).

Many agro-fibres have been applied for dye removal from aqueously developed solutions under various conditions, including oil palm fibre, coir pith, sugarcane fibre, rice husk, and banana peeling (Annadurai *et al.* 2002; Demirbas 2009; Parab *et al.* 2009). Biochar (activated carbon) manufactured from sugar beet root fibre was successful in removing heavy metals, as reported by Inyang *et al.* (2012). The findings showed that activated carbon produced from agro-waste fibres has a sorption potential equivalent to commercially available activated carbon.

4.6 THE ECONOMIC IMPACT OF FARM AND FORESTRY WASTES VALOURIZATION

Valourization of waste resources is a global phenomenon. Farm waste valourization has added impetus to the sustainability and viability of farm resources in certain key applications in the economy of nations. The water treatment sector is utilizing a significant volume of farm-based materials to meet the needs of the clientele.

Forest economics were initially concerned exclusively with the timber production process. Today, however, the scope of this subject has expanded to take account of the multi-functional role that forests have in delivering not only economic benefits to society but also environmental, social, and cultural benefits (Kabongo 2013). Several writers have treated these sub-heads on their merits, depending on the materials of interest. Adding value to forest waste has introduced another angle of forest economics, which is the focus of this section.

(i) Foresters have been exposed to the possibilities of utilizing the forest waste resources for economic gains, thereby giving them a wider latitude to make an informed judgement on forest expansion.
(ii) Selection of the tree crops that will provide the needed resources on maturity.
(iii) A wider scope of employment potentials to meet the demand of forestry.
(iv) Intensive reduction of forest waste resources as the former wastes are upgraded to useful economic materials.
(v) A wider range of forestry products.

4.7 CONCLUSION

Agricultural waste can serve as a potential candidate for biosorbent production since it is currently underutilized. The use of agro-waste to treat polluted water can increase the value of agro-waste and, in turn, minimize global agro-waste management issues. Agro-materials are mostly made up of lignin and cellulose, and other polar lignin functional groups, which are responsible for the chemistry of biosorption and the removal of contaminants from wastewater. Agricultural waste adsorbents have demonstrated promising potential for wastewater treatment and can serve as a cost-effective alternative.

4.8 REFERENCES

Abd Rahman, N.S., Yhaya, M.F., Azahari, B. and Ismail, W.R. 2018. Utilisation of natural cellulose fibres in wastewater treatment. *Cellulose*, 25 (9): 4887–4903.

Abdolali, A., Ngo, H.H., Guo, W., Zhou, J.L., Zhang, J., Liang, S., Chang, S.W., Nguyen, D.D., Liu, Y. 2017. Application of a breakthrough biosorbent for removing heavy metals from synthetic and real wastewaters in a lab-scale continuous fixed-bed column. *Bioresource Technology*, 229, 78–87.

Afroze, S. and Sen, T.K. 2018. A review on heavy metal ions and dye adsorption from water by agricultural solid waste adsorbents. *Water, Air, and Soil Pollution*, 229, 1–50.

Ahmad, A., Chong, M., Bhatia, S. and Ismail, S. 2006. Drinking water reclamation from palm oil mill effluent (POME) using membrane technology. *Desalination*, 191 (1–3): 35–44.

Ahmad, T., Rafatullah, M., Ghazali, A., Sulaiman, O. and Hashim, R. 2011. Oil palm biomass—Based adsorbents for the removal of water pollutants—A review. *Journal of Environmental Science and Health, Part C*, 29 (3): 177–222.

Ajmal, N., Saraswat, K., Bakht, M.A., Riadi, Y., Ahsan, M.J. and Noushad, M. 2019. Cost-effective and eco-friendly synthesis of titanium dioxide (TiO2) nanoparticles using fruit's peel agro-waste extracts: Characterization, in vitro antibacterial, antioxidant activities. *Green Chemistry Letters and Reviews*, 12 (3): 244–254.

Anirudhan, T. and Unnithan, M.R. 2007. Arsenic (V) removal from aqueous solutions using an anion exchanger derived from coconut coir pith and its recovery. *Chemosphere*, 66 (1): 60–66.

Annadurai, G., Juang, R.S. and Lee, D.J. 2002. Factorial design analysis for adsorption of dye on activated carbon beads incorporated with calcium alginate. *Advances in Environmental Research*, 6 (2): 191–198.

Barakat, M. 2011. Adsorption and photodegradation of Procion yellow H-EXL dye in textile wastewater over TiO2 suspension. *Journal of Hydro-environment Research*, 5 (2): 137–142.

Bharathi, K. and Ramesh, S. 2013. Removal of dyes using agricultural waste as low-cost adsorbents: A review. *Applied Water Science*, 3 (4): 773–790.

Blecken, G.T., Zinger, Y., Deletić, A., Fletcher, T.D. and Viklander, M. 2009. Impact of a submerged zone and a carbon source on heavy metal removal in stormwater biofilters. *Ecological Engineering*, 35 (5): 769–778.

Bledzki, A. and Gassan, J. 1999. Composites reinforced with cellulose-based fibres. *Progress in Polymer Science*, 24 (2): 221–274.

Brandão, P.C., Souza, T.C., Ferreira, C.A., Hori, C.E. and Romanielo, L.L. 2010. Removal of petroleum hydrocarbons from aqueous solution using sugarcane bagasse as adsorbent. *Journal of Hazardous Materials*, 175 (1–3): 1106–1112.

Cakbentra. 2017. Biomass industrial innovative projects: Coconut shell charcoal and palm kernel shell charcoal production process. Available: https://biomassproject.blogspot.com/2017/02/coconut-shell-charcoal-and-palm-kernell.html (Accessed 28/05/2021).

Carpenter, A.W., de Lannoy, C.F. and Wiesner, M.R. 2015. Cellulose nanomaterials in water treatment technologies. *Environmental Science & Technology*, 49 (9): 5277–5287.

Chen, Yun-Nen, Chai, Li-Yuan and Shu, Yu-De 2008, December. Study of arsenic(V) adsorption on bone char from aqueous solution. *Journal of Hazardous Materials*, 160 (1): 168–172. doi:10.1016/j.jhazmat.2008.02.120. PMID 18417278.

Choy, S.Y., Nagendra-Prasad, K.M., Wu, T.Y., Raghunandan, M.E., Ramanan, R.N. 2014. Utilization of plant-based natural coagulants as future alternatives towards sustainable water clarification. *Journal of Environmental Science*: 2178–2189.

Cicek, N. 2003. A review of membrane bioreactors and their potential application in the treatment of agricultural wastewater. *Canadian Biosystems Engineering*, 45: 6.37–36.37.

Das, C. and Gebru, K.A. 2017. Preparation and characterization of CA– PEG– TiO2 membranes: Effect of PEG and TiO2 on morphology, flux and fouling performance. *Journal of Membrane Science and Research*, 3 (2): 90–101.

Demirbas, A. 2009. Agricultural based activated carbons for the removal of dyes from aqueous solutions: A review. *Journal of Hazardous Materials*, 167 (1–3): 1–9.

Deydier, Eric, Guilet, Richard and Sharrock, Patrick 2003, July. Beneficial use of meat and bone meal combustion residue: An efficient low-cost material to remove lead from aqueous effluent. *Journal of Hazardous Materials*, 101 (1): 55–64. doi:10.1016/S0304-3894(03)00137-7. PMID 12850320.

Etim, U., Umoren, S. and Eduok, U. 2016. Coconut coir dust as a low-cost adsorbent for the removal of cationic dye from aqueous solution. *Journal of Saudi Chemical Society*, 20: S67–S76.

Fan, Y., Yang, R., Lei, Z., Liu, N., Lv, J., Zhai, S., Zhai, B. and Wang, L. 2016. Removal of Cr(VI) from aqueous solution by rice husk derived magnetic sorbents. *Korean Journal of Chemical Engineering*, 33, 1416–1424.

Fomina, M. and Gadd, G.M. 2014. Biosorption: Current perspectives on concept, definition and application. *Bioresource Technology*, 160: 3–14.

Gönen, F. and Serin, D.S. 2012. Adsorption study on orange peel: Removal of Ni (II) ions from aqueous solution. *African Journal of Biotechnology*, 11 (5): 1250–1258.

Gu, D., Andreev, K. and Dupre, M.E. 2021. Major trends in population growth around the world. *China CDC Weekly*, 3 (28): 604–613.

Gupta, V.K., Carrott, P.J.M., Ribeiro Carrott, M.M.L. and Suhas. 2009. Low-cost adsorbents: Growing approach to wastewater treatment—a review. *Critical Reviews in Environmental Science and Technology*, 39 (10): 783–842.

Hamada, H.M., Jokhio, G.A., Yahaya, F.M., Humada, A.M. and Gul, Y. 2018. The present state of the use of palm oil fuel ash (POFA) in concrete. *Construction and Building Materials*, 175: 26–40.

Hendriksz, V. 2017. Sustainable textile innovations: Banana fibres. Available: https://fashionunited.uk/news/fashion/sustainable-textile-innovations-banana-fibres/2017082825623 (Accessed 28/05/2021).

Hossain, M., Ngo, H., Guo, W. and Setiadi, T. 2012. Adsorption and desorption of copper (II) ions onto garden grass. *Bioresource Technology*, 121: 386–395.

Hosseinabady, B.T., Ziarati, P., Ballali, E. and Umachandran, K. 2018. Detoxification of heavy metals from leafy edible vegetables by agricultural waste: Apricot pit shell. *Journal of Environmental & Analytical Toxicology*, 8.

Igwe, J.C. and Abia, A. 2007. Adsorption isotherm studies of Cd (II), Pb (II) and Zn (II) ions bioremediation from aqueous solution using unmodified and EDTA-modified maize cob. *Eclética Química*, 32 (1): 33–42.

Inyang, M., Gao, B., Yao, Y., Xue, Y., Zimmerman, A.R., Pullammanappallil, P. and Cao, X. 2012. Removal of heavy metals from aqueous solution by biochars derived from anaerobically digested biomass. *Bioresource Technology*, 110: 50–56.

Ioannou, L.A., Puma, G.L. and Fatta-Kassinos, D. 2015. Treatment of winery wastewater by physicochemical, biological and advanced processes: A review. *Journal of Hazardous Materials*, 286: 343–368.

Juang, R.S. and Shiau, R.C. 2000. Metal removal from aqueous solutions using chitosan-enhanced membrane filtration. *Journal of Membrane Science*, 165 (2): 159–167.

Juang, R.S., Wu, F.C. and Tseng, R.L. 2002. Characterization and use of activated carbons prepared from bagasses for liquid-phase adsorption. *Colloids and Surfaces A: Physicochemical and Engineering Aspects*, 201 (1–3): 191–199.

Kabongo, J.D. (2013). *Waste Valorization. Encyclopedia of Corporate Social Responsibility, 2701–2706.* doi:10.1007/978-3-642-28036-8_680.

Kong, W., Ren, J., Wang, S. and Chen, Q. 2014. Removal of heavy metals from aqueous solutions using acrylic-modified sugarcane bagasse-based adsorbents: Equilibrium and kinetic studies. *BioResources*, 9 (2): 3184–3196.

Krishnan, V.N. 2013. Synthesis and characterization of cellulose nanofibers from coconut coir fibers. *IOSR Journal of Applied Chemistry*, 6: 18–23.

Lin, C.H., Gung, C.H., Sun, J.J. and Suen, S.Y. 2014. Preparation of polyethersulfone/plant-waste-particles mixed matrix membranes for adsorptive removal of cationic dyes from water. *Journal of Membrane Science*, 471: 285–298.

Marshall, W.E. and Champagne, E.T. 1995. Agricultural byproducts as adsorbents for metal ions in laboratory prepared solutions and in manufacturing wastewater. *Journal of*

Environmental Science and Health. Part A: Environmental Science and Engineering and Toxicology, 30: 241–261.

Mohammed, M., Shitu, A., Tadda, M. and Ngabura, M. 2014. Utilization of various Agricultural waste materials in the treatment of Industrial wastewater containing heavy metals: A review. *International Research Journal of Environmental Sciences*, 3 (3): 62–71.

Mohan, D., Singh, K.P. and Singh, V.K. 2006. Trivalent chromium removal from wastewater using low cost activated carbon derived from agricultural waste material and activated carbon fabric cloth. *Journal of Hazardous Materials*, 135 (1–3): 280–295.

Mohd-Asharuddin, S., Othman, N., Mohd Zin, N.S. and Tajarudin, H.A. 2017. A chemical and morphological study of cassava peel: A potential waste as coagulant aid. In: *MATEC Web of Conferences*. EDP Sciences, p. 06012.

Moon, C.J. and Lee, J.H. 2005. Use of curdlan and activated carbon composed adsorbents for heavy metal removal. *Process Biochemistry*, 40 (3–4): 1279–1283.

Mukhopadhyay, S. and Fangueiro, R. 2009. Physical modification of natural fibers and thermoplastic films for composites—a review. *Journal of Thermoplastic Composite Materials*, 22 (2): 135–162.

Oshida, Y. 2013. Oxidation and oxides. In: *Bioscience and Bioengineering of Titanium Materials*. Elsevier, pp. 87–115.

Osorio, L., Trujillo, E., Van Vuure, A.W., Lens, F., Ivens, J. and Verpoest, I. 2010. The relationship between the bamboo fibre microstructure and mechanical properties. In: *Proceedings of Proceedings 14th European conference on composite materials (on cd-rom)*. Budapest University of Technology and Economics. Department of Polymer . . . .

Owen, G., Bandi, M., Howell, J. and Churchouse, S. 1995. Economic assessment of membrane processes for water and waste water treatment. *Journal of Membrane Science*, 102: 77–91.

Parab, H., Sudersanan, M., Shenoy, N., Pathare, T. and Vaze, B. 2009. Use of agro-industrial wastes for removal of basic dyes from aqueous solutions. *Clean—Soil, Air, Water*, 37 (12): 963–969.

Park, H.G., Kim, T.W., Chae, M.Y. and Yoo, I.K. 2007. Activated carbon-containing alginate adsorbent for the simultaneous removal of heavy metals and toxic organics. *Process Biochemistry*, 42 (10): 1371–1377.

Peresin, M.S., Habibi, Y., Zoppe, J.O., Pawlak, J.J. and Rojas, O.J. 2010. Nanofiber composites of polyvinyl alcohol and cellulose nanocrystals: Manufacture and characterization. *Biomacromolecules*, 11: 674–681.

Powell, J.M. and Unger, P.W. 1997. Alternatives to crop residues for sustaining agricultural productivity and natural resource conservation. *Journal of Sustainable Agriculture*, 11 (2–3): 59–83.

Rafiee, E., Shahebrahimi, S., Feyzi, M. and Shaterzadeh, M. 2012. Optimization of synthesis and characterization of nanosilica produced from rice husk (a common waste material). *International Nano Letters*, 2 (1): 29.

Ramavandi, B. 2014. Treatment of water turbidity and bacteria by using a coagulant extracted from Plantago ovate. *Water Resource Industry*: 36–50.

Rehman, R., Farooq, S. and Mahmud, T. 2019. Use of agro-waste Musa acuminata and Solanum tuberosum peels for economical sorptive removal of emerald green dye in ecofriendly way. *Journal of Cleaner Production*, 206: 819–826.

Rezaee, A., Godini, H., Dehestani, S. and Khavanin, A. 2008. Application of impregnated almond shell activated carbon by zinc and zinc sulfate for nitrate removal from water. *Journal of Environmental Health Science & Engineering*, 5 (2): 125–130.

Ricordel, S., Taha, S., Cisse, I. and Dorange, G. 2001. Heavy metals removal by adsorption onto peanut husks carbon: Characterization, kinetic study and modeling. *Separation and Purification Technology*, 24 (3): 389–401.

Rowell, R.M., Lange, S.E. and Jacobson, R.E. 2000. Weathering performance of plant-fiber/thermoplastic composites. *Molecular Crystals and Liquid Crystals Science and Technology. Section A. Molecular Crystals and Liquid Crystals*, 353 (1): 85–94.

Sahin, H.I. 2020. The potential of using forest waste as a raw material in particleboard manufacturing | Şahin | BioResources. *Bioresources.com*, 15.

Siebert, S., Burke, J., Faures, J.M., Frenken, K., Hoogeveen, J., Döll, P. and Portmann, F.T. 2010. Groundwater use for irrigation – a global inventory. *Hydrology and Earth System Sciences*, 14: 1863–1880.

Singh, K.P., Mohan, D., Sinha, S., Tondon, G. and Gosh, D. 2003. Color removal from wastewater using low-cost activated carbon derived from agricultural waste material. *Industrial & Engineering Chemistry Research*, 42 (9): 1965–1976.

Siqueira, G., Bras, J. and Dufresne, A. 2010. Luffa cylindrica as a lignocellulosic source of fiber, microfibrillated cellulose and cellulose nanocrystals. *BioResources*, 5 (2): 727–740.

Tavker, N., Gaur, U.K. and Sharma, M. 2019. Highly active agro-waste-extracted cellulose-supported CuInS2 nanocomposite for visible-light-induced photocatalysis. *ACS Omega*, 4: 11777–11784.

Tavker, N., Gaur, U.K. and Sharma, M. 2020. Agro-waste extracted cellulose supported silver phosphate nanostructures as a green photocatalyst for improved photodegradation of RhB dye and industrial fertilizer effluents. *Nanoscale Advances*, 2: 2870–2884.

Teli, M. and Valia, S.P. 2013. Acetylation of banana fibre to improve oil absorbency. *Carbohydrate Polymers*, 92 (1): 328–333.

Thyagarajan, D., Barathi, M. and Sakthivadivu, R. 2013. Scope of poultry waste utilization. *Journal of Agriculture and Veterinary Science*, 6 (5): 29–35.

UNESCO. 2015. Water for a sustainable world: Facts and figures, WWDR2015Facts_Figures_ENG_web.pdf (unesco.org) (Accessed 28/05/2021).

United Nations Department of Economic and Social Affairs. 2020. World population: Past, Present, and Future. Available: https://worldstatistics.live/population (Accessed 24/05/2021).

Vara, S. 2012. Screening and evaluation of innate coagulants for water treatment: A sustainable approach. *International Journal of Energy and Environmental Engineering*, 3: 1–11.

Vieira, R.H. and Volesky, B. 2000. Biosorption: A solution to pollution? *International Microbiology*, 3 (1): 17–24.

Wang, J., Zheng, Y. and Wang, A. 2013. Preparation and properties of kapok fiber enhanced oil sorption resins by suspended emulsion polymerization. *Journal of Applied Polymer Science*, 127 (3): 2184–2191.

Wang, Z., Guo, H., Shen, F., Yang, G., Zhang, Y., Zeng, Y., Wang, L., Xiao, H. and Deng, S. 2015. Biochar produced from oak sawdust by Lanthanum (La)-involved pyrolysis for adsorption of ammonium (NH4+), nitrate (NO3–), and phosphate (PO43–). *Chemosphere*, 119: 646–653.

Wong, K., Lee, C., Low, K. and Haron, M. 2003. Removal of Cu and Pb by tartaric acid modified rice husk from aqueous solutions. *Chemosphere*, 50 (1): 23–28.

Wulandari, W.T., Rochliadi, A. and Arcana, I.M. 2016. Nanocellulose prepared by acid hydrolysis of isolated cellulose from sugarcane bagasse. *IOP Conference Series: Materials Science and Engineering*, 107: 12045.

Yang, T.C., Wu, T.L., Hung, K.C., Chen, Y.L. and Wu, J.H. 2015. Mechanical properties and extended creep behavior of bamboo fiber reinforced recycled poly (lactic acid) composites using the time—temperature superposition principle. *Construction and Building Materials*, 93: 558–563.

Ziarati, P., Moradi, D. and Vambol, V. 2020. Bioadsorption of heavy metals from the pharmaceutical effluents, contaminated soils and water by food and agricultural waste: A short review. *Labour Protection Problems in Ukraine*, 36: 3–7.

5 Sustainable Production of Nanomaterials from Agricultural Wastes

Mahen C. Perera, Ryan Rienzie and Nadeesh M. Adassooriya

CONTENTS

DOI: 10.1201/9781003245773-5

5.1 INTRODUCTION

With the rising population, the production and processing of crops have exponentially increased to meet consumer demand. Therefore, agricultural wastes have become a major global environmental concern. Haphazard disposal and combustion of these wastes are becoming major challenges to sustaining a healthy environment for living beings. In most countries, certain agricultural wastes are utilized as a fuel to fulfill part of their energy requirement due to the low cost. It is well known that in addition to the natural degradation process of organic wastes, the combustion of agricultural wastes can cause detrimental effects to the environment, such as the generation of greenhouse gases that lead to global warming (Adebisi *et al.*, 2017).

Moreover, apart from the ecological concerns, disposal of such organic wastes without utilization leads to a huge nutritional and economical loss for a country. Organic wastes are produced by various sources throughout the food lifecycle, ranging from agricultural operations to domestic operations. These wastes include seeds, peels, shells, bran, pomace, etc. (Lu *et al.*, 2019; Kringel *et al.*, 2019), which are highly rich in bioactive substances such as carotenoids, vitamins, oils, polyphenols, flavor compounds, fibers, and many other valuable compounds (Baiano, 2014; Kumar *et al.*, 2020b). As proof, peels of avocadoes, grapes, lemons, jackfruit seeds, and mangoes contain 15% higher phenolic content when compared with fruit pulp (Kumar *et al.*, 2020b). Although in the last two decades agricultural wastes were not considered an economic loss to food processing industries, recently, effective utilization of agricultural wastes has drawn great attention from researchers due to the global concerns regarding hunger, environmental degradation, and economic impacts of food wastes (Ghosh *et al.*, 2017). Renewable biomass as an alternative raw material has recently been considered a sustainable solution to overcome the problems associated with environmental degradation (Nomura and Terwilliger, 2019). Furthermore, the utilization of agricultural wastes is a sustainable approach to improve human health with the aid of functional foods containing the aforementioned bioactive substances extracted from organic wastes (Sagar *et al.*, 2018).

Nowadays, various strategies have been investigated to utilize agricultural wastes which are a valuable, low-cost, and benign resource, and the purpose of those strategies is to optimize the benefits from agricultural wastes and thereby minimize the impact on the environment (Ghosh *et al.*, 2017). In this regard, nanotechnology plays a significant role. Nanotechnology is an interdisciplinary field that offers a greater opportunity to use agricultural wastes and convert them into valuable nanomaterials such as green synthesized metal nanoparticles, which have an array of uses.

5.2 GREEN SYNTHESIS OF NANOMATERIALS FROM AGRICULTURAL WASTES: AN OVERVIEW

Synthesis of nanomaterials follows two major strategies: namely, top-down and bottom-up strategies. Reduction of bulk materials into nano size is practiced under the top-down strategy, while in the bottom-up strategy nucleation is the basis, i.e., formation through atom or molecules (Baig *et al.*, 2021). Certainly, in-between approaches are also plausible during the processes of synthesis (Gregorczyk *et al.*, 2016; Baig *et al.*, 2021). Within the scopes of these two strategies, three main methods, namely physical, chemical, and biological methods, are being employed. The strategy behind green synthesis as a biological method is bottom-up, while in chemical synthesis top-down or a strategy in between these two is applicable. Employing green methods has fetched wider attention within the scientific community due to their number of advantages.

Green methods involve the use of plants, microbes such as bacteria, yeasts, and fungal species. The use of agricultural waste materials when synthesizing nanomaterials is considered as one of the green solutions, as it provides many basic advantages including simplicity, very minimal or negligible toxic contamination, and reduced or minimum energy consumption, thus owing to their cost-effective nature. Agricultural wastes can be either plant-derived or animal-derived; most of the research studies have been conducted using plant-derived waste materials; thus, studies on the synthesis of nanomaterials based on animal-derived agricultural wastes are few. The main residual matter generated during cultivation, harvesting, and postharvest processing of crops contains lignocellulosic biomass. It has been revealed that almost 50% of the food waste generated by households is derived from fresh fruits and vegetables (De Laurentiis *et al.*, 2018). For instance, 8–20 million tons of total waste is generated during the production process of melon in the world (Rolim *et al.*, 2020), while the amount of pomace alone generated during the process of processing of apples is known to be 27000–1000000 Mt (Lyu *et al.*, 2020). Most of the studies carried out in terms of generating nanomaterials have been employed using postharvest biomass like rinds and peels, which are direct wastes. Obviously, plant-mediated agricultural wastes contain numerous phytochemical groups including alkaloids, phenylpropanoids, polyketides, and terpenoids. Such compounds act as strong reducing and capping agents. Green chemistry principles are based on some principles that target improvement of the reaction efficiency, efficient utilization of energy, usage of safe chemicals to reduce toxic contamination, minimizing the generation of waste materials during the synthesis process, utilization of renewable resources to make aforesaid safe chemicals, the focus of biodegradation of nontoxic products, less-hazardous nature of synthesized chemicals that are produced through these safer pathways, avoidance of usage of derivatives such as stabilizers, prevention of the release of harmful substances, usage of safer solvents, usage of catalysis as a way to improve the reaction efficiency, and minimization of causing accidents (Matharu and Lokesh, 2019; Bandeira *et al.*, 2020).

This chapter broadly discusses the effective utilization of agricultural wastes to produce nano materials while eliminating the high cost incurred with sophisticated industrial procedures and harmful chemical reactions. Moreover, the chapter discusses potential applications of such nanomaterials in various fields while stressing the importance of environmental sustainability.

5.3 SYNTHESIS OF METAL NANOPARTICLES FROM AGRICULTURAL WASTES

Biological synthesis processes are direct and involve a minimum number of steps due to their simple nature. Using a metal salt or a precursor solution and a reductant extracted from a low-cost or waste material, it would be sufficient to complete the process. However, the quality of the final product depends on several variables, including the chemical composition of the reductant used, the concentration of precursor used, pH, reaction time, and temperature of the mixture. Despite its eco-friendly nature, large-scale production of nanomaterials using green methods is a challenge and mostly done at a laboratory scale (Bandeira *et al.*, 2020). Table 5.1 summarizes the aspects related to the synthesis of metal nanoparticles using agricultural wastes.

5.3.1 Silver Nanoparticles (AgNPs)

Extracts generated based on water or organic solvents using various plant parts have yielded AgNPs ranging between 4 and 100 nm, and spherical shape was one of the most prominent shapes of the generated nanoparticles (Lateef *et al.*, 2016; Govindarajan *et al.*, 2017). Dang *et al.* (2017) also synthesized AgNPs with size ranges from 70 nm up to 600 nm with cubic, spherical, and triangular shapes, using waste banana plant stems. Aqueous extract of *Cassia fistula* fruit pulp was used by Fouad *et al.* (2018) to synthesize AgNPs. The size of the AgNPs synthesized by fruit pulp extract of *C. fistula* was in the range of 148–938 nm. Baran (2019) synthesized AgNPs using *Cydonia oblonga* leaf extract. The nanoparticles had cubic crystal structure, spherical shape, 40 to 60 nm in size, with uniform dispersion and high stability. Ndikau *et al.* (2017) also synthesized AgNP's aqueous extract of *Citrullus lanatus* fruit rind as the reductant and the capping agent. They were observed to be stable, spherical shapes, with an average diameter of 17.96 nm.

5.3.2 Zinc Oxide Nanoparticles (ZnONPs)

Through a work done by Fatimah (2018), ZnONPs were synthesized and the mean particle size of the formed nanoparticles was 17.16 nm and had bandgap energy of 3.18 eV. ZnO nanoparticles were produced by waste thyme (*Thymus vulgaris* L.) extracted by Abolghasemi *et al.* (2019). The shape of nanoparticles is almost regular with cubical, rectangular, radial hexagonal, and rod in morphology, while the average size of ZnO nanoparticles was estimated at 10–35 nm. Moreover, Shabaani *et al.* (2020) also synthesized ZnONPs with sizes ranging from 18–27 nm using *Eriobutria japonica* leaf extract.

5.3.3 Gold Nanoparticles (AuNPs)

Biosynthesis of AuNPs using *Croton Caudatus* Geisel extract was demonstrated by Kumar *et al.* (2019b). The generated nanoparticles were 20–50 nm in size with a spherical shape. *Cannabis sativa* (hemp) was also used to generate AuNPs (12–18 nm) in a study done by Singh *et al.* (2018). Moreover, *Scutellaria barbata* was also employed by Wang *et al.* (2019) to produce AuNPs with an average size of 154 nm, and using leaf extracts of *Sphaeranthus indicus* spherical shaped AuNPs with a mean particle size of 25 nm were synthesized. (Balalakshmi *et al.*, 2017).

5.3.4 Platinum Nanoparticles (PtNPs)

Through a study, Pt nanoparticles were generated using an aqueous extract of sugarcane bagasse by Ishak *et al.* (2021). Spherical and mild agglomerated NPs were observed therein. The size of individual nanoparticles was 5 ± 2 nm. Additionally, electron microscopic images showed a thin biomolecule layer surrounding the PtNPs. Moreover, they had higher antioxidant activity compared to the commercially available PtNPs. Pt nanoparticles (22 nm) were also synthesized using leaf extracts of *Xanthium strumarium* (Kumar *et al.*, 2019a), which is a weed species.

5.3.5 Iron-Based Nanoparticles (FeNPs and FeONPs)

Pan *et al.* (2020) also synthesized FeNPs with a mean diameter of 10.6 nm using red peanut skin extract. They dried the synthesized FeNPs under different conditions, namely vacuum drying and air drying. Vacuum-dried samples at 40 and 60°C had much smaller nanoparticles in the size range of 1–18 nm, while air-dried ones at 25°C resulted in slightly aggregated particles within the range of 25–77 nm. Peel extracts of *Citrus paradise* were used to synthesize nanoparticles with an average size of 28–32 nm with a spherical shape and aggregating in nature (Kumar *et al.*, 2020a).

5.3.6 Copper-Based Nanoparticles (CuNPs)

CuNPs synthesized using *Punica granatum* seeds extract were 40–80 nm in size, spherical in shape, and well dispersed and homogeneous (Nazar *et al.*, 2018). Meanwhile, stem extract of the holoparasitic plant (*Orobancheaegyptiaca*) was also used to synthesize nanoparticles of less than 50 nm (Akhter *et al.*, 2020). Moreover, highly crystal, face-centered, cubic structured, zero-valent CuNPs having an average size of 34.4 nm were synthesized by Nguyen *et al.* (2020) using cocoa pod extract.

5.3.7 Titanium Dioxide Nanoparticles (TiO_2NPs)

The size distribution for TiO_2NPs synthesized with lemon peel extract was 80–140 nm. Bicycle chain-like agglomerated aggregates clustered and uniformly dispersed all over the surface of the nanoparticles (Nabi *et al.*, 2020). Roopan *et al.* (2012) used *A. squamosa* peel extract to synthesize TiO_2NPs with a mean size of 23 nm. Goutam *et al.* (2018) also used *Jatrophacurcas* leaf extract to produce spherical-shaped anatase TiO_2NPs.

TABLE 5.1

Some Aspects Related to the Processes Employed in Representative Studies on Synthesizing Nanomaterials

	Agricultural waste type	Nanoparticle species	Preparation of waste solution	Reaction procedure and indications	Reference
1	Pomogranate (*Punica granatum*) fruit fleshy pericarp	Ag	Extraction at 8% w/v, 70°C/ 2 h and centrifugation at 3000 rpm/ 5 mm.	5 mL of extract was mixed with 50 mL of $AgNO_3$ and keeping for 30 min at 26 ± °C centrifugation at 10000 rpm/15 min. Formation of nanoparticles was indicated through changing the color into pale brown.	Govindappa *et al.*, 2021
3	Watermelon (*Citrullus lanatus*) rind extract	Ag	Ripe watermelon fruit was used to obtain rinds which were cut into small pieces (5 mm × 10 mm) and crushed using a blender. Then it was transferred into a 1000 mL conical flask and diluted with 400 mL of distilled water. Then it was kept in a water bath 80°C/10 min followed by filtering.	50 mL of 0.001 M $AgNO_3$ solution was added to 50 mL watermelon rind extract in a 250 mL conical flask. The pH of the reaction mixture was adjusted to 5.92 to 10 using 0.1 M NaOH solution. Then it was placed in a shaking water bath at 80°C until the color change occurred after 35 min. Formation of nanoparticles was indicated through changing the color from light-green to yellowish-brown.	Ndikau *et al.*, 2017
4	Orange (*Citrus* X *sinensis*) peel extract	Ag	Orange peels were dried at 100°C/48 h and ground to a fine powder. Then 1g of dry orange peel powder was added to 40 mL and stirred. Resultant mixture was placed in a plasmochemical reactor. Then it was treated with CNP discharge for 5 min.	Reacting orange peel extracts with $AgNO_3$ with concentration of 0.25–6.0 Mm/L. 40 mL of the orange peel extract was added to 40 mL $AgNO_3$ solution under stirring during 0.1 min. Obtained mixture was heated at 75°C. The AgNPs obtained were centrifuged at 5000 rpm/5 min, while adjusting the pH using 0.1 N NaOH and 0.1 M HCl. Brown color of the final reaction mixture indicated the formation of NPs.	Skiba and Vorobyova, 2019
6	Quince (*Cydonia oblonga*) leaf extract	Ag	25 g of *Cydonia oblonga* leaves with reduced size was mixed with 500 mL distilled water and boiled at 85°C followed by filtering and cooling down to room temperature.	500 mL of 1 mM $AgNO_3$ aqueous solution was mixed with 125 mL of plant extract and added to 1000 mL flask under room temperature. The mixture was centrifuged 7500 rpm/10 min and the supernatant was removed. The resulted solid component (AgNPs) was dried at 65°C and stored in the dark.	Baran, 2019

	Agricultural waste type	Nanoparticle species	Preparation of waste solution	Reaction procedure and indications	Reference
7	diced stems of banana (*Musa acuminata*)	Ag	1 g of the diced banana stems was added to a glass beaker containing 100 mL of water, then was heated at 75°C for 1 h and slowly cooled down to room temperature (22°C). After cooling, the aqueous mixture was filtered and then stored until use.	1 mL aqueous solution of 0.1 M $AgNO_3$ was mixed with varying quantities of diced stems of banana extract (1, 5, 10, 15, and 20 mL). After mixing for 2 minutes, the mixtures were left alone under room temperature (22°C).	Dang *et al.*, 2017
9	Waste macadamia (*Macadamia* sp.) nut shells	Ag	100 mL of Milli-Q® water was boiled and then allowed to simmer at 95°C/ 2 h soaking period. Cleaned Macadamia nut shells (10 g) was added to the simmering water until the appearance changed to a milky brown color. Then it was cooled down. The liquid was then filtered three times using 0.22 µm syringe that gave a pale-yellow appearance.	Varying amounts of extract (1, 5,10,15, 20, and 25 mL) were added to an aqueous 1 mL solution of $AuCl_4^-$ (500 ppm). The reaction mixtures were then left overnight for 22 h.	Dang *et al.*, 2019
10	Corn (*Zea mays*) cobs	Ag and Au NPs	10 g dry corncob pieces of 0.4-2.0 cm were boiled with distilled water 100 mL/2.0 h. Filtration under reduced pressure and storage at 4–10°C for further studies.	1 mL of corncob extract was mixed with 20 mL of 1.5 mM $AgNO_3$ and after some time, the mixture was centrifuged at 4000 rpm for 30 min. Recovered the nanoparticles after keeping at 90°C overnight	Doan *et al.*, 2020
12	Saffron (*Crocus sativus*) leaf extract	ZnO	Washed with deionized water, oven dried at 50°C /2 days and then crushed to obtain a fine powder. About 5 g of the powder was dissolved 100 mL deionized water, blended for 60 min at 70°C, and centrifuged at 6000 rpm/20 min. Supernatant was preserved in an airtight container in a refrigerator.	1. Drop wise addition of 5 M NaOH solution (5 M) into 0.01 M $Zn(CH_3CO_2)_2{\cdot}2H_2O$ (0.01 M) while stirring until the pH becomes 12 during a period of 30 min. 2. Drop wise addition of extract for 2 h under constant stirring at 70°C Centrifugation at 6000 rpm/30 min and washing with deionized water and 100% ethanol. Finally oven dried at 50°C. Milky and cloudy appearance of the final mixture indicated the formation of NPs.	Rahaiee *et al.*, 2020
13	Waste thyme (*Thymus vulgaris* L.)	ZnO	Waste thyme extract was prepared by mixing waste thyme powder and 100 mL deionized water in a 250 mL glass tube. The solution was heated in a water bath at 70–75°C/ 20 min and cooled down and filtration.	2.5 mM zinc acetate was prepared by using deionized water. $Zn(CH_3CO_2)_2$ and waste thyme extract were mixed in clean sterilized tubes at 2:1, 1:5, 1:10 ratio (V/V). Followed by the mixture was heated while stirring at 90°C. 3 h. Precipitate was dried in at 80°C for 6 h and then powdered.	Abolghasemi *et al.*, 2019

(Continued)

TABLE 5.1
Continued

	Agricultural waste type	Nanoparticle species	Preparation of waste solution	Reaction procedure and indications	Reference
15	Fresh banana (*Musa* sp.) peel extract	CuO	20 g of banana peels were mixed with 75 mL double distilled water and boiled at 70-80°C/ 20 min. Then the yellow color extract was filtered and was stored in the refrigerator.	30 mL of the extract was boiled up to 70-80°C by using magnetic stirrer. Then 1 g of Cu $(NO_3)_2.3H_2O$ was added to extract and the resulting greenish solution was boiled until it formed a brown paste. The paste was further heated in a ceramic crucible followed by heating in furnace at 400°C/ 2 h. Brown color of the final reaction mixture indicated the formation of AgNPs.	Aminuzzaman *et al.*, 2017
17	Dried sugarcane (*Saccharum officinarum*) bagasse, pineapple (*Ananas comosus*) peel and banana (*Musa* sp.) peel extracts	Pt	Oven dried at 60°C for several days and crusting to a fine powder. Then boiling of powder with deionized water 60°C/30 min, cooling and filtration.	10 mL of the extract was reacted with 90 mL of 1 mM Pt ion solution. Then conduction further steps as follows viz. Refluxation: 95°C; Sonication: 30 min; Centrifugation 13400 rpm/ 10 min washing with DI water and drying at 110°C/6 h. Blackish nature of the reaction mixture confirmed the formation of PtNPs.	Kamari and Ghorbani, 2020
19	Whole plant extract of *Catharanthus roseus*	Co	Whole plants of *Catharanthus roseus* were collected and thoroughly washed and shade dried for 25–30 days at room temperature. Then they were ground into fine powder and 50 grams of the powder was added to 300 mL of 30% methanol solution in a 500 mL flask and mixed. Then mixture was heated for 30 minutes under continuous stirring at 70°C. Finally, the mixture was cooled and filtered using fluted filter paper.	15 mL of the plant extract was added in drops to 50 mL of 0.03 M aqueous solution of $CoCl_2$ with constant stirring at 80°C. Co^{2+} were reduced to Co^{o} within 30-60 minutes as color of reaction mixture changed that indicated the formation of CoNPs. Then the colored solution was centrifuged at 5000 rpm/40 min, and they were washed thoroughly, dried in an oven, and stored.	Zaib *et al.*, 2020

5.3.8 Potential Applications of Metal Nanoparticles Synthesized Using Agricultural Wastes

5.3.8.1 Removal of Dyes from Wastewater

ZnO synthesized using *Eucalyptus* spp. (irregular in shape and size less than 40 nm) was tested on the removal of Congo red and malachite green; maximum removal was achieved at pH 6.0 and pH 8.0 respectively. The maximum adsorption capacity of ZnONPs was 48.3 mg/g for Congo red dye and 169.5 mg/g for malachite green dye. Moreover, ZnONPs was reusable in the process (Chauhan *et al.*, 2020). Degradation of methylene blue under UV irradiation was demonstrated using the TiO_2NPs generated by palm trunk mince extract. It was observed that methylene blue absorption decreases with the increased irradiation time, while optimum degradation occurred with the shape of the nanowire of TiO_2 (Aziz *et al.*, 2020). Employing the CoNPs synthesized with *Catharanthus roseus* extract demonstrated the degradation ability of anionic dye alizarin red S under room temperature at pH range 2–12 with and without surfactant. Overall, it showed a 91–93% of dye degradation percentage. With increased pH (higher than pH of 6), the degradation percentage was decreased. Maximum decolorization of dye could be observed between pH 4 to 6 while the optimum pH value was found to be 6 (Zaib *et al.*, 2020). Aqueous extract of Mediterranean cypress (*Cupressus sempervirens*) was used by Ebrahiminezhad *et al.* (2018) to generate zero-valent FeNPs.

Methyl orange was applied to study the decolorization efficiency, and was found to be 95% in a 6 h process. Moreover, FeNPs synthesized using red peanut extract were tested in terms of removing Cr^{6+} from contaminated water. Here, FeNPs obtained under vacuum drying at 60°C performed the best, removing 100% of Cr^{6+}, from a 10 mg/L aqueous solution of Cr^{6+} in 1 min (Pan *et al.*, 2020). The Fe/Ni bimetallic nanoparticles synthesized by Weng *et al.* (2018) were tested in terms of removing methyl orange. FeNPs alone removed around 29% of methyl orange, while Fe/Ni bimetallic nanoparticles could remove around 99.6% of methyl orange.

5.3.8.2 Removal of Heavy Metals from Water

Biosorption of Ni and Pb ions from water using agricultural residual biomasses chemically modified with TiO_2 was demonstrated by Herrera-Barros *et al.* (2018). For Ni^{2+}, the removal yields were 81.51% and 86.66% using CP-TiO_2 and YP-TiO_2 biosorbents, indicating they are highly efficacious in removing pollutants. Polymerization of polyaniline nanofibers chains was performed using clay and silica nanoparticles generated using rice husk by Diab *et al.* (2020). These composites were used to remove Zn^{2+} and Cr^{3+} ions from wastewater. The researchers found that the removal efficiency was higher with the composites than polyaniline nanofibers alone. Removal efficiencies were 51% and 92% respectively for Zn^{2+} and Cr^{3+}. Moreover, FeNPs synthesized using *Eucalyptus* leaf extracts were used to remove Cr^{6+} from an aqueous solution. Here, 84.6% of total chromium and 98.6% of Cr^{6+} were removed. The removal efficiency of Cr^{6+} was still relatively high (55.7%) after using it four times (Jin *et al.*, 2018).

5.3.8.3 Antimicrobial Activity

The cell wall composition matters in the penetration of nanomaterials into the bacterial cells. For instance, Gram-positive bacteria have rigid cell wall structures that are difficult to penetrate, while Gram-negative bacteria have thinner cell walls that make penetration easy (Shrivastava *et al.*, 2007; Dang *et al.*, 2017). The AgNPs synthesized were tested against *Escherichia coli* and *Staphylococcus epidermis* for their antimicrobial activity, where *E.coli* showed the maximum inhibition of growth. They found that banana extracts had no antibacterial effect against these two bacterial species (Dang *et al.*, 2017). AgNPs synthesized by Baran (2019) showed strong antimicrobial effects against *E. coli*, *S. aureus,* and *Candida albicans,* with minimum inhibitory concentrations of 55.2, 153.5, and 38.3 µg/mL respectively, and these were compared with certain antibiotics where it was revealed that the synthesized nanoparticles were effective over those antibiotics and 1 mM AgNO3 solution. Much lower antibacterial activity of AgNPs has been reported through various other studies that range within 11.6–100 µg/mL. Moreover, AgNPs synthesized by Ahmed *et al.* (2018) showed antimicrobial activity against both *S. aureus,* which is a Gram-positive, and *E. coli* Gram-negative bacteria species, while the bioactivity assay using human lung cell lines revealed cytotoxicity. Moreover, their antioxidant activity is also evident. Apart from this, green synthesized ZnONPs also showed their antibacterial nature over *E. coli*, *S. aureus*, and *Pseudomonas aeruginosa* bacteria (Fatimah, 2018). Antibacterial properties of AuNPs were also demonstrated by Dang *et al.* (2019), revealing the inhibition of growth of *E. coli* and *S. epidermis.* Sinsinwar *et al.* (2018) produced AgNPs using coconut shell powder extract and tested them against human pathogenic microbial species like *S. aureus, Listeria monocytogenes, E. coli, and Salmonella typhimurium*. The highest inhibition was shown against *S. aureus* (15 mm), followed by *E. coli* (13 mm), *S. typhimurium* (13 mm), and *L. monocytogenes* (10 mm), with minimum inhibitory concentrations of 26, 53, 106, and 212 µg/mL respectively. This growth suppression was found to be effective over ampicillin. Kumar *et al.* (2019b) demonstrated the antibacterial and antifungal properties of the green synthesized AuNPs, and they found AuNPs to be effective in terms of their suppressive ability toward the growth of the microbes. TiO_2 nanoparticles synthesized using palm trunk mince extract were also effective against *E. coli,* with a minimum inhibitory concentration of 200 µg/mL and an inhibition zone of 28 mm. It was formed in hexagonal nanobelts, nanorods, and nanowires (Aziz *et al.*, 2020). ZnONPs synthesized using an aqueous extract of ash derived from *Musa balbisiana* Colla pseudostem also showed efficacy against various bacterial species, namely *E. coli*, *S. aureus*, *Bacillus subtilis*, and *P. aeruginosa.* (Basumatari *et al.*, 2021). *Catharanthus roseus* extract mediated synthesis of CoNPs that were spherical in shape; the size of 27.08 nm also revealed antibacterial efficacy against *B. subtilis* JS 2004 and *E. coli* ATCC 25922; however, it was lower than rifampicin, the standard antibiotic used in the study (Zaib *et al.*, 2020). Using aqueous root extract of *Salvadora persica* plant as a reducing agent Arshad *et al.* (2021) synthesized AgNPs (10–70 nm) in spherical and rod-shaped. They confirmed the growth suppressing efficacy against *S. epidermidis* and *E. coli* with inhibition zones of 20 mm for *S. epidermidis* and 15 mm at a rate of 40 µg/well.

5.4 SYNTHESIS OF NON-METAL NANOMATERIALS FROM AGRICULTURAL WASTES

5.4.1 Extraction and Synthesis of Nanosilica

Most of the crop wastes like rice husk and wheat straw are rich in silica; thus, there is increasing attention to the extraction of silica nanoparticles from agricultural wastes (Naddaf *et al.*, 2019). Nanostructures possess a wide range of applications; they exhibit enhanced physical and chemical properties due to their high surface area to volume ratio and quantum effect (Mor *et al.*, 2017; Naddaf *et al.*, 2019). Several studies have recently reported on using crop wastes to generate silica nanoparticles.

5.4.1.1 Extraction of Silica Nanoparticles from Rice Husk

Because rice is a staple diet in most Asian countries, around 770 million tons of rice husk are produced each year. It is estimated that approximately 220kg of rice husk is obtained from 1000 kg of paddy (Mor *et al.*, 2017). Combustion of rice husk produces 25% of rice husk ash (RHA) of initial total weight, and RHA causes environmental problems like pollution and certain health problems in humans (Bazargan *et al.*, 2014; Mor *et al.*, 2017). There are research studies in which silica nanoparticles from RHA were green synthesized using the sol-gel method. The flow chart illustrates the preparation of silica nanoparticles using RHA (Figure 5.1).

The principal reactions which occur during this process are:

$$NaOH + SiO_2 \rightarrow Na_2SiO_3$$

$$Na_2SiO_3 + H_2O + HCl \rightarrow Si(OH)_4 + NaCl$$

$$\equiv Si\text{-}OH + \equiv Si\text{-}OH \rightarrow \equiv Si\text{-}O\text{-}Si \equiv + H_2O$$

During the acidification, silicon hydroxide condenses and polymerizes and is linked by siloxane bonds (-Si-O-Si-) (Zulkifli *et al.*, 2013).

5.4.1.2 Extraction of Silica Nanoparticles from Sugarcane Wastes

Sugarcane is utilized in the sugar and ethanol industries, and sugarcane bagasse, which is a fibrous residue, is obtained as a waste product. Bagasse consists of a high percentage of silica (Asagekar and Joshi, 2014; Mupa *et al.*, 2015). About 54 million tons of sugarcane bagasse are annually produced globally, and sugarcane bagasse ash contains 64.38% of silica (Kadir and Maasom, 2013; Kharade *et al.*, 2014). The combustion of sugarcane waste releases greenhouse gases, which lead to global warming and health problems (Mohd *et al.*, 2017; Rovani *et al.*, 2018). The utilization of sugarcane bagasse to produce silica nanoparticles is a green approach, and one of those research projects on "green synthesis of silica nanoparticles using sugarcane bagasse" is discussed here (Mohd *et al.*, 2017). It has been done in two consecutive processes, namely, silica extraction and the precipitation process. Extraction is carried out as explained in Figure 5.1.

During precipitation process, sodium silicate solution is precipitated by using nitric acid. Resulting precipitate is furnaced at 600°C for 30 min to obtain silica nano powder.

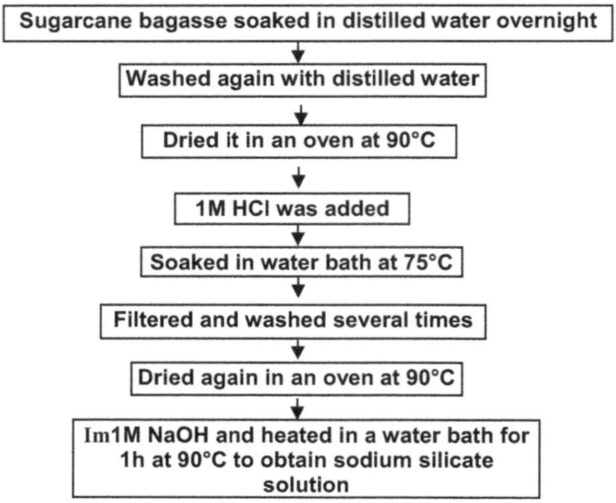

FIGURE 5.1 Silica extraction.

5.4.1.3 Extraction of Silica Nanoparticles from Olive Stones

Olive is one of the major crops in the Mediterranean region where it is used to extract olive oil, which plays an important role in their daily diet (Naddaf *et al.*, 2019; Valvez *et al.*, 2021). Production has increased with the evolution of technology. At the same time, the production of olive waste has been increased. Further, it has been estimated that 0.6 tons of olive waste is produced from 1 ton of olives. For the world harvest in 2005/2006, 2.58 and 1.73 million tons of olive stones were obtained from the olive oil and table olive industries, respectively (Rodríguez *et al.*, 2008). It has been revealed that olive pits and olive pressings contain 31% and 21% of silica, respectively (Lancaster, 2015). A process of extraction of silica nanoparticles using olive stones obtained from table olives is discussed later (Naddaf *et al.*, 2019).

Finally, white or grayish-colored silica was obtained. In order to obtain bright white silica, this silica was subjected to sintering processes by heating at 900°C for 2 h under atmospheric conditions. The silica yield of this experiment was calculated to be around 15%.

5.4.1.4 Extraction of Silica Nanoparticles from Cassava Periderm (CP)

Nigeria had the highest percentage of cassava production in the world in 2009, generating more than 3.4 million tons of cassava peeling, which contains the cassava periderm as a crop waste (Adebisi *et al.*, 2018). Ash of cassava periderm contains 6.10% of silicon, and 38% of that is silica (Adebisi *et al.*, 2017). Adebisi *et al.* (2018) investigated the utilization of cassava periderm as a source of silica nanoparticles, in which silica nanoparticles were produced using a modified sol-gel method and extraction process of silica. Through this process, silica nanoparticles were obtained having diameters within the range of 3.12–50.75 nm.

Adebisi *et al.* (2018) introduced a method for producing silicon nanoparticles by a reduction method from cassava periderm. Herein, the aforementioned process was followed to produce silica. After that, silica and magnesium powder were ground and mixed in a 1:1 ratio. To obtain silicon nanoparticles, pallets of the mixture of magnesium and silica were formed using 4% PVA, and then the pallets were heated at 650°C for 30 minutes to reduce silica into silicon. The reaction that occurred during this process is:

$$SiO_2 + Mg \rightarrow Si + MgO$$

Silica nanoparticles and silicon nanoparticles obtained were within 2–79 nm and 12–164 nm, respectively.

5.4.1.5 Potential Uses of Nanosilica

A number of uses for nanosilica exist in the building materials industry. Nanosilica-incorporated asphalt binder has been found to improve rutting resistance, thus increasing resistance to permanent deformation. Moreover, it is highly resistant to oxidative aging while exhibiting good storage stability at higher temperatures (Bhat and Mir, 2019). Nanosilica-enhanced blended cement mortar and concrete have been shown to develop mechanical properties of mortar and concrete at a young age of curing with improved microstructural arrangements (Raheem *et al.*, 2021). The experimental results showed that nanosilica has remarkable effects on the mechanical and transport properties of lightweight aggregate concrete, even in small amounts, with a significant improvement in strength and a reduction of transport properties in specimens when increasing the nanosilica content (Elrahman *et al.*, 2019).

5.4.2 Extraction Methods of Nanocellulose (NC)

Cellulose is the most abundant natural polymer on Earth, and it is a renewable, biodegradable, and non-toxic material. Therefore, cellulose has gained significant interest. Cellulose is the major structural component of the plant cell walls (Teeri *et al.*, 2007). Nowadays, interest has turned toward the extraction of nanocellulose using plant materials with the advancement of nanotechnology. NC can be produced either by a top-down method, which is chemical or physical disintegration, or a bottom-up method in which microorganisms are used (Dufresne, 2013). NC is categorized into three types: cellulose nanocrystals, cellulose nanofibrils, and bacterial cellulose (Panchal *et al.*, 2019). Agricultural wastes from the most abundantly produced crop species can be utilized for this extraction of nanocellulose. For instance, approximately 1 trillion tons of maize are produced annually (FAO Statistics, 2015), while 45 million tons of maize husk are produced for 640 million tons of maize (Pordesimo *et al.*, 2005). Maize husk contains 43.14% of α-cellulose (Smyth *et al.*, 2017).

5.4.2.1 Extraction Methods of Nanocellulose and Cellulose Nanocrystals

Smyth *et al.* (2017) presented a protocol to extract cellulose nanocrystals (CNC) from maize husk. In this study, CNCs were produced by the acid hydrolysis method, and before that alkali and leaching were carried out to remove other chemical substances from cellulose. In the acid hydrolysis method, the amorphous component of cellulose is degraded and the crystalline component is retained, confirming CNCs. Procedures for alkali and leaching treatment and the acid hydrolysis method, which is used for maize husk (MH), are mentioned here. The whole process can be divided into two main steps: alkali treatment followed by leaching (Figure 5.2) and acid hydrolysis (Figure 5.3). Isolation of CNCs was successfully achieved by this approach with a 15.61% yield percentage.

About 100 million of citrus is produced annually; oranges had the majority among the citrus crops produced in 2004. A large part of that is used for industrial extraction of juice, which leads to large amounts of wastes like segment membrane and peel, which contains 9.21% of cellulose (Rivas *et al.*, 2008). Mariño *et al.* (2015)

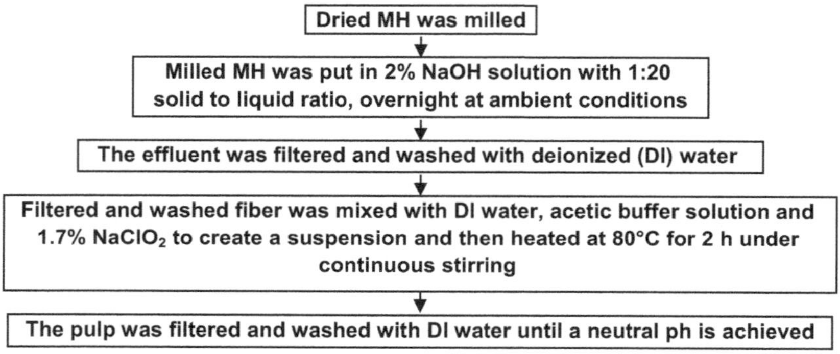

FIGURE 5.2 Process alkali treatment followed by leaching.

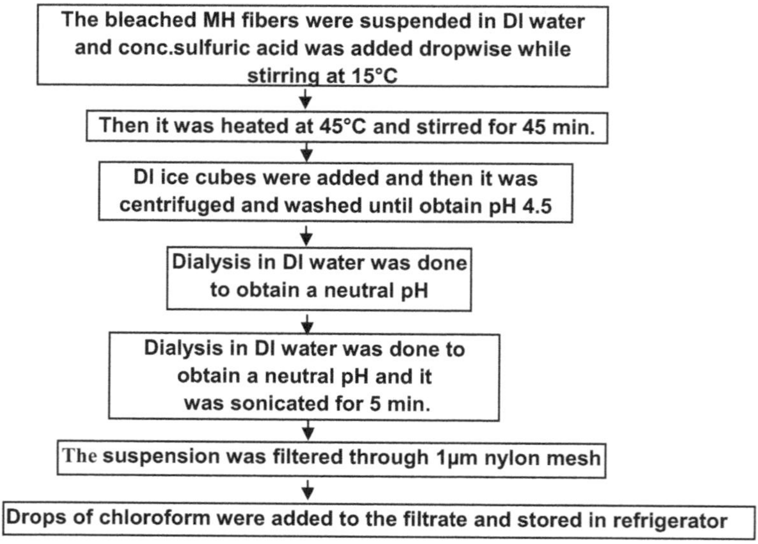

FIGURE 5.3 Process alkali treatment followed by acid hydrolysis.

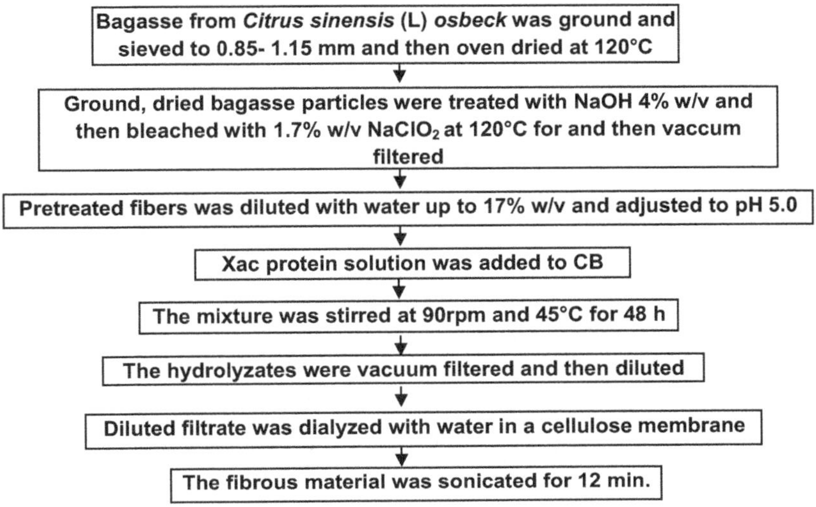

FIGURE 5.4 Process of extraction of nanocellulose.

investigated the utilization of citrus-waste biomass (CB) to produce nanocellulose by enzymatic hydrolysis process, in which enzymes produced *Xanthomonas axonopodispv. citri* (Xac) strain 306 (IBSBF 1594), a bacterium isolated from infected citrus fruits. The process of extraction of nanocellulose is described under Figure 5.4. Nanofibers with 55% crystallinity, average diameter of 10 nm, and length of 458 nm were obtained from this approach. Xac protein solution, which contains enzymes, was produced following the procedures, which were described by Awan *et al.* (2013).

5.4.2.2 Potential Uses of Nanocellulose

Nanocellulose has been widely researched in terms of its efficient adsorbent properties that aid in the remediation of heavy metals, dissolved organic substances, dyes, oil, and effluents (Shak *et al.*, 2018; Mahfoudhi and Boufi, 2017). Moreover, its photocatalytic ability and membrane technological applications have been widely studied. Among the products that have been developed using these properties, nanocellulose emulsifiers, wood adhesives, wastewater treatment, as well as new evolving biomedical applications, are vital (Trache *et al.*, 2020). Recently, this includes the development of nanocellulose-based films that can be used in touchscreens, solar cells, and nanogenerators (Fang *et al.*, 2019).

5.4.3 Extraction of Activated Carbon Nanoparticles (ACNPs)

Carbon NPs are activated mostly by using physical and chemical activation (Lakshmi *et al.*, 2018). Nowadays, biowaste materials have gained significant attention in the production of ACNPs as a green approach; this is due to the fact that it does not use harmful materials and generates less waste after the reaction (Duan *et al.*, 2015).

5.4.3.1 Extraction Methods of Activated Carbon Nanoparticles (ACNPs)

Agricultural waste materials, such as sugarcane bagasse, rice husks, ground nut shells, and various wood bark species, are widely researched to synthesize ACNPs. Finally, it was dried at 110°C for 24 hrs. Rice husks have also been used to synthesize ACNPs (Sekaran *et al.*, 2013). First, rice husks were precarbonized at 400°C to obtain mesoporous activated carbon, and then chemical activation was done with phosphoric acid at 85°C for 4 h. Then it was heated at 700, 800, and 900°C at 5°C per minute, and finally washed with hot water and dried at 110°C (Sekaran *et al.*, 2013). Ground nut shell is used in the textile industry as a fuel and in the pharmaceutical and food processing industries (Lakshmi *et al.*, 2018). It has medicinal and nutritional value (Sujatha *et al.*, 2016). Subsequently, the soaked dried shells were crushed and ground and subjected to pyrolysis at 550, 750, and 950°C for 1 h in a nitrogen atmosphere to obtain ACNPs. Sandalwood is an indigenous plant in India (Lakshmi *et al.*, 2018) and is widely used in beauticulture and Ayurveda medicine (Vachirayonstein *et al.*, 2006; Shamsi *et al.*, 2014; Dey *et al.*, 2014).

5.4.3.2 Potential Uses of Activated ACNPs

ACNPs are used for energy storage devices like high-performance supercapacitors, and air filtering devices, water purification, and gas purification (Pelgrift and Friedman, 2013; Ijaola *et al.*, 2013; Alhassan *et al.*, 2018). Moreover, ACNPs derived from biowaste have been widely used as antimicrobial agents.

5.5 FUTURE PERSPECTIVES

In a milieu where the world is emphasizing sustainable technologies, nanotechnology coupled with green technology can bring promising outcomes over conventional methods, which are not benign. It is clear that an array of agricultural wastes can effectively be utilized in the synthesis of nanomaterials, and the technology can be transferred to future generations if the drawbacks are successfully addressed through detailed studies. Such drawbacks include the higher time-consuming nature, controlling the size and shape of the nanomaterials, monodispersity, and durability over chemical and physical methods. Therefore, efforts related to large-scale production are not adequate, and large-scale production targeting industrial needs should be emphasized. Moreover, more research studies have to be done with other waste types which have not yet gained considerable attention. Aside from the lack of knowledge about the toxicity of green-synthesized nanomaterials, waste management during green processes is a major concern that has an impact on scalar production.

5.6 REFERENCES

Abolghasemi, R., Haghighi, M., Solgi, M. and Mobinikhaledi, A., 2019. Rapid synthesis of ZnO nanoparticles by waste thyme (*Thymus vulgaris* L.). *International Journal of Environmental Science and Technology*, 16(11):6985–6990.

Adebisi, J.A., Agunsoye, J.O., Bello, S.A., Ahmed, I.I., Ojo, O.A. and Hassan, S.B., 2017. Potential of producing solar grade silicon nanoparticles from selected agro-wastes: A review. *Solar Energy*, 142:68–86.

Adebisi, J.A., Agunsoye, J.O., Bello, S.A., Haris, M., Ramakokovhu, M.M., Daramola, M.O. and Hassan, S.B., 2018. Extraction of silica from cassava periderm using modified sol-gel method. *Nigerian Journal of Technological Development*, 15(2):57–65.

Ahmed, S., Kaur, G., Sharma, P., Singh, S. and Ikram, S., 2018. Fruit waste (peel) as bio-reductant to synthesize silver nanoparticles with antimicrobial, antioxidant and cytotoxic activities. *Journal of Applied Biomedicine*, 16(3):221–231.

Akhter, G., Khan, A., Ali, S.G., Khan, T.A., Siddiqi, K.S. and Khan, H.M., 2020. Antibacterial and nematicidal properties of biosynthesized Cu nanoparticles using extract of holoparasitic plant. *SN Applied Sciences*, 2(7):1–6.

Alhassan, M., Andrew, I., Auta, M., Umaru, M., Garba, M.U., Isah, A.G. and Alhassan, B., 2018. Comparative studies of CO_2 capture using acid and base modified activated carbon from sugarcane bagasse. *Biofuels*, 9(6):719–728.

Aminuzzaman, M., Kei, L.M. and Liang, W.H., 2017, April. Green synthesis of copper oxide (CuO) nanoparticles using banana peel extract and their photocatalytic activities. In *AIP Conference Proceedings* 1828(1):020016. AIP Publishing LLC.

Arshad, H., Sami, M.A., Sadaf, S. and Hassan, U., 2021. *Salvadora persica* mediated synthesis of silver nanoparticles and their antimicrobial efficacy. *Scientific Reports*, 11(1):1–11.

Asagekar, S.D. and Joshi, V.K. 2014. Characteristics of sugarcane fibres. *Indian Journal of Fibre and Textile Research*, 39:180–184.

Awan, A.T., Tsukamoto, J. and Tasic, L., 2013. Orange waste as a biomass for 2G-ethanol production using low-cost enzymes and co-culture fermentation. *RSC Advances*, 3(47):25071–25078.

Aziz, W.J., Ghazai, A.J., Abd, A.N. and Habubi, N.F., 2020, November. Synthesis of TiO_2NPs with agricultural waste for photocatalytic and antibacterial applications. In *Journal of Physics: Conference Series* 1660(1):012063. IOP Publishing.

Baiano, A., 2014. Recovery of biomolecules from food wastes—A review. *Molecules*, 19(9):14821–14842.

Baig, N., Kammakakam, I. and Falath, W., 2021. Nanomaterials: A review of synthesis methods, properties, recent progress, and challenges. *Materials Advances*, 2(6):1821–1871.

Balalakshmi, C., Gopinath, K., Govindarajan, M., Lokesh, R., Arumugam, A., Alharbi, N.S., Kadaikunnan, S., Khaled, J.M. and Benelli, G., 2017. Green synthesis of gold nanoparticles using a cheap *Sphaeranthus indicus* extract: Impact on plant cells and the aquatic crustacean *Artemia nauplii*. *Journal of Photochemistry and Photobiology B: Biology*, 173:598–605.

Bandeira, M., Giovanela, M., Roesch-Ely, M., Devine, D.M. and da Silva Crespo, J., 2020. Green synthesis of zinc oxide nanoparticles: A review of the synthesis methodology and mechanism of formation. *Sustainable Chemistry and Pharmacy*, 15:100223.

Baran, M.F., 2019. Synthesis, characterization and investigation of antimicrobial activity of silver nanoparticles from *Cydonia oblonga* leaf. *Applied Ecology and Environmental Research*, 17(2):2583–2592.

Basumatari, M., Devi, R.R., Gupta, M.K., Gupta, S.K., Raul, P.K., Chatterjee, S. and Dwivedi, S.K., 2021. *Musa balbisiana* Colla pseudostem biowaste mediated zinc oxide nanoparticles: Their antibiofilm and antibacterial potentiality. *Current Research in Green and Sustainable Chemistry*, 4:100048.

Bazargan, A., Gebreegziabher, T., Hui, C.-W. and McKay, G., 2014. The effect of alkali treatment on rice husk moisture content and drying kinetics. *Biomass and Bioenergy*, 70:468–475. doi:10.1016/j.biombioe.2014.08.018.

Bhat, F.S. and Mir, M.S., 2019. Performance evaluation of nanosilica-modified asphalt binder. *Innovative Infrastructure Solutions*, 4(1):1–10.

Chauhan, A.K., Kataria, N. and Garg, V.K., 2020. Green fabrication of ZnO nanoparticles using *Eucalyptus* spp. leaves extract and their application in wastewater remediation. *Chemosphere*, 247:125803.

Dang, H., Fawcett, D. and Poinern, G.E.J., 2017. Biogenic synthesis of silver nanoparticles from waste banana plant stems and their antibacterial activity against *Escherichia coli* and *Staphylococcus epidermis. International Journal of Research in Medical Sciences*, 5(9):3769–3775.

Dang, H., Fawcett, D. and Poinern, G.E.J., 2019. Green synthesis of gold nanoparticles from waste macadamia nut shells and their antimicrobial activity against *Escherichia coli* and *Staphylococcus epidermis. International Journal of Research in Medical Sciences*, 7(4).

De Laurentiis, V., Corrado, S. and Sala, S., 2018. Quantifying household waste of fresh fruit and vegetables in the EU. *Waste Management*, 77:238–251. doi:10.1016/j.wasman.2018.04.001.

Dey, P., Karuna, D.S. and Bhakta, T., 2014. Medicinal plants used as anti-acne agents by tribal and non-tribal people of Tripura, India. *American Journal of Phytomedicine and Clinical Therapeutics*, 2(5):556–570.

Diab, M.A., Attia, N.F., Attia, A. and El-Shahat, M.F., 2020. Green synthesis of cost-effective and efficient nanoadsorbents based on zero and two dimensional nanomaterials for Zn^{2+} and Cr^{3+} removal from aqueous solutions. *Synthetic Metals*, 265:116411.

Doan, V.D., Luc, V.S., Nguyen, T.L.H., Nguyen, T.D. and Nguyen, T.D., 2020. Utilizing waste corn-cob in biosynthesis of noble metallic nanoparticles for antibacterial effect and catalytic degradation of contaminants. *Environmental Science and Pollution Research*, 27(6):6148–6162.

Duan, H., Wang, D. and Li, Y., 2015. Green chemistry for nanoparticle synthesis. *Chemical Society Reviews*, 44(16):5778–5792.http://dx.doi.org/10.1039/C1GC15386B.

Dufresne, A., 2013. Nanocellulose: A new ageless bionanomaterial. *Materials Today* 16(6):220–227. https:// doi.org/10.1016/j.mattod.2013.06.004.

Ebrahiminezhad, A., Taghizadeh, S., Ghasemi, Y. and Berenjian, A., 2018. Green synthesized nanoclusters of ultra-small zero valent iron nanoparticles as a novel dye removing material. *Science of the Total Environment*, 621:1527–1532.

Elrahman, M.A., Chung, S.Y., Sikora, P., Rucinska, T. and Stephan, D., 2019. Influence of nanosilica on mechanical properties, sorptivity, and microstructure of lightweight concrete. *Materials*, 12(19):3078.

Fang, Z., Hou, G., Chen, C. and Hu, L., 2019. Nanocellulose-based films and their emerging applications. *Current Opinion in Solid State and Materials Science*, 23(4):100764.

FAO statistics, 2015. Food and agricultural organization of the United Nations statistics division, www.fao.org/faostat/en/.

Fatimah, I., 2018. Biosynthesis and characterization of ZnO nanoparticles using rice bran extract as low-cost templating agent. *Journal of Engineering Science and Technology*, 13(2):409–420.

Fouad, H., Hongjie, L., Hosni, D., Wei, J., Abbas, G., Ga'al, H. and Jianchu, M., 2018. Controlling *Aedes albopictus* and *Culex pipienspallens* using silver nanoparticles synthesized from aqueous extract of *Cassia fistula* fruit pulp and its mode of action. *Artificial cells, Nanomedicine, and Biotechnology*, 46(3):558–567.

Ghosh, P.R., Fawcett, D., Sharma, S.B. and Poinern, G.E., 2017. Production of high-value nanoparticles via biogenic processes using aquacultural and horticultural food waste. *Materials*, 10(8):852.

Goutam, S.P., Saxena, G., Singh, V., Yadav, A.K., Bharagava, R.N. and Thapa, K.B., 2018. Green synthesis of TiO_2 nanoparticles using leaf extract of *Jatropha curcas* L. for photocatalytic degradation of tannery wastewater. *Chemical Engineering Journal*, 336:386–396.

Govindappa, M., Tejashree, S., Thanuja, V., Hemashekhar, B., Srinivas, C., Nasif, O., Pugazhendhi, A. and Raghavendra, V.B., 2021. Pomegranate fruit fleshy pericarp mediated silver nanoparticles possessing antimicrobial, antibiofilm formation, antioxidant, biocompatibility and anticancer activity. *Journal of Drug Delivery Science and Technology*, 61:102289.

Govindarajan, M., AlQahtani, F.S., AlShebly, M.M., Benelli, G., 2017. One-pot and eco-friendly synthesis of silver nanocrystals using *Adiantum raddianum*: Toxicity against mosquito vectors of medical and veterinary importance. *Journal of Applied Biomedicine*, 15:87–95.

Gregorczyk, K. and Knez, M., 2016. Hybrid nanomaterials through molecular and atomic layer deposition: Top down, bottom up, and in-between approaches to new materials. *Progress in Materials Science*, 75:1–37.

Herrera-Barros, A., Tejada-Tovar, C., Villabona-Ortiz, A., Gonzalez-Delgado, A. and Fornaris-Lozada, L., 2018. Effect of pH and particle size for lead and nickel uptake from aqueous solution using cassava (*Manihot esculenta*) and yam (*Dioscorea alata*) residual biomasses modified with titanium dioxide nanoparticles. *Indian Journal of Science and Technology*, 11(21):1–7.

Ijaola, O.O., Ogedengbe, K. and Sangodoyin, A.Y., 2013. On the efficacy of activated carbon derived from bamboo in the adsorption of water contaminants. *International Journal of Engineering Inventions*, 2(4):29–34.

Ishak, N.A.I.M., Kamarudin, S.K., Timmiati, S.N., Karim, N.A. and Basri, S., 2021. Biogenic platinum from agricultural wastes extract for improved methanol oxidation reaction in direct methanol fuel cell. *Journal of Advanced Research*, 28:63–75.

Jin, X., Liu, Y., Tan, J., Owens, G. and Chen, Z., 2018. Removal of Cr (VI) from aqueous solutions via reduction and absorption by green synthesized iron nanoparticles. *Journal of Cleaner Production*, 176:929–936.

Kadir, A.A. and Maasom, N., 2013. Recycling sugarcane bagasse waste into fired clay brick. *International Journal of Zero Waste Generation*, 1(1):21–26.

Kamari, S. and Ghorbani, F., 2020. Extraction of highly pure silica from rice husk as an agricultural by-product and its application in the production of magnetic mesoporous silica MCM–41. *Biomass Conversion and Biorefinery*, 1–9. https://doi.org/10.1007/s13399-020-00637 w.

Kharade, A.S., Suryavanshi, V.V., Gujar, B.S. and Deshmukh, R.R., 2014. Waste product bagasse ash from sugar industry can be used as stabilizing material for expansive soils. *International Journal of Research in Engineering and Technology*, 3(3):506–512.

Kringel, D.H., Dias, A.R.G., Zavareze, E.D.R. and Gandra, E.A., 2019. Fruit wastes as promising sources of starch: Extraction, properties, and applications. *Starch-Stärke*, 72(3–4):1900200.

Kumar, B., Smita, K., Galeas, S., Sharma, V., Guerrero, V.H., Debut, A. and Cumbal, L., 2020a. Characterization and application of biosynthesized iron oxide nanoparticles using *Citrus paradisi* peel: A sustainable approach. *Inorganic Chemistry Communications*, 119:108116.

Kumar, H., Bhardwaj, K., Sharma, R., Nepovimova, E., Kuča, K., Dhanjal, D.S., Verma, R., Bhardwaj, P., Sharma, S. and Kumar, D., 2020b. Fruit and vegetable peels: Utilization of high value horticultural waste in novel industrial applications. *Molecules*, 25(12):2812.

Kumar, P.V., Kala, S.M.J. and Prakash, K.S., 2019a. Green synthesis derived Pt-nanoparticles using *Xanthium strumarium* leaf extract and their biological studies. *Journal of Environmental Chemical Engineering*, 7(3):103146.

Kumar, P.V., Kala, S.M.J. and Prakash, K.S., 2019b. Green synthesis of gold nanoparticles using *Croton Caudatus* Geisel leaf extract and their biological studies. *Materials Letters*, 236:19–22.

Lakshmi, S.D., Avti, P.K. and Hegde, G., 2018. Activated carbon nanoparticles from biowaste as new generation antimicrobial agents: A review. *Nano-Structures and Nano-Objects*, 16:306–321.

Lancaster, L.C., 2015. *Innovative Vaulting in the Architecture of the Roman Empire: 1st to 4th Centuries CE*. Cambridge University Press.

Lateef, A., Akande, M.A., Azeez, M.A., Ojo, S.A., Folarin, B.I., Gueguim-Kana, E.B., Beukes, L.S., 2016. Phytosynthesis of silver nanoparticles (AgNPs) using miracle fruit plant (*Synsepalum dulcificum*) for antimicrobial, catalytic, anticoagulant, and thrombolytic applications. *Nanotechnology Reviews*, 5(6):507–520.

Lu, Z., Wang, J., Gao, R., Ye, F. and Zhao, G., 2019. Sustainable valorisation of tomato pomace: A comprehensive review. *Trends in Food Science and Technology*. doi:10.1016/j.tifs.2019.02.020.

Lyu, F., Luiz, S.F., Azeredo, D.R.P., Cruz, A.G., Ajlouni, S. and Ranadheera, C.S., 2020. Apple pomace as a functional and healthy ingredient in food products: A review. *Processes*, 8(3):319.

Mahfoudhi, N. and Boufi, S., 2017. Nanocellulose as a novel nanostructured adsorbent for environmental remediation: A review. *Cellulose*, 24(3):1171–1197.

Mariño, M., Lopes da Silva, L., Durán, N. and Tasic, L., 2015. Enhanced materials from nature: Nanocellulose from citrus waste. *Molecules*, 20(4):5908–5923.

Matharu, A.S. and Lokesh, K., 2019. Green Chemistry Principles and Global Drivers for Sustainability–An Introduction. *Green Chemistry for Surface Coatings, Inks and Adhesives: Sustainable Applications*:1–17.

Mohd, N.K., Wee, N.N.A.N. and Azmi, A.A., 2017, September. Green synthesis of silica nanoparticles using sugarcane bagasse. In *AIP Conference Proceedings* 1885(1):020123. AIP Publishing LLC.

Mor, S., Manchanda, C.K., Kansal, S.K. and Ravindra, K., 2017. Nanosilica extraction from processed agricultural residue using green technology. *Journal of Cleaner Production*, 143:1284–1290.

Mupa, M., Hungwe, C.B., Witzleben, S., Mahamadi, C. and Muchanyereyi, N., 2015. Extraction of silica gel from *Sorghum bicolour* (L.) moench bagasse ash. *African Journal of Pure and Applied Chemistry*, 9(2):12–17.

Nabi, G., Ain, Q.U., Tahir, M.B., Nadeem Riaz, K., Iqbal, T., Rafique, M., Hussain, S., Raza, W., Aslam, I. and Rizwan, M., 2020. Green synthesis of TiO_2 nanoparticles using lemon peel extract: Their optical and photocatalytic properties. *International Journal of Environmental Analytical Chemistry*, 102(2):434–442.

Naddaf, M., Kafa, H. and Ghanem, I., 2019. Extraction and characterization of Nano-silica from olive stones. *Silicon*, 12(1):185–192.

Nazar, N., Bibi, I., Kamal, S., Iqbal, M., Nouren, S., Jilani, K., Umair, M. and Ata, S., 2018. Cu nanoparticles synthesis using biological molecule of *Punica granatum* seeds extract as reducing and capping agent: Growth mechanism and photo-catalytic activity. *International Journal of Biological Macromolecules*, 106:1203–1210.

Ndikau, M., Noah, N.M., Andala, D.M. and Masika, E., 2017. Green synthesis and characterization of silver nanoparticles using *Citrullus lanatus* fruit rind extract. *International Journal of Analytical Chemistry*, 2017:8108504.

Nguyen, P.A., Nguyen, A.V.P., Dang-Bao, T., Phan, H.P., Van Nguyen, T.T., Tran, B.A., Huynh, T.L.D., Hoang, T.C. and Nguyen, T., 2020. Green synthesis of copper nanoparticles using Cocoa pod extract and its catalytic activity in deep oxidation of aromatic hydrocarbons. *SNApplied Sciences*, 2(11):1–13.

Nomura, K. and Terwilliger, P., 2019. Self-dual Leonard pairs. *Special Matrices*, 7(1):1–19.

Pan, Z., Lin, Y., Sarkar, B., Owens, G. and Chen, Z., 2020. Green synthesis of iron nanoparticles using red peanut skin extract: Synthesis mechanism, characterization and effect of conditions on chromium removal. *Journal of Colloid and Interface Science*, 558:106–114.

Panchal, P., Ogunsona, E. and Mekonnen, T., 2019. Trends in advanced functional material applications of nanocellulose. *Processes*, 7(1):10.

Pelgrift, R.Y. and Friedman, A.J., 2013. Nanotechnology as a therapeutic tool to combat microbial resistance. *Advanced Drug Delivery Reviews*, 65(13–14):1803–1815.

Pordesimo, L.O., Hames, B.R., Sokhansanj, S. and Edens, W.C., 2005. Variation in corn stover composition and energy content with crop maturity. *Biomass and Bioenergy*, 28(4):366–374.

Rahaiee, S., Ranjbar, M., Azizi, H., Govahi, M. and Zare, M., 2020. Green synthesis, characterization, and biological activities of saffron leaf extract-mediated zinc oxide nanoparticles: A sustainable approach to reuse an agricultural waste. *Applied Organometallic Chemistry*, 34(8):e5705.

Raheem, A.A., Abdulwahab, R. and Kareem, M.A., 2021. Incorporation of metakaolin and nanosilica in blended cement mortar and concrete-A review. *Journal of Cleaner Production*, 125852.

Rivas, B., Torrado, A., Torre, P., Converti, A. and Domínguez, J.M., 2008. Submerged citric acid fermentation on orange peel autohydrolysate. *Journal of Agricultural and Food Chemistry*, 56(7):2380–2387.

Rodríguez, G., Lama, A., Rodríguez, R., Jiménez, A., Guillén, R. and Fernández-Bolanos, J., 2008. Olive stone an attractive source of bioactive and valuable compounds. *Bioresource Technology*, 99(13):5261–5269.

Rolim, P.M., Seabra, L.M.A.J. and de Macedo, G.R., 2020. Melon by-products: Biopotential in human health and food processing. *Food Reviews International*, 36(1):15–38.

Roopan, S.M., Bharathi, A., Prabhakarn, A., Rahuman, A.A., Velayutham, K., Rajakumar, G., Padmaja, R.D., Lekshmi, M. and Madhumitha, G., 2012. Efficient phyto-synthesis and structural characterization of rutile TiO_2 nanoparticles using *Annona squamosa* peel extract. *Spectrochimica Acta Part A: Molecular and Biomolecular Spectroscopy*, 98:86–90.

Rovani, S., Santos, J.J., Corio, P. and Fungaro, D.A., 2018. Highly pure silica nanoparticles with high adsorption capacity obtained from sugarcane waste ash. *ACS Omega*, 3(3):2618–2627.

Sagar, N.A., Pareek, S., Sharma, S., Yahia, E.M. and Lobo, M.G., 2018. Fruit and vegetable waste: Bioactive compounds, their extraction, and possible utilization. *Comprehensive Reviews in Food Science and Food Safety*, 17:512–531.

Sekaran, G., Karthikeyan, S., Gupta, V.K., Boopathy, R. and Maharaja, P., 2013. Immobilization of *Bacillus* sp. in mesoporous activated carbon for degradation of sulphonated phenolic compound in wastewater. *Materials Science and Engineering: C*, 33(2):735–745.

Shabaani, M., Rahaiee, S., Zare, M. and Jafari, S.M., 2020. Green synthesis of ZnO nanoparticles using loquat seed extract; Biological functions and photocatalytic degradation properties. *LWT-Food Science and Technology*, 134:10133.

Shak, K.P.Y., Pang, Y.L. and Mah, S.K., 2018. Nanocellulose: Recent advances and its prospects in environmental remediation. *Beilstein Journal of Nanotechnology*, 9(1):2479–2498.

Shamsi, T.N., Parveen, R., Afreen, S., Azam, M., Fatma, T., Haque, Q.M.R. and Fatima, S., 2014. *In-vitro* antibacterial and antioxidant activities of sandalwood (*Santalum Album*). *Austin Journal of Biotechnology and Bioengineering*, 1:2.

Shrivastava, S., Bera, T., Roy, A., Singh, G., Ramachandrarao, P. and Dash, D., 2007. Characterization of enhanced antibacterial effects of novel silver nanoparticles. *Nanotechnology*, 18:103–12.

Singh, P., Pandit, S., Garnæs, J., Tunjic, S., Mokkapati, V.R., Sultan, A., Thygesen, A., Mackevica, A., Mateiu, R.V., Daugaard, A.E. and Baun, A., 2018. Green synthesis of gold and silver nanoparticles from *Cannabis sativa* (industrial hemp) and their capacity for biofilm inhibition. *International Journal of Nanomedicine*, 13:3571.

Sinsinwar, S., Sarkar, M.K., Suriya, K.R., Nithyanand, P. and Vadivel, V., 2018. Use of agricultural waste (coconut shell) for the synthesis of silver nanoparticles and evaluation of their antibacterial activity against selected human pathogens. *Microbial Pathogenesis*, 124:30–37.

Skiba, M.I. and Vorobyova, V.I., 2019. Synthesis of silver nanoparticles using orange peel extract prepared by plasmochemical extraction method and degradation of methylene blue under solar irradiation. *Advances in Materials Science and Engineering*, 2019:8306015.

Smyth, M., García, A., Rader, C., Foster, E.J. and Bras, J., 2017. Extraction and process analysis of high aspect ratio cellulose nanocrystals from corn (*Zea mays*) agricultural residue. *Industrial Crops and Products*, 108:257–266.

Sujatha, E.R., Dharini, K. and Bharathi, V., 2016. Influence of groundnut shell ash on strength and durability properties of clay. *Geomechanics and Geoengineering*, 11(1):20–27.

Teeri, T.T., Brumer III, H., Daniel, G. and Gatenholm, P., 2007. Biomimetic engineering of cellulose-based materials. *Tends in Biotechnology*, 25(7):299–306.

Trache, D., Tarchoun, A.F., Derradji, M., Hamidon, T.S., Masruchin, N., Brosse, N. and Hussin, M.H., 2020. Nanocellulose: From fundamentals to advanced applications. *Frontiers in Chemistry*, 2020:8.

Vachirayonstein, T., Sirotamarat, S., Balachandra, K. and Saifah, E., 2006. Cytotoxicity and inhibitory activity on hepatitis B surface antigen secretion from PLC/PRF/5 cells of medicinal plant extracts. *The Journal of Pharmaceutical Sciences*, 30(1–2):1–7.

Valvez, S., Maceiras, A., Santos, P. and Reis, P.N., 2021. Olive stones as filler for polymer-based composites: A review. *Materials*, 14(4):845.

Wang, L., Xu, J., Yan, Y., Liu, H., Karunakaran, T. and Li, F., 2019. Green synthesis of gold nanoparticles from *Scutellaria barbata* and its anticancer activity in pancreatic cancer cell (PANC-1). *Artificial Cells, Nanomedicine, and Biotechnology*, 47(1):1617–1627.

Weng, X., Cai, W., Lan, R., Sun, Q. and Chen, Z., 2018. Simultaneous removal of amoxicillin, ampicillin and penicillin by clay supported Fe/Ni bimetallic nanoparticles. *Environmental Pollution*, 236:562–569.

Zaib, M., Shahzadi, T., Muzammal, I. and Farooq, U., 2020. *Catharanthus roseus* extract mediated synthesis of cobalt nanoparticles: Evaluation of antioxidant, antibacterial, hemolytic and catalytic activities. *Inorganic and Nano-Metal Chemistry*, 50(11):1171–1180.

Zulkifli, N.S.C., Ab Rahman, I., Mohamad, D. and Husein, A., 2013. A green sol–gel route for the synthesis of structurally controlled silica particles from rice husk for dental composite filler. *Ceramics International*, 39(4):4559–4567.

6 Domestic and Agricultural Wastes

Environmental Impact and Their Economic Utilization

Opeyemi A. Oyewo, Sam Ramaila, Lydia Mavuru, Damian C. Onwudiwe, Felicia O. Afolabi, Paul Musonge, Donald Tyoker Kukwa and Oluwasayo E. Ogunjinmi

CONTENTS

DOI: 10.1201/9781003245773-6

6.1 INTRODUCTION

Wastes are considered unwanted resources in any aspect of processing and yet wastes generation is unavoidable. It was reported that on average, waste generated for one person per annum is 0.74 kg but varies from 0.11 to 4.54 kg globally (Serge Kubanza and Simatele, 2020). There are different kinds of environmental waste, such as liquid and solid wastes, which could be confirmed through different perspective and policy. The generation of solid waste is the inevitable consequence of all processes where materials are used such as agricultural and domestic practices. This is because the extraction of raw materials, manufacturing of products, consumption and waste management all generate wastes (Lagerkvist and Dahlén, 2012). For example, the world generates 2.01 billion tonnes of municipal solid waste annually, and although only 16 percent of the world's population was considered, most developed countries produce about 35 percent of the wastes worldwide (Gautam and Agrawal, 2021).

Overall, there is a relationship between a country's revenue and waste generation. In fact, waste generation reflects the production and consumption patterns of economic activities in a country. Therefore, the higher the income, the more waste is generated and vice-versa (Sarath et al., 2015). As such, waste generation, disposal and management related issues are a serious concern for developing countries. It has also been reported that a reduction in the volume of waste generated per unit of gross domestic product (GDP) is an indication of the economy's move towards less material-intensive production patterns (Marchettini et al., 2007).

Although waste management is based on sources and compositions, recycling, incineration, landfills, chemical-physical and biological treatment remain the existing methods of waste disposal (Marchettini et al., 2007). Despite the efficiency of these methods, there is a direct relationship between waste generation and income, because the operational costs associated with these techniques are also increasing. Thus, there is a need to accelerate the implementation of an eventual waste transformation policy in order to develop more valuable materials that could be utilized in another application from wastes (Leal Filho et al., 2016).

Consequently, this chapter provides a comprehensive analysis of domestic and agricultural waste generation, disposal and management. The sources and effects of these wastes on human health and the environment are also reviewed. Wastes transformation routes with reference to their major application, achievements and limitations are critically reviewed in this study. An investigation on the previous and novel methods of reversing the causes of environmental damage by domestic and agricultural wastes is discussed. A suitable way to eradicate environmental pollution, future perspectives and acceptance of the proposed techniques are analysed in this chapter.

6.2 GENERATION OF DOMESTIC WASTES

Domestic wastes (DW) can be classified into two different forms, such as liquid and solid wastes. However, both forms constitute an important portion in municipal solid waste (MSW). The increase in population obviously increases the amount of domestic wastes of different kinds in the waste stream globally. Separation of solid from liquid waste from the place it is generated is a crucial task to promote recycling and a circular economy (Tassie Wegedie, 2018). DW could be dangerous to the environment if not treated or managed efficiently. However, the composition of DW is very important in an effective design for a solid waste management plan in a community. Although the quantity and composition of DW depend on its source, they vary significantly depending on socio-economic groups in the community. As such, the richer the household or community, the higher the DW generation rate, followed by the middle- and lower-income groups. Therefore, DW generation is directly proportional to family size and wealth. For example, Suthar and Singh (2015) reported the statistics of about 144 households and their DW quantification which were analyzed or characterized based on individual blocks. The average DW quantity in households was found to be 267.17 g per day. The kitchen waste appeared to be the major constituent, which created 80% of total weight of DW in the estimated area; only 1% of glass/ceramic scrap was reported, 2% of cardboard, 6% paper, 7% polythene and plastic, and other miscellaneous items such as rubber, cloths and dirt.

6.2.1 Solid and Liquid Domestic Wastes (DW)

The DW that are firm and stable in shape can be regarded as solid wastes. These wastes are mostly found in gardens, kitchens, cattle sheds and so on. Solid waste is not limited to any physically solid wastes; some solid wastes could emanate from gaseous materials, liquid or semi-solid. Most DW in rural areas is organic and non-toxic due to the environment-friendliness. Liquid waste is generated from household activities involving the use of water. Therefore, water that is not fit for house use or human consumption is considered to be wastewater or liquid waste. It can be subcategorized as industrial, household or domestic wastes, which could be originated from hotels, hostels, kitchens and commercial complexes.

The matrix of liquid waste depends on the source. For instance, in urban areas, the main sources are households, commercial establishments and industries, but kitchen waste has been found to be of maximum weight in rural areas. Yoada et al. (2014) reported that about 93.1% and 77.8% food debris and plastic materials respectively were disposed as wastes per annum. Also, about 61.0% of households use community bins or had waste picked up at their houses by wastes collectors companies. Therefore, statistics reported that about 39.0% polluted their environment which includes gutters, streets and holes with their waste products. This thus contributes to diseases such as malaria and diarrhoea in humans due to improper waste management.

6.2.2 Possible Pollutants in Domestic Wastewater

As reported by Syed (2006), materials regulated by the Resource Conservation and Recovery Act (RCRA) are solid wastes. Also, details of Environmental Protection Agency (EPA) regulations define the type of materials that could be regarded as solid or hazardous wastes. Therefore, the first step to take in the process of EPA regulation is the proper identification of the type of wastes recovered. DW are generally pollutants and if not managed accordingly, they become more harmful to humans and the environment as they decay, especially in sewage. The presence of high concentrations of ammonium nitrate, nitrogen, phosphorus and high conductivity usually found in sewage could pose a serious risk to the environment. It was reported that inputs of metals and organic contaminants to the urban wastewater system (WWTS) occur from three generic sources: domestic, commercial and urban runoff.

Although wastes are already subjected to regulation under different statutes, some hidden pollutants require special attention. The commonly found DW pollutants include oxygen-demanding wastes, sediments, synthetic organic chemicals, pathogenic organisms, inorganic chemicals, microplastics, radioactive substances, oil and plant nutrients. In another study, sampling was done from four different locations (Zhou et al., 2019), and the outcome revealed that perfluoroalkyl substances (PFASs), personal care products (PPCPs) and pharmaceutical substances in the effluent were found in higher concentration of pollutants above their discharge limit. Zn was also found in higher concentrations among the heavy metals present in the effluent. However, the estimated emission flux of these pollutants was ordered as HMs > PPCPs > PFASs through effluent discharge and sludge disposal.

6.2.3 Health and Environmental Effects of Domestic Wastes

Inadequate storage, collection or treatment of DW poses a serious health risk to humans and the environment owing to wrong disposal of these products of no intrinsic value. It was reported that most of the DW consists of about 69.1% organic wastes, hazardous materials of 1.6% and other wastes materials in small quantities. Some hidden or uncollected domestic waste endangers the life of the people more than others by causing health hazards as well as environment pollution. In fact, these DW are considered one of the biggest sources of environmental pollution as land is polluted by DW dumped upon it that makes the soil infertile.

The increase in wastes generated, the more emissions produced and the side effects are created accordingly. As such, human health effects majorly caused by wastes or dusts include cancer, asthma, cardiovascular disease, birth defects/cancer, low birth weight, and many infectious diseases. Waste management systems try as much as possible to reduce the negative impacts on health and environment, as well as the economy. However, many developing countries are seriously facing waste management problems in collection, transportation and disposal of DW.

Inadequate DW management facilities in Ghana resulted in indiscriminate disposal and unsanitary environments, which posed health risks to urban residents (Boadi and Kuitunen, 2005). Boadi and his coworkers compiled an extensive report on waste management and disposal practices in the Accra Metropolitan Area, Ghana, which is over 80% of the general Ghana population. It was reported that only 13.5% of people were served with door-to-door DW collection, while the rest of the people disposed of their waste in open spaces. This inadequate waste management in this part of the world increased the presence of houseflies in the kitchen and the environment. The presence of houseflies in the kitchen during cooking is correlated with the incidence of childhood diarrhoea and cholera. Alternative DW management involving burning and burying is also associated with incidence of respiratory health symptoms among adults and children. Consequently, poor DW management is the major cause of environmental pollution, which results in breeding grounds for pathogenic organisms and the spread of infectious diseases.

6.3 SOURCES OF AGRICULTURAL WASTES

Agricultural wastes can be solely generated from agricultural operations or activities. These practices include rotten farm produce, fruits/vegetables peels, poultry/chicken farming, slaughterhouses wastes and harvest wastes. Agricultural wastes can be defined as residues produced from the growing and processing of agricultural products such as vegetables, fruits, meat, dairy products, poultry and crops. They are not major products obtained from production and processing of agricultural products which can be beneficial to man. However, the economic worth of these non-products is lower than the cost required for the collection, transportation and processing for effective usage.

The quantity and composition of agricultural wastes vary with location and depend on the geographical and cultural nature of a particular country or region and the size of land available for agricultural use. Agricultural wastes, referred to as agro-wastes, comprise crop waste (sugarcane bagasse, corn stalks, fruits and vegetables and prunings), food processing residue (canned maize is 20%, while waste is 80%), animal waste (manure and animal carcasses) and toxic and hazardous waste (pesticides, herbicides and insecticides). Agricultural wastes are increasing daily; therefore they contribute significantly to the world's total waste generation. Increase in agricultural production led to an increase in the quantities of animal wastes, crop residues and agro-industrial by-products. Agricultural wastes are likely to increase significantly globally as the developing countries ensure continuous farming systems. It has been estimated that agricultural wastes produced yearly are about 998 million tonnes (Obi et al., 2016).

6.3.1 Impact and Health Effects of Agricultural Wastes on the Environment

Agricultural development directly increases agricultural wastes (AW) from various applications of intensive farming techniques and also from the misuse or mismanagement of chemicals applied in cultivation. This obviously negatively affects human and animal health globally, especially in rural environments. The health hazards associated with AW depend on the source of these wastes, as harmful pollutants found in AW come from different sources.

6.3.1.1 Effects of Wastes Generated from Cultivation Activities

The tropical climate favours the growth of many crops. However, it also encourages the generation and development of weeds and insects. This results in high demand for pesticides and insecticides to get rid of the insects and also guard against the spread of epidemic diseases which commonly lead to the misuse of pesticides and insecticides by farmers. After applying pesticides, the majority of the bottles and packaging containing these pesticides are left on the fields or in the ponds. The Plant Protection Department (PPD) estimated that about 1.8% of the chemicals remain in the bottles and/or packaging after use (Pan et al., 2021). These wastes have the prospect to generate unpredictable environmental effects such as food poisoning, unsafe food hygiene and polluted farmland due to their tendency to last a long time and the toxicity of the chemicals. In addition, the discarded or unused pesticides packages containing left-overs from the original pesticide result in severe environmental effects, as they can be stored or buried in such a way that contents can leak or find their way into the environment through osmosis, thereby resulting in dire environmental consequences.

Fertilizers play a major role in agricultural production as they help to maintain productivity and viability of plants. Inorganic fertilizer is costly and generates high productivity. However, many farmers apply too much fertilizer to their crops. The serious effect of such surplus application of fertilizer is the fact that it excessively increases annual agricultural yield. The absorption rate of fertilizer compounds (nitrogen, phosphorus and potassium) depends on the land features, type of plant and the method of application (Sengupta et al., 2015). As such, a certain proportion is retained in the soil and a proportion enters into the ponds, lakes and/or rivers due to surface run-off or the irrigation method adopted, which leads to pollution of surface water. A proportion enters the groundwater while a proportion evaporates or de-nitrates, resulting in air pollution. Figure 6.1 displays the agricultural solid wastes being dumped in open space.

6.3.1.2 Effects of Wastes Generated from Livestock Production

Wastes generated from livestock activities comprise solid wastes such as manure and organic materials; wastewater is generated from cage wash, urine, bathing of animals and from the sanitation of slaughterhouses; air pollutants include CH_4 and H_2S and

FIGURE 6.1 Agricultural solid wastes dumped in an open space.

unpleasant odours. Therefore, wastes generated from livestock production can cause serious environmental problems to the surroundings, especially in countries where rearing livestock within residential areas is allowed. Unpleasant odours are the most common source of air pollution emanating from digestion processes of waste generated, the putrefaction process of organic matter in manure, animal urine and wasted foods. The severity of the smell depends on the number of animals, ventilation, humidity and temperature. The quantity of NH_3, CH_4 and H_2S produced differs within the stages of the digestion process and is also subject to organic matter, the food components, microorganisms and the conditions of the animal's health. This untreated waste source is capable of producing greenhouse gases, on top of the negative impact on soil fertility and contamination of the water bodies. Livestock wastes consists of 75–95% water while the remaining constituents are organic and inorganic matter, various species of microorganisms and parasite eggs (Obi et al., 2016). Those germs and contaminated substances affect the environment negatively and can also cause deadly diseases in humans.

6.3.1.3 Effects of Wastes Generated from Aquaculture

The increasing growth in aquaculture has resulted in increased feed utilization for enhanced production. The quantity of feed utilized is an important element used for estimating the amount of wastes generated. One of the main wastes generated in aquaculture is metabolic waste, which can be dissolved and/or suspended.

Approximately 30% of the feed used in a well-managed farm usually turns into solid waste. The feeding rate of livestock depends on the ambient temperature. The feeding rate of livestock increases with increased ambient temperature, which results in increased generated waste. Consistent water flow patterns in the livestock production units are essential for proper waste management, because a well-patterned flow will reduce the disintegration of faeces and enhance the rapid settling and concentration of the settleable solids.

6.4 TRANSFORMATION OF DOMESTIC AND AGRICULTURAL WASTES

To broaden the scope of this topic, the adjective 'domestic' is replaced with the noun 'household', with the understanding that domestic waste is generated by family units, foster homes, social groups, organizations, DW administrative units, businesses, military formations and political entities. Household waste (HSW) is typically generated from sources where various human activities occur. The typical household wastes are composed of a range of materials whose composition varies according to the income, lifestyles and the grade of industrialization, institutionalization and commercialization of the community and its consumers. In the United States, the average household solid waste (HSW) composition and categorization is food (15%), paper (27%), glass (4%), metals (9%), plastics (13%), rubber, leather, and textiles (9%), wood (6%), yard trim (14%), and other products (3%) (Abdel-Shafy and Mansour, 2018). In China, on the other hand, losses were recorded as 62.64% (food), 10.41% (paper), 15.06% (plastics), 6.21% (fabrics), 1.35% (green), 1.85% (glass), 0.56% (metal), and 2.15% (metal) (Fei et al., 2018).

Agricultural waste is divided into primary and secondary wastes, which are generated as a result of various agricultural operations. Primary agricultural waste is harvest-related waste such as residues (including stalks and stubble, leaf litter and seed pods), weeds, sawdust, forest waste and animal waste (including poultry and livestock droppings). The secondary agricultural waste is postharvest-related and/or processing waste such as peels (banana, orange, mango, cassava), husk (rice, sorghum, corn), corncob, palm bunch, palm kernel shell, waste from arable land (manure, fertilizer run-off, pesticides), poultry house waste (feathers, eggshell) and slaughterhouse waste (blood, bones, hoofs, skin) (Seadi and Holm-Nielsen, 2004).

HSW can be categorized into biodegradable waste and non-biodegradable waste. The biodegradable waste consists of food wastes like vegetable and fruit peels, dead plants and livestock, poultry, garden waste, paper materials and bioplastics that can serve as food for some microorganisms within an amount of time. These materials usually consist of fats, starches, proteins, acids, aldehydes and alcohols when dissolved (Zheng et al., 2013). Non-biodegradable wastes cannot be broken down by natural organisms and thus cause pollution. Non-biodegradable waste does not decompose or dissolve in natural environments. Consumer items such as plastic bags, shopping bags, plastic containers and plastic water bottles are examples. Fibres, shoes, metals, hazardous substances and pesticides are among the others.

There are some negative impacts of non-biodegradable wastes that could be disastrous such as drains clogging, water bodies polluted, destruction of soil, and death of cattle and other animals due to air pollution, water pollution and land pollution. There are two types of non-biodegradable waste: recyclable waste and non-recyclable waste. The waste transformation aims to increase the recyclability of desired solid waste materials while attempting to add value to those that cannot be recycled. Overall, the transformation route decreases the waste's volume, weight, scale, and toxicity by using one of the two broad methods, which include biological methods and physicochemical methods. These techniques convert solid wastes into sustainable goods or energy sources that emit no harmful gases.

6.4.1 Biological Waste Conversion Methods

Biologically treated waste yields a stable output suitable for landfill disposal. Four biological processes have been identified: aerobic bio-stabilization, aerobic-in-vessel composting, anaerobic digestion (AD) and vermicomposting. Agricultural-based materials obtained from these activities usually utilize different applications, and the details of these conversion methods are as follows:

6.4.1.1 Aerobic Bio-Stabilization

This method is used to bio-stabilize the waste's organic matter in waste management, mostly in rural areas. In the open air, microbes metabolize organic waste content, reducing its volume by up to 50%. Ball et al. (2017) reported the effects of the composting method in biostabilizing the biodegradable fraction of AW obtained from one of the famous waste treatment plants in Australia. It was confirmed that about 30% and 45% reduction in oxygen consumption occurred in immature and mature compost respectively owing to the effect of biostabilization. A higher conversion rate of AW to organic carbon was also reported by Rada et al. (2005). In short, biostabilization of the biodegradable fraction of AW diverted from landfill can result in a significant reduction of greenhouse gas emissions.

6.4.1.2 Aerobic-in-Vessel Composting

This composting technique involves loading of organic materials into silos, a drum, concrete-lined trench, and other similar containers. Compost or humus is the term use for the stabilized substance or organic wastes in-vessel systems, thus allowing a higher degree of process control in the system. Cabaraban et al. (2007) reported the assessment of aerobic in-vessel composting and bioreactor landfilling using a life cycle inventory (LCI) tool. It was revealed that waste of about 9.7 MJ/kg was utilize for the recovery of enough energy from bioreactor landfilling. Therefore, collected commingled recyclables and residual wastes produce air/waterborne issues along with the environmental side effects associated with the utilization of composting and bioreactor landfilling methods.

6.4.1.3 Anaerobic Digestion (AD)

This is one of the biological processes mainly composed of gases such as carbon dioxide, methane and other gases such as biogas. Organic materials are broken down without the presence of oxygen. However, processing a separated organic-rich fraction takes much longer as it produces very little heat. In other words, biological degradation of organic wastes using microbes under controlled conditions refers to AD (Bautista Angeli et al., 2018). Postawa et al. (2021) reported an increase in biogas production efficiency by modifying the installations of a certain biogas plant technology. Also, the solutions based on mathematics have been validated using certain measurements from an existing biogas plant in Poland.

6.4.1.4 Vermicomposting

Vermicomposting is the process of composting with worms in an aerobic environment. Various species of worms, commonly red wigglers, white worms and other earthworms are often used in the decomposition process to produce a mixture of putrefying vegetable or food waste and mattresses materials. There are two main types of vermicomposting: bin vermicomposting and pile vermicomposting. The bin approach is designed for use on a small scale, such as in the kitchen or garage for home composting. The bin can be constructed from a variety of materials, including wood and plastic (Rostami, 2011). Decomposable organic matter is removed and composted in composting systems, which are close to sanitary landfilling. The end products are long-lasting and effective soil conditioners. They can be used as a fertilizer base.

6.4.2 Physiochemical Wastes Conversion Methods

6.4.2.1 Mechanical Conversion

Different sorting or separation aggregates are used in mechanical waste treatment to recycle single materials for reuse from the waste stream of household solid waste. Size reduction, classification, isolation, and compaction are some of the processes involved. Mechanical Biological Treatment (MBT) is a catch-all phrase for automated mechanical processes used in composting and anaerobic digestion (AD) plants. MBT is a residual waste management method that combines mechanical and biological processes (Department for Environment, Food & Rural Affairs, 2013). MBT has the advantage of being able to be programmed to meet a variety of goals, including the following: (i) Waste pre-treatment; (ii) Mechanical sorting of non-biodegradable and biodegradable household solid waste going to landfill into materials for recycling and/or energy recovery as refuse-derived fuel (RDF); (iii) Reducing the dry mass of biodegradable HSW before landfilling and reducing the biodegradability of BMW before landfilling; (iv) Stabilization into a compost-like output (CLO) for land use; (v) Combustible biogas for energy recovery; and/or (vi) Drying materials to achieve a high calorific organic-rich fraction for use as RDF (Fei et al., 2018). Transformation of wastes through mechanical milling to obtained valuable water treatment products have been reported extensively (Nkalane et al., 2019; Oyewo, 2018; Oyewo et al., 2019, 2020).

6.4.2.2 Thermal Conversion Techniques

Thermal treatment is a remedial technique that involves heating of solid materials such as sediments, dirt or sludge to improve mobility and make it easier to remove organic pollutants. Thermal treatment is a method of sanitizing wastes by applying heat. The primary function of thermal treatment is to convert waste into a stable and usable end-product, thereby reducing the amount of waste that must be disposed of in landfills. Incineration is the most common process for thermochemically reducing the amount of solid waste and producing electricity. Thermal therapy produces chemicals that are more likely to be harmful than their natural forms. During the operations, toxic fumes such as dioxins, furans and hazardous gases such as arsenic (As), mercury (Hg) and cadmium (Cd) are released. Chemical conversion techniques involve hazardous wastes being classified based on physiochemical properties, and also group them in categories such as toxic, ignitable, infectious, reactive, radioactive or corrosive material/wastes (Cai and Du, 2021).

6.4.3 Thermochemical Conversion of HSW and Agricultural Wastes

The degree of oxidation involved in thermochemical treatment can be graded. These methods are widely used to convert the organic matter content of household solid wastes quickly and are applied in different categories such as thermochemical conversion through combustion, thermochemical conversion through gasification and thermochemical conversion through pyrolysis. The choice of any of the aforementioned techniques depends on the desired end-product. The most common form of thermal waste treatment can be regarded as thermochemical conversion through combustion. It is also the most environmentally damaging with no pollution control systems, which allows the contaminants to escape into the atmosphere. Organic matter burns in excess air to create CO_2, CO, N_2, H_2O, slag and ash. The incineration technique involves controlled waste combustion with heat recovery for steam production, in electricity generation via steam turbines (Canabarro et al., 2013). The degradation process of organic matter into less harmful materials, which includes char, oil or gas, is termed thermochemical conversion pyrolysis. This organic matter could be recovered as 5%-10% of the total hydrocarbon fraction such as methane. Organic waste can also be thermally cracked at near-zero air supply to generate hydrogen gas (H_2), CO, C, and N_2, H_2O as well as oil (Tanger et al., 2013). Pyrolysis and gasification are refined thermal treatment methods which could be used as alternative techniques to incineration. The transformation process of AW and DW into gaseous product as an energy carrier which can be used as boiler fuel or in a gas engine refers to thermochemical conversion through gasification. As such, organic matter is gasified into H_2, CO, CO_2, CH_4, steam, slag and ash with a partial air supply (Tanger et al., 2013).

6.5 APPLICATION OF DOMESTIC AND AGRICULTURAL WASTES-BASED MATERIALS

The purpose of conversion or transformation of AW and DW is to reuse or apply as a valuable product in different establishments. In the past two decades, waste-based materials have been utilized in different applications such as, pharmaceuticals, water treatment, gasification and as a source of energy generation.

6.5.1 Application of Transformed Domestic Solid Wastes

Composting and anaerobic digestion of food waste, garden/park wastes and sludge have been used to destroy pathogens in waste and produce biogas, such as landfill gas. Waste anaerobic digestion is frequently associated with landfill gas recovery for energy. Pipelines are laid into landfills, and methane gas is collected and used to generate electricity in power plants. The biological degradation and stabilization of organic substrates under controlled thermophilic and aerobic conditions result in the production of high-quality compost, which can be used as an organic fertilizer on soil in order to improve its physical properties and capacity to retain water by supplying some organic nutrients (Pandey et al., 2015). Furthermore, recovered heat from controlled combustion of solid waste in incinerators can be used in boilers to generate steam and turn steam turbines for power plants. Gasification of solid wastes could also be used to generate carbon monoxide for the Fischer-Tropsch process, which produces hydrocarbons (Tan et al., 2016). Solid waste pyrolysis can be used to create carbon briquettes as a heating solid fuel (Bayu et al., 2020).

6.5.2 The Use of Domestic Liquid Wastes

Wastewater, fats, oils or grease (FOG), and sludges obtained from households can be harmful to the environment, especially in rural areas that depend on surface water for consumption. These liquids are either dangerous or have the potential to be hazardous to human health or the environment. Wastewater can be divided into five categories which include sewage, non-sewage, blackwater, greywater and yellow water. Sewage is an abandoned wastewater in a drainage and mostly directly from domestic sources such as homes, public restrooms, restaurants, schools, hotels and hospitals, while non-sewage is in forms of wastewater, such as floodwater, water from commercial activities such as laundries and industrial effluent. The term 'blackwater' refers to wastewater from the kitchen or toilet. All the contaminants commonly found in appliances or fixtures can be found in blackwater. It contains urine, pieces of rotting food, faeces, toilet paper and other cleaning agents or chemicals (Davis and Fisher, 2015). As a consequence, the water is extremely polluted and can be hazardous to human health. The term greywater simply means wastewater emanated from bathrooms or washing machines; it is considered blackwater excluding faeces or urine. It does consist of chemicals from cleaning agents, but it is more appropriate for treatment and reuse since it is not pathogenic. Yellow water is urine in its purest form. It is urine from particular sources that is free of pollutants present in greywater and blackwater, such as pesticides, food wastes, faeces and used toilet paper.

Wastewater can be reused. However, adequate treatment to remove all the harmful pollutants is required. The source and the proposed application of this processed water determines certain requirements and specifications to be adopted. The organic load of the water is one of these parameters, which can be calculated using the Total Organic Carbon (TOC) in the water. One of the most important criteria for controlling the discharge of household and industrial wastewater is organic load. Permissible limits are defined by laws and must be adhered to. However, controlling organic loading in upstream treatment processes provides another opportunity to improve plant efficiency (Laurens, 2017; Onukak et al., 2017).

Limit values can be adhered to, items can be saved, and plants can be secured against unexpected incidents using online controls. The most appropriate parameter for measuring organic load is TOC. In comparison to traditional methods such as COD or BOD, TOC analysis allows for shorter measuring times with no hazardous waste and is best suited for online analysis.

6.5.3 Application of Agricultural Wastes-Based Materials

Bioactive compounds are commonly found in agricultural residues, and these compounds can be used as a replacement for toxic materials in any production such as biogas, pharmaceutical and biofuel. AW can equally be utilized as animal feed, water treatment products and in pharmaceutical industries (Sadh et al., 2018). The utilization of fibrous agricultural waste in building or decoration of houses were also reported (Dungani et al., 2016). AW is environmentally friendly, abundantly available, durable and not expensive; therefore, it has a great quality in composite formation for building and production purposes. AW waste fibres have generated a lot of research in terms of replacing artificial fibres with them in order to minimize their environmental effects. As such, agricultural wastes can hereby be regarded as one of the base materials for renewable energy production. These wastes can be obtained from oil palm plantations, rice production, and through harvesting of fruits such as pineapple, sugarcane, banana and coconut. Natural fibres can also be classified as one of the major biomasses from agricultural wastes, such as cellulose production for biocomposite, automotive component, biomedical and others.

Recently, an attempt has been made to utilize AW for the synthesis of novel water treatment products such as coagulant, adsorbent, photocatalyst and membrane modules. The main component of plant wastes is cellulose; therefore, numerous investigations have been reported on the extraction of cellulose from agricultural wastes for water treatment or pharmaceutical application (Carpenter et al., 2015; Choudhury et al., 2020; Espíndola et al., 2021).

6.5.3.1 Agricultural Wastes-Based Material as a Pharmaceutical Product

The saying ‘farmer to pharma’ simply means developing new valuable products from agricultural wastes in pharmaceuticals. For example, some reactive compound found in mushroom, banana and potato peels can be used as a substitute to other chemicals in the production of health-promoting foods and beverages, skincare products and pharmaceutical products. Also, some organic matter could be processed and fermented for the production of ethanol or green oils. AW have also been used for the development of skincare, carbohydrate and pharmaceutical products. Duolikun et al. (2021) reported the important functional groups paramount in pharmaceutical application in AW, the physiochemical properties, which include non-toxicity, light weight, and biocompability of these products, make them considerably important candidates in biomedical application. In addition, AW-based materials can be used for pharmaceutical wastewater treatment. The removal of certain emerging pollutants (pharmaceuticals) from water using a charcoal-like biochar has been reported (Zhou et al., 2019).

6.5.3.2 Agricultural Wastes-Based Material as a Water Treatment Product

There is no doubt that AW, especially plants/fruits peels, is regarded as a burden in the communities; therefore, there is a need for further investigation to lessen the discomfort in society (Bhatnagar et al., 2015). However, the utilization of these wastes for the development of low-cost and sustainable water treatment products is one of the options to reuse the wastes. Bhatnagar and his co-workers reported extensively on the use of AW for various adsorbents synthesis. The outcome of this investigation confirmed that different AW-based products have affinity for different organic or inorganic pollutants. The use of AW, which include mangosteen, palm oil ash and corncob in industrial effluents treatment, were also reviewed and reported by Mohammed et al. (2014).

AW-based materials have been found as a replacement to the commercial materials in different applications, especially water treatment. Activated carbons have also been synthesized from a series of agricultural wastes using different methods and equipment such as oven, microwave or furnace. It was reported that adequate preparation techniques and investigation of various activation methods are the main factors required to achieve highly efficient activated carbon (De Gisi et al., 2016). The use of maize tassel–silver nanoparticles for activated carbon development was also reported (Omo-Okoro et al., 2017).

Manganese (Mn) is of serious concern in water pollution due to its toxicity and carcinogenic effects (Okereafor et al., 2020; Fan et al., 2017). This is due to the fact that discharge of manganese to water bodies via industrial effluents is increasing daily. In response to these challenges, the use of agricultural waste-based adsorbent in Mn removal from water were explored and was found to be highly efficient. An extensive review was also conducted on efficient Mn removal from water. The factors affecting the adsorption process and the adsorbent performances are reported in detail (Rudi et al., 2020). The removal of other contaminants of serious concern which include V, Ni, Cd (Oyewo et al., 2019), Fe, Pb, Cr and dyes (Adegoke and Bello, 2015) have also been investigated using AW-based materials.

6.6 APPLICATION OF AGRICULTURAL WASTE-BASED PRODUCT IN DOMESTIC WATER TREATMENT

It is amazing how waste is reprocessed to treat wastes. The use of AW-based materials in domestic wastewater treatment has been reported in the literature (Santos et al., 2016). For example, the use of agricultural waste as a bio-filter media to treat domestic wastewater before its disposal is considered to be one of the ways to protect the environment (Elazizy et al., 2014). Rice husk and ficus trees trimming stalks have also been used in another study as media in bio-filter reactors in a pilot plant which consists of two different units. Each unit was packed with AW of 60 cm, and individual units worked as standard and high-rate bio-filter reactors.

The peristatic pump was used to transport domestic wastewater from the top of each unit for 8 hours per day. It was reported that the removal ratios of high-rate bio-filter reactors are higher than that of standard-rate bio-filter reactors for wastewater parameter (COD, BOD) using the same type and dosage of agricultural wastes (Ghazy et al., 2016). With the use of rice husk media, the average removal ratio of COD was 24.80% and 36.84%, of BOD was 24.42% and 36.69% for standard rate and high-rate, respectively. On the other hand, with the use of ficus tree trimming stalks, the average removal ratio of COD was 27.24% and 69.48%, of BOD was 27.18% and 69.48% for standard rate and high-rate, respectively (El-Nadi et al., 2014). These values illustrated the possibility of applying both rice husk and ficus trees trimming stalks as bio-filter media.

6.7 ECONOMIC IMPACT OF WASTES UTILIZATION

Global waste management results in a significant reduction in greenhouse gas emissions, which result in the creation of millions of jobs, and also generated hundreds of billions of dollars in economic benefits. Any work generated in the waste management industry has a positive effect on the labour market, lowering unemployment rates worldwide. Solid and liquid waste mitigation systems are robust in waste management (Tan et al., 2016). Waste disposal jobs (landfilling and incineration) influence the following three recycling categories: Recycling Industries, which include the storage and processing of recyclables in order to make them useful in modern manufacturing processes; Recycling-dependent businesses, such as those that buy secondary materials from the recycling industry; and Reuse and Remanufacturing Sectors, which include businesses that reuse or remanufacture goods for their original purpose.

Solid waste piles not only degrade the environment's beauty, but they also act as a source of foul odour, posing significant health risks. The accumulation of waste leads to landscape degradation, contamination of the air, water, and soil and the extinction of terrestrial and marine life. Epidemics may sometimes break out due to the accumulation of wastes, particularly near water bodies. For example, in the United States, approximately 40% of food is wasted each year, costing an estimated $218 billion or 1.3% of GDP. Food makes up 21% of solid waste in landfills in the United States (Epa, 2020).

Sustainable solid waste management is important for both people's health and environmental conservation. Energy continues to be the world's most pressing issue for developing countries. Energy must be generated in a sustainable manner, ideally from renewable sources with low environmental impact. It is important to reuse waste resources in energy production in order to address potential waste management issues and to harness renewable energy.

Waste-to-energy has been identified as a viable option for recycling rates, reducing reliance on fossil fuels, reducing the amount of waste sent to landfills and avoiding pollution. Fei et al. (2018) used life cycle assessment scenario analyses and other methods to evaluate the energy, environmental and economic performance of MBT and other conventional solid waste treatment technologies, reporting that economic performance is one of the decisive factors affecting solid waste treatment technologies. Furthermore, the researchers demonstrated economic performance in the following order: Waste-to-Energy (WtE)>Mechanical Biological Treatment (MBT) for electricity generation>MBT for natural gas production. Although the incineration procedure has high construction and operating costs, it is best suited for use in urban areas where landfill sites are scarce (Abdel-Shafy and Mansour, 2018).

The transformation of global economic growth to green growth for sustainable development using agricultural and domestic wastes is an important way to improve the economic impact of wastes. However, municipal solid waste (MSW) management remains an environmental concern and remains one of the major constraints for green growth across the globe (Iqbal et al., 2020). MSW includes waste generated in households within a community under a municipality and waste generated from industrial or commercial sectors. The trend indicates that the amount of MSW is predicted to increase globally by 2.3 billion tonnes (Igoni et al., 2017; Das, Islam, Billah, Sarker et al., 2021).

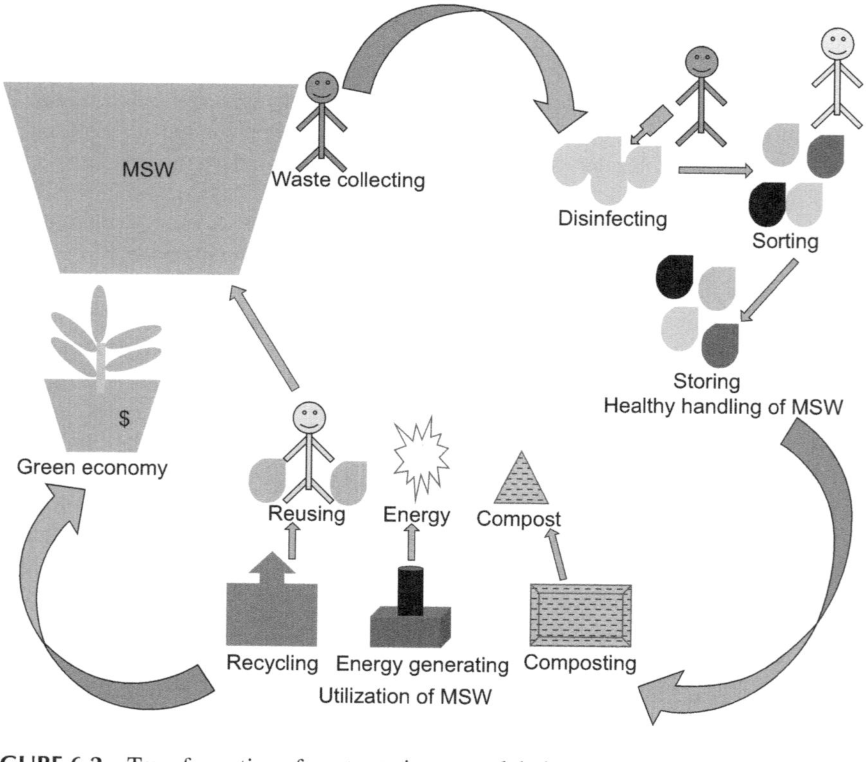

FIGURE 6.2 Transformation of wastes to improve global economic growth.

6.8 CONCLUSION AND FUTURE PERSPECTIVES

Agricultural and domestic wastes are readily available, eco-friendly, and can be utilized in different applications. AW and DW have great physico-chemical properties and can easily be modified or processed to develop novel materials. AW and DW are non-toxic, biodegradable and highly efficient in different applications, especially in pharmaceutical and water treatment. However, an appropriate management using efficient techniques coupled with the transformation of these wastes is mitigating against the challenge. The success of AW- and DW-based materials in different applications has been stated clearly. However, these materials have only been produced on a very small scale. Therefore, further investigation is required. Processing and production of valuable materials using these wastes on a very large scale or pilot plant is proposed based on this review. This will surely expand the implementation of these materials and also improve the global economy. Overall, when all procedures are relatively safe, these materials would exhibit less aggregation and offer advanced applications which have not been explored.

6.9 ACKNOWLEDGEMENT

The authors would like to acknowledge the University of Johannesburg under the Global Excellence Stature Fellowship (GES) for financial support and Durban University of Technology.

6.10 REFERENCES

Abdel-Shafy, H. I. & Mansour, M. S. M. 2018. Solid waste issue: Sources, composition, disposal, recycling, and valorization. *Egyptian Journal of Petroleum*, 27, 1275–1290.

Adegoke, K. A. & Bello, O. S. 2015. Dye sequestration using agricultural wastes as adsorbents. *Water Resources and Industry*, 12, 8–24.

Adejumo, I. & Adebiyi, O. 2020. Agricultural solid wastes: Causes, effects, and effective management, *In*: Saleh, H.M. (ed.) *Strategies of Sustainable Solid Waste Management*. IntechOpen. DOI: 10.5772/intechopen.93601.

Ball, A. S., Shahsavari, E., Aburto-Medina, A., Kadali, K. K., Shaiban, A. A. J. & Stewart, R. J. 2017. Biostabilization of municipal solid waste fractions from an Advanced Waste Treatment plant. *Journal of King Saud University—Science*, 29, 145–150.

Bautista Angeli, J. R., Morales, A., LeFloc'h, T., Lakel, A. & Andres, Y. 2018. Anaerobic digestion and integration at urban scale: Feedback and comparative case study. *Energy, Sustainability and Society*, 8, 29.

Bayu, A. B., Amibo, T. A. & Akuma, D. A. 2020. Conversion of degradable municipal solid waste into fuel briquette: Case of Jimma city municipal solid waste %J Iranian (Iranica). *Journal of Energy & Environment*, 11, 122–129.

Bhatnagar, A., Sillanpää, M. & Witek-Krowiak, A. 2015. Agricultural waste peels as versatile biomass for water purification—A review. *Chemical Engineering Journal*, 270.

Boadi, K. O. & Kuitunen, M. 2005. Environmental and health impacts of household solid waste handling and disposal practices in third world cities: The case of the Accra Metropolitan Area, Ghana. *Journal of Environmental Health*, 68**,** 32–36.

Cabaraban, M. T., Khire, M. & Alocilja, E. 2007. Aerobic in-vessel composting versus bioreactor landfilling using life cycle inventory models. *Clean Technologies and Environmental Policy*, 10**,** 39–52.

Cai, X. & Du, C. 2021. Thermal plasma treatment of medical waste. *Plasma Chemistry and Plasma Processing*, 41**,** 1–46.

Canabarro, N., Soares, J. F., Anchieta, C. G., Kelling, C. S. & Mazutti, M. A. 2013. Thermochemical processes for biofuels production from biomass. *Sustainable Chemical Processes*, 1**,** 22.

Carpenter, A., de Lannoy, C.-F. & Wiesner, M. 2015. Cellulose nanomaterials in water treatment technologies. *Environmental Science & Technology*, 49.

Choudhury, R. R., Sahoo, S. K. & Gohil, J. M. 2020. Potential of bioinspired cellulose nanomaterials and nanocomposite membranes thereof for water treatment and fuel cell applications. *Cellulose*, 27**,** 6719–6746.

Das, A. K., Islam, M. N., Billah, M. M. & Sarker, A. 2021. COVID-19 and municipal solid waste (MSW) management: A review. *Environmental Science and Pollution Research*, 28**,** 28993–29008.

Das, A. K., Islam, M. N., Billah, M. M., Sarker, A. 2021. COVID-19 and municipal solid waste (MSW) management: A review. *Environmental Science and Pollution Research*, 28, 1–16.

Davis, A. & Fisher, R. 2015. *Kitchen & Bath Sustainable Design: Conservation, Materials, Practices*. John Wiley & Sons.

De Gisi, S., Lofrano, G., Grassi, M. & Notarnicola, M. 2016. Characteristics and adsorption capacities of low-cost sorbents for wastewater treatment: A review. *Sustainable Materials and Technologies*, 9**,** 10–40.

Department for Environment, Food & Rural Affairs. 2013. *Mechanical biological treatment of municipal solid waste*. PB13890.

Dungani, R., Karina, M., Sulaeman, S., A., Hermawan, D. & Hadiyane, A. 2016. Agricultural waste fibers towards sustainability and advanced utilization: A review. *Asian Journal of Plant Sciences*, 15**,** 42–55.

Duolikun, T., Lai, C. W. & Bin Johan, M. R. 2021. 26—Agricultural waste-based bionanocomposites in tissue engineering and regenerative medicine. *In:* Ahmed, S. & Annu (eds.) *Bionanocomposites in Tissue Engineering and Regenerative Medicine*. Woodhead Publishing.

Elazizy, I., El Nadi, D. M. E. H. & Abdalla, M. A. F. 2014. Use of agricultural wastes as bio filter media in aerobic sewage treatment. *Australian Journal of Basic and Applied Sciences*, ISSN:1991-8178.

El-Nadi, M., Elazizy, I., Abdalla, M. A. F. 2014. Use of agricultural wastes as bio filter media in aerobic sewage treatment. *Australian Journal of Basic and Applied Sciences*, 8, 181–185.

Espíndola, S. P., Pronk, M., Zlopasa, J., Picken, S. J. & van Loosdrecht, M. C. M. 2021. Nanocellulose recovery from domestic wastewater. *Journal of Cleaner Production*, 280**,** 124507.

Fan, Y., Zhu, T., Li, M., He, J. & Huang, R. 2017. Heavy metal contamination in soil and brown rice and human health risk assessment near three mining areas in central China. *Journal of Healthcare Engineering*, 2017**,** 4124302. doi: 10.1155/2017/4124302

Fei, F., Wen, Z., Huang, S. & Clercq, D. D. 2018. Mechanical biological treatment of municipal solid waste: Energy efficiency, environmental impact and economic feasibility analysis. *Journal of Cleaner Production*, 178**,** 731–739.

Gautam, M. & Agrawal, M. J. C. F. C. S. 2021. Greenhouse gas emissions from municipal solid waste management: A review of global scenario. *Environmental Science*, 123–160.

Ghazy, M. R., Basiouny, M. A. & Badawy, M. 2016. Performance of agricultural wastes as a biofilter media for low-cost wastewater treatment technology. *Advances in Research*, 1–13.

Igoni, A. H., Harry, I. S. K. 2017. Environmental crises in government controlled municipal solid waste management in rivers state, Nigeria. *Journal of Environment and Earth Science*, 7, 38–50.

Iqbal, A., Liu, X. & Chen, G.-H. 2020. Municipal solid waste: Review of best practices in application of life cycle assessment and sustainable management techniques. *Science of the Total Environment*, 729, 138622.

Lagerkvist, A. & Dahlén, L. 2012. Solid Waste solid waste Generation solid waste generation and Characterization solid waste characterization. *In:* Meyers, R. A. (ed.) *Encyclopedia of Sustainability Science and Technology*. Springer New York.

Laurens, L. M. 2017. *State of Technology Review—Algae Bioenergy*. IEA Bioenergy.

Leal Filho, W., Brandli, L., Moora, H., Kruopienė, J. & Stenmarck, Å. 2016. Benchmarking approaches and methods in the field of urban waste management. *Journal of Cleaner Production*, 112, 4377–4386.

Marchettini, N., Ridolfi, R. & Rustici, M. 2007. An environmental analysis for comparing waste management options and strategies. *Waste Management*, 27, 562–571.

Mohammed, M. A., Shitu, A., Tadda, M. A. & Ngabura, M. 2014. Utilization of various Agricultural waste materials in the treatment of Industrial wastewater containing Heavy metals: A review. *International Research Journal of Environment Sciences*, 3, 62–71.

Nkalane, A., Oyewo, O. A., Leswifi, T. & Onyango, M. S. 2019. Application of coagulant obtained through charge reversal of sawdust-derived cellulose nanocrystals in the enhancement of water turbidity removal. *Materials Research Express*, 6, 105060.

Obi, F. O., Ugwuishiwu, B. O. & Nwakaire, J. N. 2016. Agricultural waste concept, generation, utilization and management. *Nigerian Journal of Technology (NIJOTECH)*, 35, 957–964.

Okereafor, U., Makhatha, M., Mekuto, L., Uche-Okereafor, N., Sebola, T., Mavumengwana, V. 2020. Toxic metal implications on agricultural soils, plants, animals, aquatic life and human health. *International Journal of Environmental Research and Public Health*, 17, 2204.

Omo-Okoro, P., Daso, A. P. & Okonkwo, O. 2017. A review of the application of agricultural wastes as precursor materials for the adsorption of per- and polyfluoroalkyl substances: A focus on current approaches and methodologies. *Environmental Technology & Innovation*, 9.

Onukak, I. E., Mohammed-Dabo, I. A., Ameh, A. O., Okoduwa, S. I. R. & Fasanya, O. O. 2017 Production and characterization of biomass briquettes from tannery solid waste. *Recycling*, 2, 17.

Oyewo, O. A. 2018. Lanthanides removal from mine water using banana peels nanosorbent. *International Journal of Environmental Science and Technology*, 15(6), 10–1274–2018.

Oyewo, O. A., Adeniyi, A., Sithole, B. B. & Onyango, M. S. 2020. Sawdust-based cellulose nanocrystals incorporated with ZnO nanoparticles as efficient adsorption media in the removal of methylene blue dye. *ACS Omega*, 5, 18798–18807.

Oyewo, O. A., Mutesse, B., Leswifi, T. Y. & Onyango, M. S. 2019. Highly efficient removal of nickel and cadmium from water using sawdust-derived cellulose nanocrystals. *Journal of Environmental Chemical Engineering*, 7, 103251.

Pan, X., Huang, T., Fang, Y., Rao, W., Guo, X., Nie, D., Zhang, D., Cao, F., Guan, X. & Chen, Z. 2021. Effect of *Bacillus thuringiensis* biomass and insecticidal activity by cultivation with vegetable wastes. *Royal Society Open Science*, 8, 201564.

Pandey, P., Ramegowda, V. & Senthil-Kumar, M. 2015. Shared and unique responses of plants to multiple individual stresses and stress combinations: Physiological and molecular mechanisms. *Frontiers in Plant Science*, 6, 723–723.

Postawa, K., Szczygieł, J. & Kułażyński, M. 2021. Innovations in anaerobic digestion: A model-based study. *Biotechnology for Biofuels*, 14, 19.

Rada, E., Ragazzi, M., Panaitescu, V. & Apostol, T. 2005. Bio-drying or bio-stabilization process? *UPB Scientific Bulletin, Series C: Electrical Engineering*, 67, 51–60.

Rostami, R. 2011. Vermicomposting. *In:* Kumar, M. S. (ed.) *Integrated Waste Management Rijeka*. IntechOpen.

Rudi, N. N., Muhamad, M. S., Te Chuan, L., Alipal, J., Omar, S., Hamidon, N., Abdul Hamid, N. H., Mohamed Sunar, N., Ali, R. & Harun, H. 2020. Evolution of adsorption process for manganese removal in water via agricultural waste adsorbents. *Heliyon*, 6, e05049.

Sadh, P. K., Duhan, S. & Duhan, J. S. 2018. Agro-industrial wastes and their utilization using solid state fermentation: A review. *Bioresources and Bioprocessing*, 5, 1.

Santos, M., Puna, J., Barreiros, A. & Matos, M. 2016. *Agricultural Wastes for Wastewater Treatment*. 4th International Conference on Sustainable Solid Waste Management, Limassol, Chipre.

Sarath, P., Bonda, S., Mohanty, S. & Nayak, S. K. 2015. Mobile phone waste management and recycling: Views and trends. *Waste Management*, 46, 536–545.

Seadi, T. & Holm-Nielsen, J. 2004. VI.1 Utilization of waste from food and agriculture. *Waste Manage*, 4.

Sengupta, S., Nawaz, T. & Beaudry, J. 2015. Nitrogen and phosphorus recovery from wastewater. *Current Pollution Reports*, 1, 155–166.

Serge Kubanza, N. & Simatele, M. D. 2020. Sustainable solid waste management in developing countries: A study of institutional strengthening for solid waste management in Johannesburg, South Africa. *Journal of Environmental Planning and Management*, 63, 175–188.

Suthar, S. & Singh, P. 2015. Household solid waste generation and composition in different family size and socio-economic groups: A case study. *Sustainable Cities and Society*, 14, 56–63.

Syed, S. 2006. Solid and liquid waste management. *Emirates Journal for Engineering Research*, 11, 19–36.

Tan, E. C. D., Talmadge, M., Dutta, A., Hensley, J., Snowden-Swan, L. J., Humbird, D. & Biddy, M. 2016. Conceptual process design and economics for the production of high-octane gasoline blendstock via indirect liquefaction of biomass through methanol/dimethyl ether intermediates. *Biofuels Bioproducts & Biorefining-Biofpr*, 10, 17–35.

Tanger, P., Field, J., Jahn, C., DeFoort, M. & Leach, J. 2013. Biomass for thermochemical conversion: Targets and challenges. *Frontiers in Plant Science*, 4, 218.

Tassie Wegedie, K. 2018. Households solid waste generation and management behavior in case of Bahir Dar City, Amhara National Regional State, Ethiopia. *Cogent Environmental Science*, 4, 1471025.

U.S. EPA. 2020. *Best practices for solid waste management: A guide for decision-makers in developing countries*. Environmental Protection Agency, EPA 530-R-20-002, pp. 1–166, 21 MB.

Yoada, R. M., Chirawurah, D. & Adongo, P. B. 2014. Domestic waste disposal practice and perceptions of private sector waste management in urban Accra. *BMC Public Health*, 14, 697.

Zheng, W., Phoungthong, K., Lü, F., Shao, L.-M. & He, P.-J. 2013. Evaluation of a classification method for biodegradable solid wastes using anaerobic degradation parameters. *Waste Management*, 33, 2632–2640.

Zhou, Y., Meng, J., Zhang, M., Chen, S., He, B., Zhao, H., Li, Q., Zhang, S. & Wang, T. 2019. Which type of pollutants need to be controlled with priority in wastewater treatment plants: Traditional or emerging pollutants? *Environment International*, 131, 104982.

7 Sustainable Utilization and Management of Agricultural and Kitchen Waste

Soumya Pandey and Neeta Kumari

CONTENTS

DOI: 10.1201/9781003245773-7

7.1 INTRODUCTION

Agriculture is the backbone of the economy of any country (Boudreault et al., 2019; Guenat et al., 2019; Patel et al., 2019), since the beginning of civilization. Exponential population growth (Seitzinger et al., 2000; Tilman et al., 2001; Misselhorn et al., 2012; Bhattacharyya et al., 2015; Franco-Ramos et al., 2016; Joksimović et al., 2020) has put forward immense pressure on the agriculture sector. According to FAO 2016, global agriculture has increased by 60% (FAO, 2015, 2019), especially in developed countries, while there has been an increase of 5% in agricultural land in the last 40 years worldwide. According to the Economic Survey Report 2019–2020, the average annual growth in India of agriculture and its associated sectors has been consistent from 2014–2015 to 2018–19 at 2.88% (Jitendra, 2020). This slow growth has mostly impacted the small and average-scale farming communities which comprise around 87% of our farmer population (Sabiiti, 2011; Erana et al., 2019). Lack of sources of irrigation and knowledge of modern equipment and cultivation practices is attributed to this slow growth. Ever since the green revolution (Duhan et al., 2017; Sufian et al., 2020), although the initial land productivity has increased, the survey now claims that there has been tremendous enhancement in soil fatigue (Koondhar et al., 2020). Soil fatigue occurs due to exploitation and excessive use of fertilizers, loss of fertile soil, erosion, etc (Pavlidis and Tsihrintzis, 2018; Wang et al., 2019). A reduction in soil fertility was observed from 13.4% in the 1970s to 4.1% in 2000. In the year 2019–20, food production is estimated to be 291.95 million tons but the food demand is projected to increase up to 345 million tons by 2030 (Sharma et al., 2019; Wang et al., 2021.

On one hand, food demands are much more than the supply (Roy et al., 2002; Bryngelsson et al., 2017; Devi et al., 2017; Hinz et al., 2020), and the major population is facing a food crisis, while on the other hand, management of agricultural and kitchen food (Sun et al., 2018; Matta et al., 2020) waste has become a global concern. Global wastage created from the food and agriculture sector causes both economic loss and environmental damage. According to FAO's report of 2013, around 1.6 billion tons per year of food end up being wasted, causing a loss of around 2.6 trillion dollars while 815 million people are suffering and are malnourished due to lack of food and nutrition (Godfray et al., 2010; FAO, 2013). One of the reasons for increased food waste is subjected to globalization (Godfray et al., 2010). From cultivation to the export of food in the form of fruit, veggies, dairy, and meat, are all subject to both loss and waste at every stage.

Loss of food is considered when food is damaged or is unconsumable in the processes of cultivation, storage, and transportation (Hamelin et al., 2014). Food is also considered waste when it is thrown away by consumers or end-users. But overall, both these processes are responsible for creating a huge pile of waste. This waste not only costs us product loss, but its improper disposal causes environmental degradation. Furthermore, even its disposal costs a lot. Waste generation in developing and developed countries are similar in quantity (that is 670 million tons and 630 million tons respectively per FAO 2013), but their type varies (OECD-FAO, 2012; FAO, 2013). Waste produced by developed countries is more on the consumer end (61%) and less in production stages (32%), whereas developing countries face more food loss in production stages and much less at the consumer end, as per the FAO 2020 report (FAO, 2020).

Food waste generation is dependent on multiple reasons like the economy of the region (Oldfield et al., 2016; Bhuyan et al., 2020), climate, available equipment and facilities, awareness among suppliers and consumers regarding food, etc. During stages of cultivation, crop failure due to natural hazards like droughts and flood or pest infestation (Kautz et al., 2017; Mátyás et al., 2018) is a common problem. For example, there was a destruction of crops due to infestation by locust swarms in Kenya, Somalia, Ethiopia, Southern Iran, Pakistan, and several north-western states of India in early 2020 (Desert Locust FAO, 2020). In Gujarat, the state of India, 25000 hectares of land were destroyed in 3 months.

Sometimes there is a surplus growth of crops, but lack of storage and transportation or loss in market demand leads to crop destruction (Butterly et al., 2015). A very recent example of food loss and wastage was seen during the coronavirus pandemic (Kumar and Kumar Singh, 2021; Singh et al., 2021), which has brought similar losses to the agriculture industry as the Great Depression of the 1930s. U.S. farmers are bound to destroy their crops, milk, eggs, and slaughter their livestock as the food supply chain has been at a standstill for many months (Garraty, 1976; Moak et al., 1994). The situation is similar in China, as their wet markets are closed and lack of feed for livestock and animals force the farmers to starve and slaughter them as their products will not sell (Lynteris, 2016; Ho and Chu, 2019). In some countries, the ready crops could not be harvested because of the lack of workers available on farms. Also, self-isolation measures became compulsory per government norms (Ciotti et al., 2020; Singh et al., 2021). Throughout history, many such incidences have occurred. Some were noted and some were not, but altogether improper food and livestock waste disposal always creates risk for diseases and environmental degradation (McMichael et al., 2006).

Agriculture waste is in the form of farm agricultural wastes, industrial agricultural wastes, livestock (meat) production and processing wastes, fertilizers and on-farm medical wastes (Delang, 2018; Mehran et al., 2021), horticultural and other chemical waste, packaging and manufacturing waste, kitchen waste, etc. (Meals and Hopkins, 2002; Rosenbaum et al., 2011; Brancoli et al., 2017). Since ancient times, the practice of disposal of agriculture and food waste was done mainly by burning crops, dumping in open land, or by flushing in water-bodies. Agricultural and kitchen waste decompose to form methane (Adhikari et al., 2006), which is 86 times more harmful a greenhouse gas than carbon dioxide (Lenka et al., 2015). These wastes are responsible for 21–37% of the annual emission of GHG. The livestock sector alone contributes to 14.5% of global GHG (Laubach and Hunt, 2018). In almost every country, especially developed and highly populated countries, these wastes constitute the majority of MSWs. Disposal of the organic part of these wastes is becoming more challenging day by day (Kato et al., 2009). As agricultural products turn to waste, they are also responsible for the loss of energy used in every stage of production throughout the supply-chain journey from farm to table (Cakar et al., 2020). They consume 25% of water and are one of the main causes of water pollution, particularly in the U.S. (Mateo-Sagasta et al., 2017)

Overall, agricultural waste is generated every year in massive quantities, and is responsible for socio-economic and environmental losses globally (Vermeulen et al., 2012; Maity et al., 2020). Other forms of waste that are produced from the agricultural sector and kitchens are chemicals, plastics, paper, bottles, and metals as lots of packaging is required for safe storing and transportation to their intended end-users (Gentil et al., 2011). In the past decade, many researchers and scientists have studied the physical, chemical, and biological properties of these wastes (Sharma et al., 2019). They found that wastes can be recycled and have a high potential to be used in other industries. They worked on the sustainable disposal of all forms of wastes produced from agriculture and kitchen industries and their utilization in other industries like construction industries (Susilawati et al., 2020) as cement mix and soil stabilizers, manufacturing industries, soap and cosmetic industries (Jayathilakan et al., 2012), energy (Sarmento et al., 2010), and biofuel industries (Tonini et al., 2016). In this chapter, insight is given on the effects of agricultural waste on our environment and advanced methods of their sustainable disposal and utilization in an eco-friendly and cost-effective manner.

7.2 TYPES OF AGRICULTURAL AND KITCHEN WASTE

As mentioned earlier, the agricultural sector produces various forms of waste in every stage starting from cultivation to storage, transportation, supermarkets, kitchen, and finally as kitchen waste (Brancoli et al., 2017). Food in the form of crops, meat, dairy, eggs, vegetables and fruits are the basic requirements for the survival of life (Gornall et al., 2010). Then why or how do these products turn into waste? In developed countries, consciousness toward food and nutrition or branding of products and their nutrition and manufacturing labels plays an important role (Liegeard and Manning, 2020). Places with a high-income population tend to choose only fresh and aesthetically good products. Hence, the supermarkets and food industries, which are the largest consumers of agricultural products (Lobley et al., 2009; Vermeulen et al., 2012), eliminate a large number of crops and other agro-products. Every year these reasons also dent the pockets of farmers. In developing countries and low-income regions, lack of poor infrastructure, lack of cold storage, improper transportation, lack of proper distribution of food in the market, and lack of proper packaging techniques and material are responsible for food loss and wastage (Al-Yaqout and Hamoda, 2002; Isaac et al., 2017). For example, in the process of milk transportation, milk is carried in small milk cans by local producers to deliver to traders or collectors by bicycle, animal, or foot. In this process, the milk is subjected to heat and many times goes rancid. Hence a large quantity of milk is discarded every day. Similarly, another form of waste is described in Table 7.1.

7.2.1 FARM WASTE

Farm wastes mostly comprise waste generated in the process of crop production (Christofoletti et al., 2013; Butterly et al., 2015; Kanders et al., 2017; Ho and Chu, 2019). Crop production leaves a huge volume of residue both on-field (field residue) and off-field (processed residue). On-field residue material consists of the waste

TABLE 7.1
Agricultural Waste, Origin, Source, and Types of Wastes Produced from Them

Classification of agricultural waste	Waste Generation/ origin	Types of Waste produced
Farm agricultural waste	Crop production, farms	Crop residue, husks, Leaf litter, etc.
Industrial agricultural waste	Commercial-scale farming and processing, breweries, wineries, fruit juice producers, non-alcoholic, coffee and tea industries, oil producers	Wood waste, paper production waste, other commercially grown crops like jute, sugarcane pulp, fruit peels from juice industries, coffee, oil mills, canned foods, tobacco processing etc.,
Livestock production, dairy, and meat processing waste	Livestock and dairy farming and processing	Bedding/litter, animal manure, carcasses, damaged feeders, water trough, hoofs, bones, feathers, peels, slurry, organs, blood, dairy, etc.,
Fertilizer waste	Infestation remedy in crops and livestock	Herbicides, pesticides, insecticides, containers, bottles, etc.,
On-farm medical waste	Veterinary care	Drugs, vaccines, syringes, bottles, wrappers, containers
Horticultural waste	Cultivation, and maintenance of plants and landscape beautification	Pruning, grass cuttings
Packaging and manufacturing waste	Food manufacturing and packaging	Plastics, cardboards, paper, tissues, wood, metals, containers, bottles, adhesive etc.
Household and commercial kitchen waste	Daily Kitchen activities, bakeries, confectioneries	Wet food waste, organic waste, plastic, wrappers, metal containers, bottles, cartons, etc.

Source: Literature survey (Flotats et al., 2009; Sharpley et al., 2011; Brancoli et al., 2017; Banks et al., 2018; Landry et al., 2018)

left after harvesting on farms, like stalks, stems, leaves, and seed pods, which could be ploughed or sometimes burned to ashes (Savant et al., 1999; Mupangwa et al., 2007; Himanshu et al., 2019). But the burning of crop residue generates greenhouse gases, increases the amount of particulate matter (PM) (Rim-Rukeh, 2014; Bhuvaneshwari et al., 2019; Kim and Yoo, 2020) and smog in the air (Solarte-Toro et al., 2018), many microbial diversities are lost in the fire (Sujatha and Sridhar, 2018), and soil fertility and health deteriorate. This causes soil degradation (Kumar and Mishra, 2015) and environmental pollution (Ngoc and Schnitzer, 2009). According to the ministry of new and renewable energy, about 350 million tons per year of agricultural waste are generated in India (Norsuraya et al., 2016; Delang, 2018). These wastes have the potential to generate 18000 MW of power yearly and produce green fertilizer (Eichler-Lóbermann et al., 2009; Devi et al., 2017) for chemical-free cultivation.

7.2.2 Industrial Agriculture Waste

Industrial-scale farming has a target to produce crops, dairy, and livestock for other industries like food processing, manufacturing, packaging, cosmetics, and pharmaceuticals. Some examples of such industries are meat, juice, paper, and jute (Jayathilakan et al., 2012; Brancoli et al., 2017; Wansink, 2018). In India alone, 67 million tons of food are wasted, including 18% of fruit and vegetables wasted every year (FAO, 2021). According to FAO 2013, the amount of food lost or wasted accounts for about 1 trillion dollars' worth and is enough to feed 2 billion people if it is conserved (FAO, 2013). According to the food sustainability index ranking (FAO, 2021), France produces the least amount of food waste and loss and stands at the top, while India has been ranked number 4 among middle-income group countries. According to FAO 2015, food waste and loss comprises 30% of cereals, 20% of dairy products, 35% of fish and seafood, 45% of fruits and vegetables, 20% of meat, 20% oilseeds and pulses, and 45% of roots and tubers (FAO, 2015). Annual losses of such huge quantities of waste also amount to a loss of other resources, including capital and labor, water, land, energy, and end up generating GHGs, air, water, and soil pollution, producing toxic leachate and contributing to climate change (FAO, 2013, 2015).

7.2.3 Livestock and Meat Processing Waste

Livestock and meat processing industries mainly consist of poultry like pigs, sheep, cows, hens, goats, and fish (Ju et al., 2005; Flotats et al., 2009; Lynch and Pierrehumbert, 2019). These animals are farmed on a large scale for commercial purposes to fulfill the meat demands of consumers and other processed food industries. They contribute more than 40% of the agriculture sector. The majority of waste produced from these industries comprises slaughter-house waste (Alvarenga et al., 2015; Wei et al., 2020), which are parts of animals that cannot be used for sales and consumption, like skin, feathers, blood, internal organs, bones, tendons, and gastrointestinal waste, manure etc.,(Darine et al., 2010). These by-products pose a lot of health and sanitation concerns. But in recent decades, these by-products have found their use in other non-food industries; for example, bone is used pharmaceutical industries (Boxall et al., 2006) and fats are used to make lards (Silva et al., 2009). Asia and Europe are the largest consumers of meat by-products (Jayathilakan et al., 2012). These wastes have a higher nutritional value and are preferred for pharmaceutical purposes as they contain high amounts of minerals, vitamins, amino acids, hormones, fatty acids, and protein. Some natural essential chemicals like riboflavin, proline, hydroxyproline, glycine, tryptophan, tyrosine, niacin, vitamin B12, B6, folacin, ascorbic acid, vitamin A, Iron, copper, manganese, phosphorous, sodium, and calcium are all found in kidneys, livers, lungs, spleens, brains, ears, and heart (Sadh et al., 2018). The government has regularized the use of these by-products only after cooking for direct consumption or proper processing for commercial processes to avoid any disease.

7.2.4 Fertilizers and Other Chemical Waste

The use of fertilizers in the form of pesticides, insecticides, and herbicides is done to improve crop productivity and control weeds and infestation of pests

(Woods et al., 2010; Pastor and Hernández, 2012; Bai et al., 2018). But the major problem occurs when most farmers do not understand the quantity and type of fertilizer to be used. Lack of soil testing labs and financial constraints mean farmers buy cheap and feasible fertilizers that they apply on crops during cultivation. Ever since the green revolution (Tartowski and Howarth, 2001; Duhan et al., 2017; Sufian et al., 2020), the use of fertilizers has increased exponentially in crop production, but with time, crop production has become static and some adverse effects are seen on natural resources like land in the form of soil fatigue (loss of fertility and nutrient content), loss of microbial diversities and nutrient pollution in waterbodies, resulting in eutrophication and algal bloom. Overall, unused chemicals, bottles, spray, and tubes generate waste and environmental pollution. During the process of application, fertilizers also spread in the air, causing air pollution (Ju et al., 2005; Seppälä et al., 2006; Makarenko and Budak, 2017).

7.2.5 On-Farm Medical Waste

On-farm wastes consist of needles, syringes, medicines, bottles, vaccines, stock of infectious agents, culture samples, plates, research animals. animal carcasses, body parts, and bedding that may pose some serious threat to humankind, and exposure to any form of life could result in fatal diseases (Boxall et al., 2006). Hence, it is especially important to decontaminate these wastes before disposal. But the majority of hospitals dump these wastes in regular MSWs, throw them in rivers or nearby waterbodies (Al-Yaqout and Hamoda, 2002), or burn the waste in open air, causing some serious health and environmental risks.

7.2.6 Packaging and Processing Waste

Another major type of agricultural waste is generated from the packaging and processing industries (Brancoli et al., 2017; Wansink, 2018). Demand for clean and aesthetically rich agricultural products has increased the demand for proper and attractive packaging of products. Furthermore, in the processed food industry, ready-made food packaging is quite common (Akande and Olorunnisola, 2018). From fresh products to dairy, meat, fruits, juices, baked foods, cooked foods, frozen foods, and fried foods all are subject to packaging and labeling. The use of plastics, glass, and wood is practiced to conserve food for safe long-distance transportation and distribution (Sun et al., 2018). These types of packaging increase the shelf value of products. The waste generated from this sector consists of organic food, peels, fiber residues, plastic, wood, glass, bottles, containers, tetra packs or cartons, and rubber. But real food wastage occurs when manufacturing and expiration dates are mentioned by the manufacturing companies on the product (Ho and Chu, 2019). Due to lack of awareness about the concepts of these mentioned periods on labels, food gets dumped as waste (Brancoli et al., 2017). Also, excess chemicals used in packaging the fresh products have started to show short- and long-term health risks like obesity, cancer, and heart diseases.

7.2.7 Kitchen Waste

Household and commercial kitchens are end-users of the majority of food produced, be it raw fresh products or processed food (Brancoli et al., 2017; Tonini et al., 2018). These wastes are defined as the organic matter left from commercial and household kitchens. Some kitchen waste is produced out of necessity like fruit and veggie peels, whereas a huge quantity of food products gets wasted due to unplanned buying and excessive storage for future use. Densely populated places produce a huge quantity of kitchen waste every day (Pandey and Tyagi, 2012). Disposal of these wastes pose a major problem, and they comprise most municipal solid wastes. The huge amount of moisture content makes it difficult to incinerate the waste. These wastes are rich in organic molecules like carbohydrates, proteins, lipids, etc., while non-organic waste (Valenzuela and Cervantes, 2020) consists of materials like plastics, paper, metals, and glass. These organic wastes are used to produce biofuel and fertilizers. China produces 30 million tons of kitchen waste, while the U.S. produces 80 billion pounds of food waste (FAO, 2021).

7.3 EFFECTS OF AGRICULTURAL AND KITCHEN WASTE ON THE ENVIRONMENT

All these varying types of agricultural waste cause economic losses, loss of resources and energy, and create excessive loads on MSWs and landfills and environmental pollution. The detailed impact of these wastes is presented here (Hargreaves et al., 2008).

7.3.1 Effect on Soil Resources

The majority of agricultural and kitchen waste ends up as municipal solid waste (Azeez et al., 2011; Calleja-Cervantes et al., 2015). The residues and wastes generated from agricultural industries are believed to be rich in organic matter, have high nutrient content, and possess less toxicity (Ingrao et al., 2018; Erana et al., 2019. If these wastes are utilized in agriculture itself, then there would be a huge reduction in waste load on MSWs. Agricultural and kitchen waste seldom have toxic components to begin with, but slow decomposition and properties of leachate produced in open dumps and landfills are hard to predict (Essienubong et al., 2019; Mekonnen et al., 2020). Lack of management of leachates has been known to reduce crop productivity by deteriorating soil quality and fertility, causing water pollution as the toxic leachate gets transported to nearby waterbodies via soil pores (Khodary et al., 2020) or in the form of runoff (Mekonnen et al., 2020). Moreover, the crops grown in nearby land tend to biologically accumulate metals (Hargreaves et al., 2008) and other toxic chemicals which magnify in the food chain and can cause fatal diseases. Negligence in proper waste management could lead to disastrous consequences for the environment and human health. Hence it is especially important to study the impact of organic waste or agricultural waste on soil properties (Mekonnen et al., 2020).

Traditional agriculture practices have been a major reason for degradation in soil properties because of the decline in the structure of soil and loss in its organic matter due to ploughing and tillage (Mupangwa et al., 2007). Change in soil structure results in closing of pores, and the soil infiltration and moisture content decrease significantly. But the addition of food waste to soil causes a reduction in bulk density and increases the porosity, which leads to increased water holding capacity and macro-aggregation of soil (Bot and Benites, 2005; Adugna, 2016). The organic content of food waste helps in water retention as they improvise the pore structure to hold water, whereas the increased infiltration causes efficient use of irrigation water and precipitation (Gülser and Candemir, 2015). The fertility of any type of soil is dependent on its organic matter content. The use of organic waste in the soil in form of compost or fertilizers has shown tremendous improvement in soil fertility and other hydraulic and physical factors like density, porosity, moisture content, pore size, shear strength, and particle structural stability (Adugna, 2016). Moreover, the rate of improvement of soil fertility and its associated factors are more in loamy soil than clay after organic waste application.

Various environmental parameters like moisture availability of soil, soil temperature, soil density, pore size, aeration, and nutrients and microbes present in the soil, along with the type of components in organic waste, mainly its C:N ratio and total nitrogen content, decide its rate of decomposition in soil. If the carbon-nitrogen ratio is higher, waste decomposition occurs at a slower rate (Kumar et al., 2010; Wang et al., 2015). An increase in organic matter of soil due to the application of organic waste in the form of organic fertilizers helps crop production in many ways. The application of organic fertilizers results in enhancement of soil nutrients, provides carbon and nitrogen for better survival and functioning of soil microbes, enhances the seed germination processes, reduces soil erosion by holding the topsoil, and retains soil moisture. Furthermore, numerous studies support the fact that organic content increases more with the long-term application of compost rather than fresh organic waste content (de Nobile et al., 2021)

The application of organic compost made up of agriculture and kitchen waste in gardening and small-scale agriculture has gained huge popularity (Bouzouidja et al., 2020). But unmonitored and excessive use of organic compost may cause nutrient saturation in soil. Also, vegetables and fruits utilized in fertilizer-making should be monitored for the number of chemical fertilizers present in them or the processes of their cleaning, to avoid any sort of metal or heavy metal pollution in soil, plants, and water (Yoshida et al., 2015). This process of waste utilization also reduces the load on landfills and other disposal processes. It is believed to be a cost-effective, eco-friendly, and feasible process of waste disposal.

Other forms of organic fertilizers used in agriculture are sewage sludge and animal manures, especially in arid and semiarid regions (Nascimento et al., 2021). Although these types of organic waste fertilizers tend to improve soil fertility and crop productivity and growth, these fertilizers need constant monitoring (Novara et al., 2020). The use of sewage sludge as fertilizer has resulted in heavy metal toxicity in soil and plants and hence their application was ceased (Wang et al., 2019). However, they are still popular in low-income countries as cheap sources of fertilizer and an easy way to decompose sludge waste. But this comes with many environmental risks and health problems in the form of metal-organic compound contamination and disease from pathogen contact (Aktar et al., 2010).

7.3.2 Effect on Water Resources

Water is the elixir of life, but the sources of fresh water are limited. The sustainability and functioning of any form of life requires water. In the past few decades, eutrophication and algal bloom have posed a great threat to water quality. In South Africa, the problems of eutrophication, cyanobacterial blooms producing cyanotoxins, and macrophyte growth of species like water hyacinth, red water fern, water lettuce, and weeds have gained attention since the late 1970s. USA freshwaters suffer from cultured eutrophication, which causes a loss of around 2.2 billion dollars annually. In China, eutrophication has become the most problematic water quality issue. According to literature (Issaka and Ashraf, 2017; Wu et al., 2017; Teng et al., 2021), most of the lakes of China are in eutrophic (44%) and hypo trophic states (22%), including Taihu, Chaohu and Dianchi lake (Dong et al., 2010; Gao and Zhang, 2010). This has resulted in socio-economic instability as well heavy environmental damage. Incidences of red tide, which happened every five years in the 1960s, has now turned into 90 red tides a year (He et al., 2021).

In the global phosphorous pollution ranking, India ranks second by dumping 8% or around 1.5 billion kg of phosphorous into freshwaters (Mateo-Sagasta et al., 2017). China tops the charts with 30% input and the USA ranks third with 7% input of phosphorous in freshwater systems. Huang He River, followed by the Indus and Ganges River basins are listed as the most polluted rivers of the world affected by nutrient pollution (Mateo-Sagasta et al., 2017; Bowes et al., 2020). The flow from sewage sludge dumped around 0.97 billion kg of phosphorous in Indian freshwater resources, while the contribution from the agricultural field was 38% in the form of erosion, runoff, and leaching, with an increase of 27% from 2002 to 2010 (FAO, 2021).

Eutrophication is a phenomenon where waterbodies get nutrient enrichment from various point and non-point sources. This causes major ecosystem structural changes and degradation in the quality of water. Increased algae growth and other aquatic weeds and plants and deterioration of aquatic biomes and fish species are the results of eutrophication. Eutrophication affects lakes of Asia (54%), North America (48%), Africa (28%), Europe (53%), and South America (41%), as per a survey of world lakes from ILE committee (Ma et al., 2011; Mehan et al., 2016; FAO, 2021). One of the major contributors to this water quality problem is fertilizers, wastes, and excess nutrient flushing from the farms into river bodies. The unplanned use of fertilizers in farms result in nutrient saturation of farm soils. Excess nutrients are washed away with or without soil particles, along with irrigation water or precipitation runoff. Another reason for increased eutrophication is the discharge of sewage sludge and greywater from farm soil into waterbodies (Hait and Tare, 2012; Hossain et al., 2017). The number of nutrients, organic and inorganic compounds, and heavy metals all get deposited into waterbodies and they support the growth of algae, phytoplankton, zooplanktons, bacteria, and fungi which are collectively called algal bloom (Rabalais, 2002; Amin et al., 2017; Caruana et al., 2020). Growth of these particulate substances causes loss of dissolved oxygen and hence make the respiration of other animals difficult. After the consumption of dissolved oxygen or when the lake is at hypoxic levels, the aquatic animals die (Howarth et al., 2002).

Further deterioration occurs when the anaerobic condition persists, creating the environment for the growth of toxin-producing bacteria (Caruana et al., 2020). These toxins are harmful to all forms of life. Furthermore, the algal bloom blocks the movement of light and thus leads to aquatic dead zones and loss of aquatic biomes. Since serious deterioration occurs in water quality, the water becomes unfit for any human or animal consumption (Ouyang et al., 2018; Solarte-Toro et al., 2018). Additionally, the stench produced from the waterbody is hard to mask by any cleaning process like chlorination. Also, the amount of chemical toxicity produced in water corrodes the pipes in supply systems and hinders the flow of water (Makarenko and Budak, 2017). For example, in Wuxi, China, 2 million people did not have access to potable drinking water for more than 7 days, as the main source of water, the Taihu lake, suffered from algal bloom (Gao and Zhang, 2010).

Also, the cyanobacteria that grow and create red tides are known for their toxic and poisonous nature even in low concentration in water (Moreno-Jiménez et al., 2020). These sources of contaminated water can be fatal to humans, animals, and livestock in ingestion. For example, shellfish poisoning killing adults and blue baby syndrome (Dolan et al., 2016; Caruana et al., 2020) in infants are commonly seen. Also, these blooms form a mat that floats and can also affect the land surface.

7.3.3 Pollution from Agricultural Waste

The process of agricultural waste disposal is also responsible for producing greenhouse gases that cause air pollution. The GHGs are released in the process of burning as well as decomposition of agricultural waste. For example, the burning of crop residues produces CO, NO_2, N_2O, and smoke (Bhuyan et al., 2020). These gases act as a contaminant and are associated with the formation of ozone and nitric acid, leading to acid deposition and rains causing risk to all forms of life and pollution of resources (Ling et al., 2010).

Animal waste, like faeces, manure, urine, respiration and decaying gases, cause pollution in the environment (Sharpley et al., 2011). These wastes pose a greater threat when an animal population is concentrated in a small space and has less space for manure disposal. The gases emerging from decay, fermentation of animal bodies (Maiti et al., 2016), and their respiration when alive are lost in the air, but the solid and liquid animal waste formed gets converted from organic matter to microbial biomass and other liquid and gaseous products (Devi et al., 2017). Also, animal waste such as manure is used as organic fertilizers and conditioners for soil (Hargreaves et al., 2008), but they also tend to pollute the soil and water resources.

In agriculture, aquaculture also plays a major role (Obi et al., 2016). Deterioration in water quality often adversely impacts the aquatic biomes. Fishermen depend on natural or artificial waterbodies like lakes, ponds, rivers, and seas for their fish farming, but nutrient pollution and waste deposition leading to eutrophication and algal blooms kill the fish communities and other sea animals. Even if they do not kill the fish, toxins produced by algal bloom gets stored in the muscles of fish and other sea animals (Karaboga and Akay, 2009; Angela et al., 2015; Caruana et al., 2020). These toxins are mostly poisonous and tend to biologically accumulate and magnify in the food chain (Delang, 2018). Hence, upon the consumption of contaminated seafood, humans may suffer from fatal diseases. Many such cases have been reported in various states of India, like Punjab, Jammu & Kashmir, and Mathura. Bulk dying of fish has been reported due to heavy deterioration in water quality.

7.3.4 Need for Better Agricultural and Kitchen Waste Management

Disposal of kitchen and agricultural waste is a complicated task as it requires a lot of planning and processes, which include labor, capital, transportation, and land resources. These wastes increase the load of MSWs, or when dumped openly or burned cause various environmental, land, and water pollution (Calleja-Cervantes et al., 2015; Amritha and Kumar, 2019). Food loss or wastage of any form causes huge economic losses. Lack of awareness about the physical, chemical, and biological properties of agricultural waste and their potential utility in other sectors from construction to biofuel causes further economic losses for farmers (Lobley et al., 2009). Additionally, if the 4Rs (that is, Reduce, Reuse, Recycle and Recover) (Ray and Adhya, 2020) are not practiced, then the amount of load is extremely high and disposing of so much load even in landfills creates many problems. Currently, where there is a serious lack of land resources and pollution is at its peak while the economy is fluctuating, it is important to apply the best, most efficient, cost-effective and most eco-friendly management practices for agricultural waste disposal (Sharma et al., 2019).

7.4 SUSTAINABLE UTILIZATION AND MANAGEMENT OF AGRICULTURAL AND KITCHEN WASTE

Traditionally, waste is anything that does not return value. Commonly, anything that one does not use or is unwanted ends as trash or garbage or waste. However, according to science, what one defines as waste could be beneficially utilized by someone else (Duque-Acevedo et al., 2020). The main objective of any waste management is to extract the maximum economic benefits in an efficient and eco-friendly manner. Whatever the method, it needs to be feasible, affordable, and simple so that the maximum population can apply it. Also, waste management practices should aim at reducing the waste load on MSWs and landfills (Hargreaves et al., 2008). With these motives, many management practices have been researched, started, and followed to date. For example, in India, sacred temple flowers generate a huge amount of waste managing, which has been a huge issue as a lot of religious and sentimental values are attached to it. Many start-ups like "Phool", "Greenwave", "Holy Waste", etc., have taken the initiative to make incense sticks and cones, soaps, and compost out of these sacred flowers (Holywaste—Making Waste Beautiful, 2020; GreenWave, 2020). This has helped in the reduction of flower waste as well as control river pollution and given financial stability to many low-income and rural populations, especially women (Obi et al., 2016). Furthermore, plenty of laws have been developed by every country to manage the agricultural and kitchen waste produced. In countries like the USA, South Korea, and China and other mostly developed countries, segregation of waste (Ahn et al., 2006) is done at the core; that is, dividing wastes into biodegradables and non-biodegradables and giving it to collectors for further processing. But these techniques may not be successfully applied in developing countries, where the population is huge and no strict actions are taken against illegal waste disposal. Besides, landfill and incineration plants are concepts addressed in the past few decades (Adejumo and Adebiyi, 2021).

7.4.1 Understanding the Waste Characteristics

In many Asian countries like Japan, South Korea, China, and Taiwan, food waste is directly converted to feed for animals. The remaining waste is either converted into compost or valorized into energy resources. In other countries, like Hong Kong, concepts of the 4Rs (Refuse, Reduce, Reuse, and Recycle) (Ma and Wu, 2018) and landfill disposal are a much-preferred option. In India, although the most common practice has been open dumping, in recent times, composting, vermicomposting, incineration, waste compaction, and biogas generation have been actively practiced. Many new concepts of waste disposal and utilization have come into practice worldwide, but before applying any process, one should be aware of the characteristics and properties of the type of waste (Yusuf, 2017). Agricultural wastes are divided into solid, liquid, and gas. The amount of gaseous waste produced from respiration, decomposition of dead bodies and dumped food waste, burning of crop residues, etc, dissolves in the air (Ortiz et al., 2008). The stench produced from these processes is quite uncomfortable for the human population residing nearby. Burning fumes and decomposition release a lot of methane into the air, which is a major greenhouse gas, along with carbon dioxide and carbon monoxide. The waste produced from animals and livestock comprises manure, fecal, urine, and bedding (Devi et al., 2017).

Animal wastes tend to have high moisture content; for example, in poultry manure, the amount of moisture is found to be 75%, while in swine manure the moisture content could be up to 85%. Manure is complex organic and nutrient-rich waste (Hamelin et al., 2014). The biological chemical properties present in manure makes it an excellent organic fertilizer, as around 75% of food nutrients supplied to the livestock come out as manure (Citak and Sonmez, 2010). Even the liquid portion of farm waste comprises 50% N (nitrogen) and 75% K (Potassium) while the solid portion is rich in P (phosphorous). But before application, they need to be treated or converted into compost, as a direct application of manure can be responsible for many environmental problems (Obi et al., 2016).

Food waste can be both in raw form (that is, fruits, vegetables, dairy, meat, eggs, juices, alcoholic and non-alcoholic beverages) or cooked form (Lemes et al., 2016). All types of waste have different compositions, but almost all of them are easily degradable, have excess moisture or water content, have organic compounds like fats, protein, and lipids, and have less pH and high solubility. They provide higher energy if the food mass is dry or volatile. Fruits and vegetables especially have a low C/N ratio (Gülser and Candemir, 2015) and are also rich in lignocellulose material, which makes them resistant to microbes. Decomposing the food waste through different processes creates different issues due to its varying characteristics.

Plastic waste is generated as agricultural and kitchen waste in many forms (Goel, 2017), from cultivation practices to every step of storage, packaging, transportation, and arrival in supermarkets (Horowitz et al., 2018). Almost all the food we buy is in some form of plastic containers, bags, or packets. Milk cartons, juices, sauces, curd, bakery products, confectioneries, grains, and cooked meals all are being sold in some sort of plastic (Styles et al., 2016). Packaging food and crops for better shelf life and easier and damage-free transportation has helped in the progressive distribution of food, but has generated an immense amount of plastic waste (Eriksson et al., 2016). Also, since the 1950s, the use of plastic mulching in cultivation has helped in improving the productivity of the land by holding the soil moisture, regulating soil temperature and other soil properties, but again the use of plastic in soil is damaging the quality of soil in the long run (Blumenstein et al., 2012; Dong et al., 2019). Plastic toxins stay in the soil for centuries and the residue is around 300kg/hectare, causing white pollution (Banerjee et al., 2013; Liu et al., 2014). The porosity and aeration of soils are hindered due to excessive plastics, as well as microbial diversities, due to lack of aeration thus impacting soil fertility (Banerjee et al., 2013).

Other than mulching, plastics are also used in micro-irrigation, which is form of sprinklers and drip irrigation systems (Mall et al., 2017). Geomembranes are used in creating artificial ponds or pond liners in places where water is scarce (Tirkey et al., 2016). Some plants are grown by creating greenhouse farming techniques, which also include plastics to create the greenhouse effect. Use of excess plastic made up of a wide range of polymers like Polypropylene (PP), Ethelene-Vinyl acetate, and PVC pour carcinogenic chemicals into the soil which indirectly get accumulated on plants grown in these soils, and end up in human and animal bodies after consumption. These plastics once used are hard to reuse (Mupangwa et al., 2007; Dong et al., 2019). For example, the mulching plastic films used in China are found to be less than 8 microns; because of this thinness, most of the time the plastic films get torn and are impossible to recover and reuse; however, the mulching sheets or wraps in other countries like the US, India, and Europe have more thickness, around 15–20 microns, and hence can be reused. Plastic waste generated from all over the world amounted to around 6300MT in 2015 and this is expected to increase up to 12000 MT of plastic by 2050 (Awasthi et al., 2017; Issac and Kandasubramanian, 2021; Tsai, 2021).

Among the different types of agricultural and kitchen waste, the majority portion comprises food waste, animal waste, crop residue (Ju et al., 2005; Sharpley et al., 2011; Hamelin et al., 2014), and then other forms of waste like plastics, cardboards, papers, and bottles (Riber et al., 2009; Arfania and Asadzadeh, 2015; Mehran et al., 2021). Most of these non-organic wastes are recycled or reused after processing. They have different collection processes. For example, in India, there is a separate set of workers who collect bottles, metals, cardboards, books, and papers weekly or monthly at a certain rate (per kg) and pay money in return to the customers. They are often known as "Kabadiwala or ragpickers". This whole process is very motivating for the population as they get a certain return value for their waste. Some of the waste like jute bags and containers are also reused for other purposes in households (Hamelin et al., 2014). For example, baling twines are commonly used to hold hay bales together, but could be reused for many other tying purposes in different household activities (Knutson et al., 2011). Similarly, the concept of vertical indoor gardening and self-watering plant growing systems utilize plastic containers and bottles for growing plants. This adds to the greenery and creates an aesthetically sound environment (Bouzouidja et al., 2020).

7.5 ADVANCES IN RECYCLING AND DECOMPOSITION OF WASTE

The biggest problem faced by the world is waste management. Exponential population growth is the main stem of these problems. Although the concept of a circular economy is being prioritized all over the world, agricultural and kitchen waste cannot be easily controlled or monitored, as many factors control the type and amount of waste generated, along with their successful management and disposal (Oldfield et al., 2016; Goel, 2017). Starting from population density, the economy of the region, climate, infrastructure, types of agriculture, awareness about food and nutrition, developments in technology and techniques of cultivation to availability of markets, branding, global connectivity of export and import, all these factors decide the fate of agricultural sectors. Overall, many common and recent techniques have been used to dispose and manage the generated agricultural and kitchen waste (Corrado and Sala, 2018). Some of these are discussed here.

7.5.1 The 4 Rs: Refuse, Reduce, Recycle, Reuse

The easiest and most convenient way of controlling waste load is to refuse the items if not required. From food to equipment, this could be applied to all (Kazerooni Sadi et al., 2012). For example, during the sale season most people buy food in bulk quantities and store it, but most of the time they are unable to consume all of it and most of these foods end up wasted (First Nations of Quebec and Labrador Sustainable Development Institute, 2008). On the other hand, bulk buying creates a food shortage in many cases, and the needy ones do not get to buy their necessities. Moreover, instead of buying many small packets of any food item, one can buy one large bag and reduce the amount of plastic waste. According to statistics, around 40% of perfectly edible food is wasted, estimating up to 125 to 160 billion pounds in the US every year (US EPA, 2017). Hence, planning before buying is a good way to control both situations.

The next step is to reduce agricultural and kitchen waste. Waste prevention is a much-appreciated approach in waste management, as what was never created does not need to be disposed of. For example, the excess packaging from manufactured products creates unnecessary plastic waste. The use of plastics in buying local groceries could be replaced with permanent jute bags that could be reused and have better durability for carrying a lot of weight. Also, most people do not prefer local vegetables if they are not aesthetically sound or if the fruits are too ripe. These concepts increase food waste.

Furthermore, recycling is another option. Most of our inorganic products in the agriculture sector could be reused. In the US, the amount of food wasted is highest compared to all other countries, while China and India generate most household food waste. Frequently thrown away items are bread (240 million slices/year), milk (5.9 million glasses), potatoes (5.8 million), cheese, and apples (1.3 million) (Scholz et al., 2015). Using different processes and cooking techniques, one could utilize these products. Other than food waste, containers, plastics, and bottles could all be washed and reused most of the time for other household work (US EPA, 2020). Paper and plastics are usually collected and sent to recycling plants.

At last, whatever we consider waste may not be a complete waste. Every object could be utilized in some other form. For example, feed bags could be reused as trash bags, weed barriers in gardens, rag cloths in homes for cleaning and scrubbing purposes. Kitchen leftovers could be used to feed strays and animals; some farmers tend to use the eggshells and other waste to feed their poultry animals as these wastes are great sources of nutrients and could also be used as manure or compost (Sharpley et al., 2011). Similarly, nylons could be used to store crops like onions and potatoes for a long time.

7.5.2 Landfill

Landfills are places or land designated for organized waste disposal. Since olden times they have been used to bury the waste by layering with soil or heaping in a place far away from the population (Eriksson et al., 2015). The most common problem was the production of toxic leachates, whose characteristics were not defined, and the generation of hazardous GHGs, especially methane (Oshita et al., 2012). It was believed that through natural processes like adsorption, rainfall, exchange of ions, and decomposition from microbes, the gases and leachate would be mitigated. In recent decades, the practice of landfills has seen new techniques; a lot of effort is being put into landfill design and management (Nandan et al., 2017; Babalola, 2020; US EPA, 2020).

Use of geomembranes, geocomposites, and geocells can confine the landfill areas or layer them to prevent the leachates produced from flowing out. Also, various gas management techniques are applied to utilize the gas generated from landfills as biofuels. The overall objective has been shifted from dumping heaps of waste to converting landfills into bioreactors and biofuel plants (Karimi, 2015). Furthermore, the study of leachate characteristics is actively done, along with monitoring the quality of soil and water around the area. The production of leachates and gases in landfills is based on the working of many components like waste composition, degree of compaction, type of soil, soil properties, availability of water resources, level of the water table, temperature, amount of moisture in waste, and age and depth of refuse or waste (Väisänen et al., 2016).

Originally, landfills were located far away from the population, as it requires a lot of precaution to maintain a landfill. Toxic leachate production, soil and water pollution, agricultural productivity failure, pest infestation, and ecological hazards are some commonly expected disasters an unattended landfill can produce (Das et al., 2002). In India, many landfill policies and laws are made; some crucial points include that within 100 m there should be no rivers; within 200 m, there should be no pond or lake; no highways should cross within 500 m; other public facilities like recreational parks and residential areas should all be 500 m away from the landfill, and airports or airbases should be at least 20 km away from landfills (University of Delhi, 2019). In some countries like Ukraine, there are sanitary protection zones developed around landfills, which cover a 500 m radius of areas to protect the population from any type of epidemic or disease breakouts (Makarenko and Budak, 2017).

But in the long run, the scarcity of lands and the growth of suburbs have forced authorities to find new methods of waste disposal, as creating new landfill spaces is not easy and maintaining them is even harder (US EPA, 2020). Some other associated problems with landfill management are that waste segregation, transportation, and storage all require both excess labor and capital. Sometimes the increased transportation cost creates a huge problem in disposal services (Lisk, 1991; Nandan et al., 2017). Many policies have been created by governments all over the world to extract the cost of storage and transportation of waste disposal, like in India, where every household pays taxes to municipal councils for the collection of wastes (Narayana, 2009; Eriksson et al., 2015).

Also, building landfills in hilly areas is avoided at all costs. Furthermore, the areas prone to earthquakes need even more precautions, as, due to seismic waves, there is a high chance that the landfill material may overstretch and tear. Also, the leachate and gas management systems may move and crack, leading to many ecological disasters like landfill fires (Castelli et al., 2013). Combustion of landfill gases could produce deadly toxic smog that has the potential to travel for kilometers and be hazardous. Moreover, the non-organic waste compacted in landfills does not dissolve or disappear for millions of years; it is always in a state of differential settlement as the organic waste keeps decomposing (Li et al., 2013). Hence the old landfill sites can be only used as recreational parks or playgrounds and not for infrastructure development. Recently, integrated waste management is being adopted everywhere globally, which supports the 4Rs in the initial phases of waste management to fuel extraction as a result (Kazerooni Sadi et al., 2012).

7.5.3 Composting

Another efficient, eco-friendly, biological, and commonly practiced technique is composting. It is a form of treatment in which the biodegradable agricultural and kitchen waste break down by naturally occurring microbial communities under aerobic conditions in a closed chamber (Adejumo and Adebiyi, 2021). It is found to be a tremendously good alternative to landfill disposal and can be practiced as home composting by every household for their everyday organic waste (Oldfield et al., 2016). Although the compost releases a lot of foul-smelling gases, it typically depends upon the type of waste undergoing composting. Decomposing farm waste might require a large space as it produces volatile organic compounds and aeration must be maintained, so waste is often mixed. The compost produced is known as humus (Sands and Edmonds, 2005; Oktaba et al., 2018) and the soil has a dark color as well as an earthy texture. It acts as organic fertilizer and is efficiently used in farms for cultivation, organic farming, and home and urban gardening (Boechat et al., 2017).

The number of organic compounds and nutrients, especially nitrogen (N-2%), phosphorous (P-0.5 to 1%), and potassium (K-2%) are present in low quantities, along with some other micronutrients like sulfur, calcium, magnesium, carbon, and hydrogen (Väisänen et al., 2016; Adejumo and Adebiyi, 2021). This amount is adequate for small-scale gardening, but crop cultivation needs it to be applied in large quantities on farms. The speed of decomposition is usually slow, so to speed up the process additional N fertilizers and manure are added to the compost batch. The N compound in compost slowly becomes available for plant absorption in small quantities, but that is how N is available to plants throughout the season for crops (Bailey and Lazarovits, 2003). Biological and chemical processes in compost formation are complex and depend on many factors like C/N ratio, pH, temperature, moisture or water content of waste, and the types of nutrients available in waste. C/N ratio is the basic determiner of the composting process. If this ratio is too high, the whole process diminishes due to lack of nitrogen; if the ratio is too low, then ammonia formation takes place, creating loss of nitrogen (Jeyabal and Kuppuswamy, 2001).

Furthermore, the calorific and nutrient value of waste also determines the process of composting. For example, food that is rich in protein gets composted faster as proteinaceous fraction supports bacterial activities, whereas fat-rich foods are not easy to decompose (Barajas-Aceves, 2016). Waste having high moisture content and low pH tends to hinder the process of decomposition by creating anaerobic conditions, and are usually dealt with by adding lime to the mixture. Compost improves the soil properties, enhances soil structure, and improves soil aggregation and stability while increasing pore space and moisture-retaining capacity (Oldfield et al., 2016). It provides all sorts of nutrients for better productivity and growth and over the years, regular application of compost as needed improves the overall soil quality (Sharma et al., 2019). As composts are chemical-free, they usually do not create any side effects.

7.5.4 Vermicomposting

Another form of composting is a mesophilic process called vermicomposting, in which worms are used for the decomposition of organic waste into nutrient-rich compounds (Sinha et al., 2002). Worms like earthworms, red wigglers, and white worms tend to consume and digest the organic waste matter and excrete the nutrient-rich by-products, called castings or vermicompost. These enhance the quality of soil and promote the growth of soil microbes and fungi, which help in process of crop growth (Atiyeh et al., 2002). The chemical composition of vermicompost or castings has exceedingly high nutrient content, which is readily available for absorption by plants (Coria-Cayupán et al., 2009; Kelova et al., 2021). The types of waste that can be decomposed through this process are crop residues like husk, straw, leaves, weeds, livestock wastes, litter, dairy, processed food, kitchen waste, and bagasse. In the process of vermicomposting, the volume of organic waste is reduced by 40–60% (Sharma et al., 2019).

There are two types of earthworm communities used for vermicomposting: the burrowing and non-burrowing kinds (Lazcano et al., 2010; Blanchet et al., 2016). Both are equally efficient in producing compost, but their functioning is different. In India, the species that have been traditionally used are Indian blue, African nightcrawler, and tiger worms (Sinha et al., 2002). Also, vermicomposting is practiced on both the commercial scale and the small scale. This is a great alternative to chemical fertilizers and helps reduce the load on landfills. Although this technique is ancient, it is now regaining popularity due to its ecological and economic benefits like soil regeneration and sustainability in cultivation while managing waste (Arancon et al., 2008). Western countries like Italy and Canada and Asian countries like India, Malaysia, Japan, and the Philippines practice large-scale vermicomposting.

7.5.5 Incineration

One of the most readily practiced waste management methods is incinerating waste. It is a process in which the waste is reduced under controlled combustion to incombustible materials like ash and gas (Eriksson et al., 2015). Also, the gases produced tend to be toxic and hazardous and hence require special treatment before they are released into the atmosphere (Narayana, 2009). This process is effective in reducing almost 90% of agricultural and organic waste and is a most hygienic method (Oldfield et al., 2016). Unlike other forms of waste management technique which require time and space and pose a risk of health issues, incineration is the best option, especially when time and land are not readily available. Also, the incinerator does not produce any smell, and it eliminates all forms of bacterial infection as high heat burns everything into ashes and the energy is utilized for another purpose (Yang et al., 2005). The disadvantage of using this method is that it produces a lot of GHGs and so is not a very eco-friendly method. Additionally, it requires lots of resources and is expensive, especially when practiced on a large scale. The waste is turned into ash and loses its nutrient value, hence it cannot be used in soil for improving fertility (Styles et al., 2016).

7.5.6 Anaerobic Digestion

Recycling MSW using the anaerobic digestion process is another effective method that is more eco-friendly than all other methods like landfills, incineration, and composting (Hamelin et al., 2014). This method helps to recycle the agricultural waste and kitchen waste into energy-efficient and value-returning products like bioethanol (Tonini et al., 2016), enzymes, compost, and bioplastics while reducing the production of GHGs. As the target is to generate biofuel (Karimi, 2015) or methane from food waste, this low-cost process has three phases: 1) enzymatic hydrolysis; 2) acid formation; and 3) gas production (España-Gamboa et al., 2011; Qin et al., 2016; Bhuvaneshwari et al., 2019). The working of this process is influenced by the nature of waste, bacterial composition and concentration, pH, and the temperature of bioreactors (Ma and Wu, 2018). The working of anaerobic digestion is presented in Figure 7.1.

The degradation of organic agricultural and food waste depends upon the composition of the matter present (Stangherlin et al., 2020). But usually, it is challenging to understand the waste's characteristics and percentage of content due to its heterogeneity (Ren et al., 2021). For anaerobic reactions to take place, it is important to control the working environment conditions as the anaerobes may be sensitive. The various factors that influence the anaerobic processes like bio-methanation are seeding (digested slurry, goat rumen fluid), temperature (35–40°C (mesophilic) and 55°C (thermophilic)), pH (6.3–7.8), C:N ratio (25:1), volatile fatty acids, ammonia content,

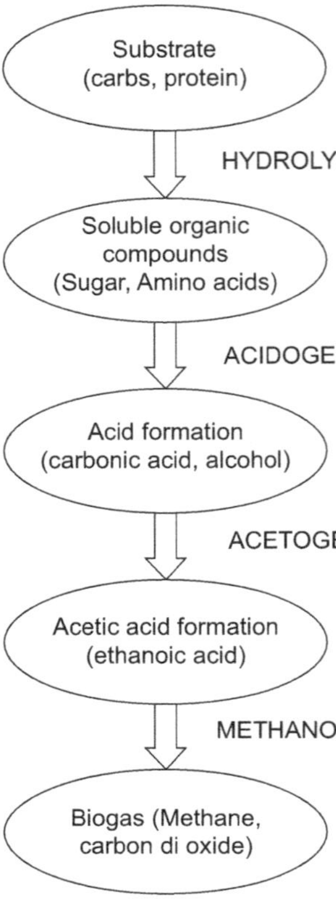

FIGURE 7.1 Process of anaerobic digestion.

Source: Literature survey (España-Gamboa et al., 2011; Tonini et al., 2018; Bhatia et al., 2020; Kapoor et al., 2020)

nutrient content, and organic loading rate. (Vimala Ebenezer et al., 2020) Overall, the process is found to be economical and more eco-friendly, though the costs of collection, segregation, and transportation need to be considered.

7.6 USE OF AGRICULTURAL AND KITCHEN WASTE IN OTHER INDUSTRIES

Other than waste decomposition methods, agricultural and kitchen wastes are also widely used in different industries like pharmaceuticals, textile, construction, cosmetics, and energy. Some of these uses are been briefly explained here:

In the construction field, agricultural waste is commonly used, from soil stabilization to making construction material, aggregate replacement, brick-making, development of lightweight material, and road construction (Adesanya and Raheem, 2009; Kazerooni Sadi et al., 2012; Susilawati et al., 2020). Some very commonly used agricultural wastes are coconut husk, bagasse ash, rice straw ash, banana fiber, ground nutshell ash, palm oil fuel ash, jute fiber, wheat straw ash, glass powder, fish bones, oyster shell (Prusty et al., 2016), raw wood, giant reed ash, cork, tobacco waste, and sawdust. For construction purposes, the waste generated is selected based on its strength, workability, mechanical properties, tensile strength (for fiber), durability, thermal conductivity, and biological and chemical properties like rate of decomposition and resistance to microbial attack (Dalgaard et al., 2008; Zheng et al., 2014; Gülser and Candemir, 2015). Agricultural waste is then selected for the desired purpose based on its compatibility with the material it is to replace or the amount of strength it has to achieve.

For example, in research, for replacement of fine aggregate, instead of sand or crushed stone, ground nutshell, waste oyster shell, and cork fines showed uniform grain size distribution, specific gravity, bulk density, water absorption, and fineness modulus. Also, the bulk density was found to be much smaller than the bulk density of fine sand, which may impact the workability, strength, and durability of mortar (Vaibhav et al., 2015; Prusty et al., 2016). Similarly, when these agricultural wastes are used in soil stabilization of problematic soils, waste is used in fiber, ash, and shell in powdered form. Recently, the field of science and technology is focused on utilizing more agricultural wastes in the construction industry, as this would solve the problem of extracting costly raw material from mining, waste management and disposal, and lowering pollution in the environment (Magar, 2020).

In the cosmetics and pharmaceutical industries, the nutrient and caloric content of waste is monitored and utilized. Since ancient times, fruit pulps, seeds, peels, and vegetables have all been known for their therapeutic, flavoring, dyes, preservative, and fragrance properties (Jayathilakan et al., 2012; Spatafora, 2012). These properties help in managing human health, skin rejuvenation, antioxidant enrichment, brain health, and haircare (Spatafora, 2012; Keles and Özdemir, 2018). Most of the food waste or organic waste is raw fruits and vegetables that may not have stood up to their aesthetic standards. It could also be that the fresh produce was not consumed before the labeled period or other similar reasons, and have been discarded as waste (Anand et al., 2019). These food wastes are rich in minerals, carbohydrates, vitamins A, B, C, K, and E, have fiber, bioactive compounds, sugar, and organic acids and are actively being used in various cosmetic and medicine industries.

For example, the pulp of apple, mango, papaya, pineapple, peel of banana, cactus, mandarin, citrus fruits, tomato and potato, and seeds, rind, pomace, and cores of other fruits are commonly used in cosmetics (Arvanitoyannis and Kassaveti, 2008; Jayathilakan et al., 2012; Barbulova et al., 2015). The nutrients present in fruits, grains, and vegetables help in fighting heart diseases, carcinogenic issues, skin pigmentation, UV rays, and artificially creating these nutrients is very harmful. Now the world prefers organic products which use fewer chemicals and are eco-friendly. Hence, use of these wastes in creating such effective products and brands has helped with waste management and saved our health and ecosystems (Barbulova et al., 2015; Anand et al., 2019).

7.6.1 Management of Waste in India

On a global front, developing countries face more challenges compared to developed countries, especially in terms of agriculture and food waste management.

The amount of waste generated by the developing countries per capita is 1.91 times less than developed countries (Thi et al., 2015). This difference is seen due to the evident higher standard of living that demands better quality, easy availability, variety, and aesthetics of food products. Higher-quality food requires more raw ingredients, hence adding more to the waste. Lack of proper facility of storage and transportation (Kumar and Goel, 2009), poor infrastructure development (Nandan et al., 2017), lack of awareness, and improper waste management practices are some of the major causes for excessive food waste generation (Ellison et al., 2019).

In India, food wastage occurs mostly in the initial stages in rural areas, while in urban areas the waste is generated in later stages. India has rainfed agriculture (Petare et al., 2016). Around 58% of the Indian population is dependent on the agriculture sector as their primary livelihood. As per the Indian Brand Equity Foundation (IBPE) 2021, the value added from agriculture and its associated sectors adds up to US $276.37 billion in the fiscal year 2020. The food demands of the markets are driven by the immense population growth and increased external demands for agricultural products (Mekonnen et al., 2020). In India, waste generation occurs equally in rural and urban areas, but the amount and type of waste generated have distinctive features. Rural areas have more organic and biodegradable waste, while urban areas have somewhat more non-biodegradable waste (Dung et al., 2014). Moreover, segregation of wet and dry waste is not actively adopted by the population, despite the government's rules and awareness programs (Nandan et al., 2017). Lack of segregation of waste results in the increased amount of organic waste in landfills. After decomposition, it produces a huge amount of greenhouse gases, adding to climate change. Unfortunately, in developing countries, a large number of the population struggles to meet their daily nutritional demands. Hence, it is crucial to adopt feasible food waste management strategies.

In recent years, many studies have pointed out the change in consumers' approach toward food waste management. A study conducted in restaurants in Mumbai, India showed that waste management was dealt with at micro levels rather than making large-scale plans. Most of the restaurants were found practicing minimum extra preparation of food (10%–20%) (Bharucha, 2018). Some had redistribution of food that may be surplus for them. Some have a completely disposable policy (Bharucha, 2018; Corrado et al., 2019). This has helped them be more successful and appreciated in the restaurant business. Chauhan 2020 studied food waste management with the application of technology. Technology can help monitor the generation and management of food waste in supply chain processes. The use of an automated programmable logic controller can help in human error reduction; enterprise resource planning can help in an efficient logistics network; the barcode can help in identifying the expiration dates on products (Chauhan, 2020). Some easily applicable ways to reduce food waste are redistribution of food, use of coolers, use of small metal silos, storing food in plastic crates, food labels with dates, and campaigns to improve consumer awareness.

7.7 SUMMARY

The purpose of this review study was to present insights on the effects, sustainability, and utilization of agricultural and kitchen waste on the environment and economy. Agriculture and food sectors are one of the most important and productive sectors of any country. From cultivation to one's kitchen waste production, every step consumes energy, labor, capital, and natural resources. In developing countries, the loss of products or waste generation is more related to the supply chain, while in developed countries it is more on the consumer end. Waste generation is more dependent on consumer behavior, income of the population, nutritional awareness, infrastructure development, and availability of transportation and storage facilities. Also, some environmental factors like climate and topography play important roles. Waste generated from this sector is rich in organic content and creates a huge load on MSWs. Disposal of this type of waste is most difficult as it is hard to define the waste characteristics. Unplanned and unmonitored disposal poses a lot of environmental, economic, and health risks. Many advancements in science and technology have given varying choices for waste decomposition and management, like incineration, anaerobic digestion, composting, landfills, and recycling, and can be used depending upon the available finances, land resources, tools and equipment, awareness among the population, environmental impact, waste composition, type of population, etc. Recently, many industries have started utilizing these wastes, like construction for stabilizers, cosmetics, medicine, textile, paper, and decorative materials like candles, incense sticks, and cones. These advancements have helped in creating wealth out of waste and conserving our planet.

7.8 REFERENCES

Adejumo, I. O. and Adebiyi, O. A. (2021) 'Agricultural solid wastes: Causes, effects, and effective management', in *Strategies of Sustainable Solid Waste Management*. IntechOpen. doi: 10.5772/intechopen.93601.

Adesanya, D. A. and Raheem, A. A. (2009) 'Development of corn cob ash blended cement', *Construction and Building Materials*, 23(1), pp. 347–352. doi: 10.1016/j.conbuildmat.2007.11.013.

Adhikari, B. K., Barrington, S. and Martinez, J. (2006) 'Predicted growth of world urban food waste and methane production', *Waste Management and Research*, 24(5), pp. 421–433. doi: 10.1177/0734242X06067767.

Adugna, G. (2016) 'A review on impact of compost on soil properties, water use and crop productivity', *Agricultural Science Research Journal*, 4(3), pp. 93–104. doi: 10.14662/ARJASR2016.010.

Ahn, J. W. et al. (2006) 'PCDDs/PCDFs in municipal solid waste incineration bottom ash in korea', *Geosystem Engineering*, 9(4), pp. 91–96. doi: 10.1080/12269328.2006.10541261.

Akande, O. M. and Olorunnisola, A. O. (2018) 'Potential of briquetting as a waste-management option for handling market-generated vegetable waste in Port Harcourt, Nigeria', *Recycling*, 3(2). doi: 10.3390/recycling3020011.

Alvarenga, P. et al. (2015) 'Sewage sludge, compost and other representative organic wastes as agricultural soil amendments: Benefits versus limiting factors', *Waste Management*, 40, pp. 44–52. doi: 10.1016/j.wasman.2015.01.027.

Al-Yaqout, A. F. and Hamoda, M. F. (2002) 'Report: Management problems of solid waste landfills in Kuwait', *Waste Management and Research*, 20(4), pp. 328–331. doi: 10.1177/0734247X0202000403.

Amin, M. N., Kroeze, C. and Strokal, M. (2017) 'Human waste: An underestimated source of nutrient pollution in coastal seas of Bangladesh, India and Pakistan', *Marine Pollution Bulletin*, 118(1–2), pp. 131–140. doi: 10.1016/j.marpolbul.2017.02.045.

Amritha, P. K. and Kumar, P. P. A. (2019) 'Productive landscapes as a sustainable organic waste management option in urban areas', *Environment, Development and Sustainability*, 21(2), pp. 709–726. doi: 10.1007/s10668-017-0056-0.

Anand, K. et al. (2019) *Innovation in Materials Science and Engineering, Innovation in Materials Science and Engineering*. Springer Singapore. doi: 10.1007/978-981-13-2944-9.

Angela, C. B. et al. (2015) 'Hydrological evaluation of a peri-urban stream and its impact on ecosystem services potential', *Global Ecology and Conservation*, 3, pp. 628–644. doi: 10.1016/j.gecco.2015.02.008.

Arancon, N. Q. et al. (2008) 'Influences of vermicomposts, produced by earthworms and microorganisms from cattle manure, food waste and paper waste, on the germination, growth and flowering of petunias in the greenhouse', *Applied Soil Ecology*, 39(1), pp. 91–99. doi: 10.1016/j.apsoil.2007.11.010.

Arfania, H. and Asadzadeh, F. (2015) 'Mobility of heavy metals after spiking in relation to sediment and metal properties: Leaching column study', *Journal of Soils and Sediments*, 15(11), pp. 2311–2322. doi: 10.1007/s11368-015-1166-7.

Arvanitoyannis, I. S. and Kassaveti, A. (2008) 'Fish industry waste: Treatments, environmental impacts, current and potential uses', *International Journal of Food Science and Technology*, 43(4), pp. 726–745. doi: 10.1111/j.1365-2621.2006.01513.x.

Atiyeh, R. M. et al. (2002) 'The influence of humic acids derived from earthworm-processed organic wastes on plant growth', *Bioresource Technology*, 84(1), pp. 7–14. doi: 10.1016/S0960-8524(02)00017-2.

Awasthi, A. K., Shivashankar, M. and Majumder, S. (2017) 'Plastic solid waste utilization technologies: A review', *IOP Conference Series: Materials Science and Engineering*, 263(2). doi: 10.1088/1757-899X/263/2/022024.

Azeez, J. O., Hassan, O. A. and Egunjobi, P. O. (2011) 'Soil contamination at dumpsites: Implication of soil heavy metals distribution in municipal solid waste disposal system: A case study of Abeokuta, southwestern Nigeria', *Soil and Sediment Contamination*, 20(4), pp. 370–386. doi: 10.1080/15320383.2011.571312.

Babalola, M. A. (2020) 'A benefit-cost analysis of food and biodegradable waste treatment alternatives: The case of Oita City, Japan', *Sustainability (Switzerland)*, 12(5). doi: 10.3390/su12051916.

Bai, Z. et al. (2018) 'Effects of agricultural management practices on soil quality: A review of long-term experiments for Europe and China', *Agriculture, Ecosystems and Environment*, 265(May), pp. 1–7. doi: 10.1016/j.agee.2018.05.028.

Bailey, K. L. and Lazarovits, G. (2003) 'Suppressing soil-borne diseases with residue management and organic amendments', *Soil and Tillage Research*, 72(2), pp. 169–180. doi: 10.1016/S0167-1987(03)00086-2.

Banerjee, T., Srivastava, R. K. and Hung, Y. T. (2013) 'Plastics waste management in India: An integrated solid waste management approach', in *Handbook of Environment and Waste Management: Volume 2: Land and Groundwater Pollution Control*. World Scientific Publishing Co., pp. 1029–1060. doi: 10.1142/9789814449175_0017.

Banks, C. et al. (2018) Food waste digestion: Anaerobic digestion of food waste for a circular economy. *IEA Bioenergy Task*, 37.

Barajas-Aceves, M. (2016) 'Organic Waste as fertilizer in semi-arid soils and restoration in mine sites', in *Organic Fertilizers—From Basic Concepts to Applied Outcomes*. InTech. doi: 10.5772/62665.

Barbulova, A., Colucci, G. and Apone, F. (2015) 'New trends in cosmetics: By-products of plant origin and their potential use as cosmetic active ingredients', *Cosmetics*, 2(2), pp. 82–92. doi: 10.3390/cosmetics2020082.

Bharucha, J. (2018) 'Tackling the challenges of reducing and managing food waste in Mumbai restaurants', *British Food Journal*, 120(3), pp. 639–649. doi: 10.1108/BFJ-06-2017-0324.

Bhatia, R. K. et al. (2020) 'Conversion of Waste biomass into gaseous fuel: Present status and challenges in India', *Bioenergy Research*, 13(4), pp. 1046–1068. doi: 10.1007/s12155-020-10137-4.

Bhattacharyya, R. et al. (2015) 'Soil degradation in india: Challenges and potential solutions', *Sustainability (Switzerland)*, 7(4), pp. 3528–3570. doi: 10.3390/su7043528.

Bhuvaneshwari, S., Hettiarachchi, H. and Meegoda, J. N. (2019) 'Crop residue burning in India: Policy challenges and potential solutions', *International Journal of Environmental Research and Public Health*, 16(5). doi: 10.3390/ijerph16050832.

Bhuyan, N. et al. (2020) 'Valorization of agricultural wastes for multidimensional use', in *Current Developments in Biotechnology and Bioengineering*. Elsevier, pp. 41–78. doi: 10.1016/b978-0-444-64309-4.00002-7.

Blanchet, G. et al. (2016) 'Responses of soil properties and crop yields to different inorganic and organic amendments in a Swiss conventional farming system', *Agriculture, Ecosystems and Environment*, 230, pp. 116–126. doi: 10.1016/j.agee.2016.05.032.

Blumenstein, B. et al. (2012) 'Economic assessment of the integrated generation of solid fuel and biogas from biomass (IFBB) in comparison to different energy recovery, animal-based and non-refining management systems', *Bioresource Technology*, 119, pp. 312–323. doi: 10.1016/j.biortech.2012.05.077.

Boechat, C. L. et al. (2017) 'Solid waste in agricultural soils: An approach based on environmental principles, human health, and food security', *Solid Waste Management in Rural Areas*. doi: 10.5772/intechopen.69701.

Bot, A. and Benites, J. (2005) 'The importance of soil organic matter Key to drought-resistant soil and sustained food production The importance of soil organic matter Key to drought-resistant soil and sustained food production The importance of soil organic matter Key to drought-resist', FAO report 2005. Available at: www.fao.org (Accessed: 12 October 2021).

Boudreault, M. et al. (2019) 'Comparison of sampling designs for sediment source fingerprinting in an agricultural watershed in Atlantic Canada', *Journal of Soils and Sediments*, 19(9), pp. 3302–3318. doi: 10.1007/s11368-019-02306-6.

Bouzouidja, R. et al. (2020) 'Simplified performance assessment methodology for addressing soil quality of nature-based solutions', *Journal of Soils and Sediments*. doi: 10.1007/s11368-020-02731-y.

Bowes, M. J. et al. (2020) 'Nutrient and microbial water quality of the upper Ganga River, India: Identification of pollution sources', *Environmental Monitoring and Assessment*, 192(8). doi: 10.1007/s10661-020-08456-2.

Boxall, A. B. A. et al. (2006) 'Uptake of veterinary medicines from soils into plants', *Journal of Agricultural and Food Chemistry*, 54(6), pp. 2288–2297. doi: 10.1021/jf053041t.

Brancoli, P., Rousta, K. and Bolton, K. (2017) 'Life cycle assessment of supermarket food waste', *Resources, Conservation and Recycling*, 118, pp. 39–46. doi: 10.1016/j.resconrec.2016.11.024.

Bryngelsson, D. et al. (2017) 'How do dietary choices influence the energy-system cost of stabilizing the climate?' *Energies*, 10(2). doi: 10.3390/en10020182.

Butterly, C. R. et al. (2015) 'Carbon and nitrogen partitioning of wheat and field pea grown with two nitrogen levels under elevated CO2', *Plant and Soil*, 391(1–2), pp. 367–382. doi: 10.1007/s11104-015-2441-5.

Cakar, B. et al. (2020) 'Assessment of environmental impact of FOOD waste in Turkey', *Journal of Cleaner Production*, 244. doi: 10.1016/j.jclepro.2019.118846.

Calleja-Cervantes, M. E. et al. (2015) 'Thirteen years of continued application of composted organic wastes in a vineyard modify soil quality characteristics', *Soil Biology and Biochemistry*, 90, pp. 241–254. doi: 10.1016/j.soilbio.2015.07.002.

Caruana, A. M. N. et al. (2020) 'Alexandrium pacificum and Alexandrium minutum: Harmful or environmentally friendly?' *Marine Environmental Research*, 160. doi: 10.1016/j.marenvres.2020.105014.

Castelli, F., Lentini, V. and Maugeri, M. (2013) 'Stability analysis of landfills in seismic area', pp. 1226–1239. doi: 10.1061/9780784412787.124.

Chauhan, Y. (2020) 'Food waste management with technological platforms: Evidence from indian food supply chains', *Sustainability (Switzerland)*, 12(19). doi: 10.3390/su12198162.

Christofoletti, C. A. et al. (2013) 'Sugarcane vinasse: Environmental implications of its use', *Waste Management*, 33(12), pp. 2752–2761. doi: 10.1016/j.wasman.2013.09.005.

Ciotti, M. et al. (2020) 'The COVID-19 pandemic', *Critical Reviews in Clinical Laboratory Sciences*, 57(6), pp. 365–388. doi: 10.1080/10408363.2020.1783198.

Citak, S. and Sonmez, S. (2010) 'Effects of conventional and organic fertilization on spinach (Spinacea oleracea L.) growth, yield, vitamin C and nitrate concentration during two successive seasons', *Scientia Horticulturae*, 126(4), pp. 415–420. doi: 10.1016/j.scienta.2010.08.010.

Coria-Cayupán, Y. S., De Pinto, M. I. S. and Nazareno, M. A. (2009) 'Variations in bioactive substance contents and crop yields of lettuce (lactuca sativa L.) cultivated in soils with different fertilization treatments', *Journal of Agricultural and Food Chemistry*, 57(21), pp. 10122–10129. doi: 10.1021/jf903019d.

Corrado, S. and Sala, S. (2018) 'Food waste accounting along global and European food supply chains: State of the art and outlook', *Waste Management*, 79, pp. 120–131. doi: 10.1016/j.wasman.2018.07.032.

Corrado, S. et al. (2019) 'Food waste accounting methodologies: Challenges, opportunities, and further advancements', *Global Food Security*, 20, pp. 93–100. doi: 10.1016/j.gfs.2019.01.002.

Dalgaard, R. et al. (2008) 'LCA of soybean meal', *International Journal of Life Cycle Assessment*, 13(3), pp. 240–254. doi: 10.1065/lca2007.06.342.

Darine, S., Christophe, V. and Gholamreza, D. (2010) 'Production and functional properties of beef lung protein concentrates', *Meat Science*, 84(3), pp. 315–322. doi: 10.1016/j.meatsci.2009.03.007.

Das, K. C. et al. (2002) 'Stability and quality of municipal solid waste compost from a landfill aerobic bioreduction process', *Advances in Environmental Research*, 6(4), pp. 401–409. doi: 10.1016/S1093-0191(01)00066-1.

de Nobile, F. O. et al. (2021) 'A novel technology for processing urban waste compost as a fast-releasing nitrogen source to improve soil properties and broccoli and lettuce production', *Waste and Biomass Valorization*. doi: 10.1007/s12649-021-01415-z.

Delang, C. O. (2018) 'Causes and distribution of soil pollution in China', *Environmental & Socio-economic Studies*, 5(4), pp. 1–17. doi: 10.1515/environ-2017-0016.

Devi, S. et al. (2017) 'Crop residue recycling for economic and environmental sustainability: The case of India', *Open Agriculture*, 2(1), pp. 486–494. doi: 10.1515/opag-2017-0053.

do Carmo Stangherlin, I., de Barcellos, M. D. and Basso, K. (2020) 'The impact of social norms on suboptimal food consumption: A solution for food waste', *Journal of International Food and Agribusiness Marketing*, 32(1), pp. 30–53. doi: 10.1080/08974438.2018.1533511.

Dolan, J. R. et al. (2016) 'Fiv046.Full', (June), pp. 1–11. doi: 10.1093/femsec.

Dong, H. et al. (2010) 'Impacts of environmental change and human activity on microbial ecosystems on the Tibetan Plateau, NW China', *GSA Today*, 20(6), pp. 4–10. doi: 10.1130/GSATG75A.1.

Dong, Q. et al. (2019) 'Effect of different mulching measures on nitrate nitrogen leaching in spring maize planting system in south of Loess Plateau', *Agricultural Water Management*, 213, pp. 654–658. doi: 10.1016/j.agwat.2018.09.044.

dos Santos Nascimento, G. et al. (2021) 'Soil physico-chemical properties, biomass production, and root density in a green manure farming system from tropical ecosystem, Northeastern Brazil', *Journal of Soils and Sediments*. doi: 10.1007/s11368-021-02924-z.

Duhan, J. S. et al. (2017) 'Nanotechnology: The new perspective in precision agriculture', in *Biotechnology Reports*. Elsevier B.V., pp. 11–23. doi: 10.1016/j.btre.2017.03.002.

Dung, T. N. B. et al. (2014) 'Food waste to bioenergy via anaerobic processes', *Energy Procedia*, 61, pp. 307–312. doi: 10.1016/j.egypro.2014.11.1113.

Duque-Acevedo, M. et al. (2020) 'Agricultural waste: Review of the evolution, approaches and perspectives on alternative uses', *Global Ecology and Conservation*, 22. doi: 10.1016/j.gecco.2020.e00902.

Eichler-Löbermann, B., Gaj, R. and Schnug, E. (2009) 'Improvement of soil phosphorus availability by green fertilization with catch crops', *Communications in Soil Science and Plant Analysis*, 40(1–6), pp. 70–81. doi: 10.1080/00103620802623612.

Ellison, B. et al. (2019) 'Every plate counts: Evaluation of a food waste reduction campaign in a university dining hall', *Resources, Conservation and Recycling*, 144(January), pp. 276–284. doi: 10.1016/j.resconrec.2019.01.046.

Erana, F. G., Tenkegna, T. A. and Asfaw, S. L. (2019) 'Effect of agro industrial wastes compost on soil health and onion yields improvements: Study at field condition', *International Journal of Recycling of Organic Waste in Agriculture*, 8(s1), pp. 161–171. doi: 10.1007/s40093-019-0286-2.

Eriksson, M., Strid, I. and Hansson, P. A. (2015) 'Carbon footprint of food waste management options in the waste hierarchy—A Swedish case study', *Journal of Cleaner Production*, 93, pp. 115–125. doi: 10.1016/j.jclepro.2015.01.026.

Eriksson, M., Strid, I. and Hansson, P. A. (2016) 'Food waste reduction in supermarkets—Net costs and benefits of reduced storage temperature', *Resources, Conservation and Recycling*, 107, pp. 73–81. doi: 10.1016/j.resconrec.2015.11.022.

España-Gamboa, E. et al. (2011) 'Vinasses: Characterization and treatments', *Waste Management and Research*, pp. 1235–1250. doi: 10.1177/0734242X10387313.

Essienubong, I. A., Okechukwu, E. P. and Ejuvwedia, S. G. (2019) 'Effects of waste dumpsites on geotechnical properties of the underlying soils in wet season', *Environmental Engineering Research*, 24(2), pp. 289–297. doi: 10.4491/EER.2018.162.

FAO (2013) 'Food wastage footprint, Fao'. Available at: www.fao.org/publications.

FAO (2015). Global food losses and food waste facts. *Global Food Losses and Food Waste – Extent, Causes and Prevention. Food and Agricultural Organisation*. 4807

FAO (2019). The State of Food and Agriculture: Moving Forward on Food Waste and Loss Reduction. Rome. Available at: http://www.fao.org/3/ca6030en/ca6030en.pdf

FAO (2020) *The State of Food and Agriculture 2020. Overcoming Water Challenges in Agriculture*. Fao.

FAO (2021) *Food Waste Index Report 2021*. Unep.

First Nations of Quebec and Labrador Sustainable Development Institute (2008) 'Reduce, reuse, recycle and recover waste: A 4R's guide for the first nations communities of quebec and labrador'. Available at: www.iddpnql.ca/an/documents/4rsguide.pdf.

Flotats, X. et al. (2009) 'Manure treatment technologies: On-farm versus centralized strategies. NE Spain as case study', *Bioresource Technology*, 100(22), pp. 5519–5526. doi: 10.1016/j.biortech.2008.12.050.

Franco-Ramos, O., Castillo, M. and Muñoz-Salinas, E. (2016) 'Using tree-ring analysis to evaluate intra-eruptive lahar activity in the Nexpayantla Gorge, Popocatépetl volcano (central Mexico)', *Catena*, 147, pp. 205–215. doi: 10.1016/j.catena.2016.06.045.

Gao, C. and Zhang, T. (2010) 'Eutrophication in a Chinese context: Understanding various physical and socio-economic aspects', *Ambio*, 39(5), pp. 385–393. doi: 10.1007/s13280-010-0040-5.

Garraty, J. A. (1976) 'Unemployment during the great depression', *Labor History*, 17(2), pp. 133–159. doi: 10.1080/00236567608584378.

Gentil, E. C., Gallo, D. and Christensen, T. H. (2011) 'Environmental evaluation of municipal waste prevention', *Waste Management*, 31(12), pp. 2371–2379. doi: 10.1016/j.wasman.2011.07.030.

Godfray, H. C. J. et al. (2010) 'The future of the global food system', *Philosophical Transactions of the Royal Society B: Biological Sciences. Royal Society*, pp. 2769–2777. doi: 10.1098/rstb.2010.0180.

Goel, S. (2017) 'Advances in solid and hazardous waste management', *Advances in Solid and Hazardous Waste Management*, pp. 1–371. doi: 10.1007/978-3-319-57076-1.

Gornall, J. et al. (2010) 'Implications of climate change for agricultural productivity in the early twenty-first century', *Philosophical Transactions of the Royal Society B: Biological Sciences. Royal Society*, pp. 2973–2989. doi: 10.1098/rstb.2010.0158.

GreenWave. (2020) Available at: https://www.greenwave.org/ (Accessed: 13 September 2022).

Guenat, S. et al. (2019) 'Effects of urbanisation and management practices on pollinators in tropical Africa', *Journal of Applied Ecology*, 56(1), pp. 214–224. doi: 10.1111/1365-2664.13270.

Gülser, C. and Candemir, F. (2015) 'Effects of agricultural wastes on the hydraulic properties of a loamy sand cropland in Turkey', *Soil Science and Plant Nutrition*, 61(3), pp. 384–391. doi: 10.1080/00380768.2014.992042.

Hait, S. and Tare, V. (2012) 'Transformation and availability of nutrients and heavy metals during integrated composting-vermicomposting of sewage sludges', *Ecotoxicology and Environmental Safety*, 79, pp. 214–224. doi: 10.1016/j.ecoenv.2012.01.004.

Hamelin, L., Naroznova, I. and Wenzel, H. (2014) 'Environmental consequences of different carbon alternatives for increased manure-based biogas', *Applied Energy*, 114, pp. 774–782. doi: 10.1016/j.apenergy.2013.09.033.

Hargreaves, J. C., Adl, M. S. and Warman, P. R. (2008) 'A review of the use of composted municipal solid waste in agriculture', *Agriculture, Ecosystems and Environment*, pp. 1–14. doi: 10.1016/j.agee.2007.07.004.

He, X. et al. (2021) 'Spatial-temporal distribution of red tide in coastal China', *IOP Conference Series: Earth and Environmental Science*, 783(1). doi: 10.1088/1755-1315/783/1/012141.

Himanshu, S. K. et al. (2019) 'Evaluation of best management practices for sediment and nutrient loss control using SWAT model', *Soil and Tillage Research*, 192(August 2018), pp. 42–58. doi: 10.1016/j.still.2019.04.016.

Hinz, R. et al. (2020) 'Agricultural development and land use change in India: A scenario analysis of trade-offs between UN Sustainable Development Goals (SDGs)', *Earth's Future*, 8(2), pp. 1–19. doi: 10.1029/2019EF001287.

Ho, K. S. and Chu, L. M. (2019) 'Characterization of food waste from different sources in Hong Kong', *Journal of the Air and Waste Management Association*, 69(3), pp. 277–288. doi: 10.1080/10962247.2018.1526138.

Holy Waste—Making Waste Beautiful. 2020. Available at: https://oorvi.org/holy-waste/ (Accessed: 13 September 2022),

Horowitz, N., Frago, J. and Mu, D. (2018) 'Life cycle assessment of bottled water: A case study of Green2O products', *Waste Management*, 76, pp. 734–743. doi: 10.1016/j.wasman.2018.02.043.

Hossain, M. Z. et al. (2017) 'Effect of different organic wastes on soil propertie s and plant growth and yield: A review', *Scientia Agriculturae Bohemica*, 48(4), pp. 224–237. doi: 10.1515/sab-2017-0030.

Howarth, R. W., Sharpley, A. and Walker, D. (2002) 'Sources of nutrient pollution to coastal waters in the United States: Implications for achieving coastal water quality goals', *Estuaries*, 25(4), pp. 656–676. doi: 10.1007/BF02804898.

Ingrao, C. et al. (2018) 'Food waste recovery into energy in a circular economy perspective: A comprehensive review of aspects related to plant operation and environmental assessment', *Journal of Cleaner Production*, 184, pp. 869–892. doi: 10.1016/j.jclepro.2018.02.267.

Issac, M. N. and Kandasubramanian, B. (2021) 'Effect of microplastics in water and aquatic systems', *Environmental Science and Pollution Research International*. doi: 10.1007/s11356-021-13184-2.

Isaac, O. et al. (2017) 'Evaluation of selected agricultural solid wastes on biochemical profile and liver histology of Albino rats', *Food and Feed Research*, 44(1), pp. 73–79. doi: 10.5937/ffr1701073a.

Issaka, S. and Ashraf, M. A. (2017) 'Impact of soil erosion and degradation on water quality: A review', *Geology, Ecology, and Landscapes*, 1(1), pp. 1–11. doi: 10.1080/24749508.2017.1301053.

Jayathilakan, K. et al. (2012) 'Utilization of byproducts and waste materials from meat, poultry and fish processing industries: A review', *Journal of Food Science and Technology*. Springer, pp. 278–293. doi: 10.1007/s13197-011-0290-7.

Jeyabal, A. and Kuppuswamy, G. (2001) 'Recycling of organic wastes for the production of vermicompost and its response in rice-legume cropping system and soil fertility', *European Journal of Agronomy*, 15(3), pp. 153–170. doi: 10.1016/S1161-0301(00)00100-3.

Jitendra (2020) 'Economic survey 2019–20: Agriculture growth stagnant in last 6 years, down to earth'. Available at: www.downtoearth.org.in/news/agriculture/economic-survey-2019-20-agriculture-growth-stagnant-in-last-6-years-69076 (Accessed: 31 May 2021).

Joksimović, D. et al. (2020) 'Assessment of heavy metal pollution in surface sediments of the Montenegrin coast: A 10-year review', *Journal of Soils and Sediments*, 20(6), pp. 2598–2607. doi: 10.1007/s11368-019-02480-7.

Ju, X. et al. (2005) 'Utilization and management of organic wastes in Chinese agriculture: Past, present and perspectives', *Science in China. Series C, Life sciences/Chinese Academy of Sciences*, 48 Spec No(965), pp. 965–979. doi: 10.1007/BF03187135.

Kanders, M. J. et al. (2017) 'Catch crops store more nitrogen below-ground when considering rhizodeposits', *Plant and Soil*, 417(1–2), pp. 287–299. doi: 10.1007/s11104-017-3259-0.

Kapoor, R. et al. (2020) 'Valorization of agricultural waste for biogas based circular economy in India: A research outlook', in *Bioresource Technology*. Elsevier Ltd, p. 123036. doi: 10.1016/j.biortech.2020.123036.

Karaboga, D. and Akay, B. (2009) 'A survey: Algorithms simulating bee swarm intelligence', *Artificial Intelligence Review*, 31(1–4), pp. 61–85. doi: 10.1007/s10462-009-9127-4.

Karimi, K. (ed.) (2015) *Lignocellulose-Based Bioproducts*. Springer International Publishing (Biofuel and Biorefinery Technologies). doi: 10.1007/978-3-319-14033-9.

Kato, T., Kuroda, H. and Nakasone, H. (2009) 'Runoff characteristics of nutrients from an agricultural watershed with intensive livestock production', *Journal of Hydrology*, 368(1–4), pp. 79–87. doi: 10.1016/j.jhydrol.2009.01.028.

Kautz, M. et al. (2017) 'Biotic disturbances in Northern Hemisphere forests—a synthesis of recent data, uncertainties and implications for forest monitoring and modelling', *Global Ecology and Biogeography*, 26(5), pp. 533–552. doi: 10.1111/geb.12558.

Kazerooni Sadi, M. A. et al. (2012) 'Reduce, reuse, recycle and recovery in sustainable construction waste management', *Advanced Materials Research*, 446–449(September 2018), pp. 937–944. doi: 10.4028/www.scientific.net/AMR.446-449.937.

Keles, Y. and Özdemir, Ö. (2018) 'Extraction, purification, antioxidant properties and stability conditions of phytomelanin pigment on the sunflower seeds', *International Journal of Secondary Metabolite*, pp. 140–148. doi: 10.21448/ijsm.377470.

Kelova, M. E. et al. (2021) 'Small-scale on-site treatment of fecal matter: Comparison of treatments for resource recovery and sanitization', *Environmental Science and Pollution Research*. doi: 10.1007/s11356-021-12911-z.

Khodary, S. M. et al. (2020) 'Effect of hazardous industrial solid waste landfill leachate on the geotechnical properties of clay', *Arabian Journal of Geosciences*, 13(15). doi: 10.1007/s12517-020-05699-8.

Kim, Y. J. and Yoo, G. (2020) 'Suggested key variables for assessment of soil quality in urban roadside tree systems', *Journal of Soils and Sediments*. doi: 10.1007/s11368-020-02827-5.

Knutson, C. L. et al. (2011) 'Farmer perceptions of sustainable agriculture practices and drought risk reduction in Nebraska, USA', *Renewable Agriculture and Food Systems*, 26(3), pp. 255–266. doi: 10.1017/S174217051100010X.

Koondhar, M. A. et al. (2020) 'Looking back over the past two decades on the nexus between air pollution, energy consumption, and agricultural productivity in China: A qualitative analysis based on the ARDL bounds testing model', *Environmental Science and Pollution Research*, 27(12), pp. 13575–13589. doi: 10.1007/s11356-019-07501-z.

Kumar, K. N. and Goel, S. (2009) 'Characterization of Municipal Solid Waste (MSW) and a proposed management plan for Kharagpur, West Bengal, India', *Resources, Conservation and Recycling*, 53(3), pp. 166–174. doi: 10.1016/j.resconrec.2008.11.004.

Kumar, M., Ou, Y. L. and Lin, J. G. (2010) 'Co-composting of green waste and food waste at low C/N ratio', *Waste Management*, 30(4), pp. 602–609. doi: 10.1016/J.WASMAN.2009.11.023.

Kumar, P. and Kumar Singh, R. (2021) 'Strategic framework for developing resilience in Agri-Food Supply Chains during COVID 19 pandemic', *International Journal of Logistics Research and Applications*, pp. 1–24. doi: 10.1080/13675567.2021.1908524.

Kumar, S. and Mishra, A. (2015) 'Critical erosion area identification based on hydrological response unit level for effective sedimentation control in a river basin', *Water Resources Management*, 29(6), pp. 1749–1765. doi: 10.1007/s11269-014-0909-3.

Landry, C. et al. (2018) 'Food waste and food retail density food waste and food retail density', *Journal of Food Products Marketing*, 24(5), pp. 632–653. doi: 10.1080/10454446.2018.1472697.

Laubach, J. and Hunt, J. E. (2018) 'Greenhouse-gas budgets for irrigated dairy pasture and a winter-forage kale crop', *Agricultural and Forest Meteorology*, 258, pp. 117–134. doi: 10.1016/j.agrformet.2017.04.013.

Lazcano, C. et al. (2010) 'Vermicompost enhances germination of the maritime pine (Pinus pinaster Ait.)', *New Forests*, 39(3), pp. 387–400. doi: 10.1007/s11056-009-9178-z.

Lemes, A. C. et al. (2016) 'A review of the latest advances in encrypted bioactive peptides from protein-richwaste', *International Journal of Molecular Sciences*. MDPI AG. doi: 10.3390/ijms17060950.

Lenka, S. et al. (2015) 'Contribution of agriculture sector to climate change', in *Climate Change Impact on Livestock: Adaptation and Mitigation*. Springer India, pp. 37–48. doi: 10.1007/978-81-322-2265-1_3.

Li, J. S. et al. (2013) 'Influence of leachate pollution on mechanical properties of compacted clay: A case study on behaviors and mechanisms', *Engineering Geology*, 167, pp. 128–133. doi: 10.1016/j.enggeo.2013.10.013.

Liegeard, J. and Manning, L. (2020) 'Use of intelligent applications to reduce household food waste', *Critical Reviews in Food Science and Nutrition*, 60(6), pp. 1048–1061. doi: 10.1080/10408398.2018.1556580.

Ling, D. J., Huang, Q. C. and Ouyang, Y. (2010) 'Identification of most susceptible soil parameters for latosol under simulated acid rain stress using principal component analysis', *Journal of Soils and Sediments*, 10(7), pp. 1211–1218. doi: 10.1007/s11368-010-0231-5.

Lisk, D. J. (1991) 'Environmental effects of landfills', *The, Science of the Total Environment*, 100(C), pp. 415–468. doi: 10.1016/0048-9697(91)90387-T.

Liu, E. K., He, W. Q. and Yan, C. R. (2014) '"White revolution" to "white pollution"—Agricultural plastic film mulch in China', *Environmental Research Letters*, 9(9). doi: 10.1088/1748-9326/9/9/091001.

Lobley, M., Butler, A. and Reed, M. (2009) 'The contribution of organic farming to rural development: An exploration of the socio-economic linkages of organic and non-organic farms in England', *Land Use Policy*, 26(3), pp. 723–735. doi: 10.1016/j.landusepol.2008.09.007.

Lynch, J. and Pierrehumbert, R. (2019) 'Climate impacts of cultured meat and beef cattle', *Frontiers in Sustainable Food Systems*, 3. doi: 10.3389/fsufs.2019.00005.

Lynteris, C. (2016) 'The prophetic faculty of epidemic photography: Chinese wet markets and the imagination of the next pandemic', *Visual Anthropology*, 29(2), pp. 118–132. doi: 10.1080/08949468.2016.1131484.

Ma, A. Y. W. and Wu, S. X. (2018) 'Food waste total recycling system—a novel zero effluent discharge process for converting food waste into three high market value products', *HKIE Transactions Hong Kong Institution of Engineers*, 25(1), pp. 17–28. doi: 10.1080/1023697X.2017.1406827.

Ma, X. et al. (2011) 'Assessment and analysis of non-point source nitrogen and phosphorus loads in the three gorges reservoir area of Hubei province, China', *Science of the Total Environment*, 412–413, pp. 154–161. doi: 10.1016/j.scitotenv.2011.09.034.

Magar, J. (2020) 'Application of industrial and agricultural waste for sustainable construction', *International Journal for Research in Applied Science and Engineering Technology*, 8(7), pp. 1869–1875. doi: 10.22214/ijraset.2020.30699.

Maiti, S. et al. (2016) 'Agro-industrial wastes as feedstock for sustainable bio-production of butanol by Clostridium beijerinckii', *Food and Bioproducts Processing*, 98, pp. 217–226. doi: 10.1016/j.fbp.2016.01.002.

Maity, P. P. et al. (2020) 'Do elevated CO 2 and temperature affect organic nitrogen fractions and enzyme activities in soil under rice crop?' *Soil Research*, 58(4), pp. 400–410. doi: 10.1071/SR19270.

Makarenko, N. and Budak, O. (2017) 'Waste management in Ukraine: Municipal solid waste landfills and their impact on rural areas', *Annals of Agrarian Science*, 15(1), pp. 80–87. doi: 10.1016/j.aasci.2017.02.009.

Mall, R. K., Gupta, A. and Sonkar, G. (2017) *Effect of Climate Change on Agricultural Crops, Current Developments in Biotechnology and Bioengineering: Crop Modification, Nutrition, and Food Production*. Elsevier B.V. doi: 10.1016/B978-0-444-63661-4.00002-5.

Mateo-Sagasta, J., Zadeh, S. M. and Turral, H. (2017) 'Water pollution from agriculture: A global review', *Executive Summary*, p. 35. Available at: www.fao.org/3/a-i7754e.pdf.

Matta, G. et al. (2020) 'Water quality assessment using NSFWQI, OIP and multivariate techniques of ganga river system, Uttarakhand, India', *Applied Water Science*, 10(9), pp. 1–12. doi: 10.1007/s13201-020-01288-y.

Mátyás, C. et al. (2018) 'Sustainability of forest cover under climate change on the temperate-continental xeric limits', *Forests*, 9(8), pp. 1–32. doi: 10.3390/f9080489.

McMichael, A. J., Woodruff, R. E. and Hales, S. (2006) 'Climate change and human health: Present and future risks', in *Lancet*. Elsevier Limited, pp. 859–869. doi: 10.1016/S0140-6736(06)68079-3.

Meals, D. W. and Hopkins, R. B. (2002) 'Phosphorus reductions following riparian restoration in two agricultural watersheds in Vermont, USA', in *Water Science and Technology*. IWA Publishing, pp. 51–60. doi: 10.2166/wst.2002.0203.

Mehan, S. et al. (2016) 'Climate change impacts on the hydrological processes of a small agricultural watershed', *Climate*, 4(4), pp. 1–22. doi: 10.3390/cli4040056.

Mehran, M. T. et al. (2021) 'Global plastic waste management strategies (Technical and behavioral) during and after COVID-19 pandemic for cleaner global urban life', *Energy Sources, Part A: Recovery, Utilization and Environmental Effects*, pp. 1–10. doi: 10.1080/15567036.2020.1869869.

Mekonnen, B., Haddis, A. and Zeine, W. (2020) 'Assessment of the effect of solid waste dump site on surrounding soil and river water quality in tepi town, Southwest Ethiopia', *Journal of Environmental and Public Health*. doi: 10.1155/2020/5157046.

Misselhorn, A. et al. (2012) 'A vision for attaining food security', *Current Opinion in Environmental Sustainability*, pp. 7–17. doi: 10.1016/j.cosust.2012.01.008.

Moak, S. K., Gyan-Baffour, G. and Turner, J. E. (1994) 'Market potential for u.s. small farmers in 2000a', *Journal of International Food and Agribusiness Marketing*, 6(1), pp. 59–70. doi: 10.1300/J047v06n01_04.

Moreno-Jiménez, E. et al. (2020) 'Biocrusts buffer against the accumulation of soil metallic nutrients induced by warming and rainfall reduction', *Communications Biology*, 3(1). doi: 10.1038/s42003-020-1054-6.

Mupangwa, W. et al. (2007) 'Effect of minimum tillage and mulching on maize (Zea mays L.) yield and water content of clayey and sandy soils', *Physics and Chemistry of the Earth*, 32(15–18), pp. 1127–1134. doi: 10.1016/j.pce.2007.07.030.

Nandan, A. et al. (2017) 'Recent Scenario of Solid Waste Management in India', *World Scientific News*, 66(January), pp. 56–74. Available at: www.worldscientificnews.com.

Narayana, T. (2009) 'Municipal solid waste management in India: From waste disposal to recovery of resources?' *Waste Management*, 29(3), pp. 1163–1166. doi: 10.1016/j.wasman.2008.06.038.

Ngoc, U. N. and Schnitzer, H. (2009) 'Sustainable solutions for solid waste management in Southeast Asian countries', *Waste Management*, 29(6), pp. 1982–1995. doi: 10.1016/j.wasman.2008.08.031.

Norsuraya, S., Fazlena, H. and Norhasyimi, R. (2016) 'Sugarcane bagasse as a renewable source of silica to synthesize santa barbara amorphous-15 (SBA-15)', in *Procedia Engineering*. Elsevier Ltd, pp. 839–846. doi: 10.1016/j.proeng.2016.06.627.

Novara, A. et al. (2020) 'Cover crop impact on soil organic carbon, nitrogen dynamics and microbial diversity in a mediterranean semiarid vineyard', *Sustainability (Switzerland)*, 12(8). doi: 10.3390/SU12083256.

Obi, F., Ugwuishiwu, B. and Nwakaire, J. (2016) 'Agricultural waste concept, generation, utilization and management', *Nigerian Journal of Technology*, 35(4), p. 957. doi: 10.4314/njt.v35i4.34.

OECD-FAO (2012) *Agricultural Outlook 2011–2020, Organisation For Economic Co-Operation And Development*. doi: 10.1787/agr_outlook-2011-en.

Oktaba, L., Odrobińska, D. and Uzarowicz, Ł. (2018) 'The impact of different land uses in urban area on humus quality', *Journal of Soils and Sediments*, 18(8), pp. 2823–2832. doi: 10.1007/s11368-018-1982-7.

Oldfield, T. L., White, E. and Holden, N. M. (2016) 'An environmental analysis of options for utilising wasted food and food residue', *Journal of Environmental Management*, 183, pp. 826–835. doi: 10.1016/j.jenvman.2016.09.035.

Ortiz, R. et al. (2008) 'Climate change: Can wheat beat the heat?' *Agriculture, Ecosystems and Environment*, 126(1–2), pp. 46–58. doi: 10.1016/j.agee.2008.01.019.

Oshita, K. et al. (2012)'8Th iwa symposium on waste management problems in agro-industries- agro'2011: Emission of greenhouse gases from controlled incineration of cattle manure', *Environmental Technology (United Kingdom)*, 33(13), pp. 1539–1544. doi: 10.1080/09593330.2012.683818.

Ouyang, W. et al. (2018) 'Heavy metal loss from agricultural watershed to aquatic system: A scientometrics review', *Science of the Total Environment*, 637–638, pp. 208–220. doi: 10.1016/j.scitotenv.2018.04.434.

Pandey, R. and Tyagi, A. K. (2012) 'Particulate matter emissions from domestic biomass burning in a rural tribal location in the lower himalayas in India: Concern over climate change', *Small-scale Forestry*, 11(2), pp. 185–192. doi: 10.1007/s11842-011-9177-8.

Pastor, J. and Hernández, A. J. (2012) 'Heavy metals, salts and organic residues in old solid urban waste landfills and surface waters in their discharge areas: Determinants for restoring their impact', *Journal of Environmental Management*, 95(Suppl.). doi: 10.1016/j.jenvman.2011.06.048.

Patel, S. K., Verma, P. and Shankar Singh, G. (2019) 'Agricultural growth and land use land cover change in peri-urban India', *Environmental Monitoring and Assessment*, 191(9). doi: 10.1007/s10661-019-7736-1.

Pavlidis, G. and Tsihrintzis, V. A. (2018) 'Environmental benefits and control of pollution to surface water and groundwater by agroforestry systems: A review', *Water Resources Management*, 32(1), pp. 1–29. doi: 10.1007/s11269-017-1805-4.

Petare, K. J. et al. (2016) 'Livelihood system assessment and planning for poverty alleviation: A case of rainfed agriculture in Jharkhand', *Current Science*, 110(9), pp. 1773–1783. doi: 10.18520/cs/v110/i9/1773-1783.

Prusty, J. K., Patro, S. K. and Basarkar, S. S. (2016) 'Concrete using agro-waste as fine aggregate for sustainable built environment—A review', *International Journal of Sustainable Built Environment*, 5(2), pp. 312–333. doi: 10.1016/j.ijsbe.2016.06.003.

Qin, X. et al. (2016) 'Assessment of food waste biodegradability by biochemical methane potential tests', *Energy Sources, Part A: Recovery, Utilization and Environmental Effects*, 38(24), pp. 3599–3605. doi: 10.1080/15567036.2016.1161678.

Rabalais, N. N. (2002) 'Nitrogen in aquatic ecosystems', *Ambio. Royal Swedish Academy of Sciences*, pp. 102–112. doi: 10.1579/0044-7447-31.2.102.

Ray, L. and Adhya, T. K. (2020) 'Energy, nutrient, and water resource recovery from agriculture and aquaculture wastes', in *Current Developments in Biotechnology and Bioengineering: Resource Recovery from Wastes*. Elsevier, pp. 343–362. doi: 10.1016/B978-0-444-64321-6.00018-5.

Ren, B. et al. (2021) 'Eco-friendly geopolymer prepared from solid wastes: A critical review', in *Chemosphere*. Elsevier Ltd. doi: 10.1016/j.chemosphere.2020.128900.

Riber, C., Petersen, C. and Christensen, T. H. (2009) 'Chemical composition of material fractions in Danish household waste', *Waste Management*, 29(4), pp. 1251–1257. doi: 10.1016/j.wasman.2008.09.013.

Rim-Rukeh, A. (2014) 'An assessment of the contribution of municipal solid waste dump sites fire to atmospheric pollution', *Open Journal of Air Pollution*, 3(3), pp. 53–60. doi: 10.4236/ojap.2014.33006.

Rosenbaum, R. K. et al. (2011) 'USEtox human exposure and toxicity factors for comparative assessment of toxic emissions in life cycle analysis: Sensitivity to key chemical properties', *International Journal of Life Cycle Assessment*, 16(8), pp. 710–727. doi: 10.1007/s11367-011-0316-4.

Roy, R. N., Misra, R. V. and Montanez, A. (2002) 'Decreasing reliance on mineral nitrogen—Yet more food', *Ambio. Royal Swedish Academy of Sciences*, pp. 177–183. doi: 10.1579/0044-7447-31.2.177.

Sabiiti, E. N. (2011) 'Utilising agricultural waste to enhance food security and conserve the environment', *African Journal of Food, Agriculture, Nutrition and Development*, 11(6), pp. 1–7.

Sadh, P. K., Duhan, S. and Singh Duhan, J. (2018) 'Agro-industrial wastes and their utilization using solid state fermentation: A review', *Bioresources and Bioprocessing*, 5, p. 1. doi: 10.1186/s40643-017-0187-z.

Sands, R. D. and Edmonds, J. A. (2005) 'Climate change impacts for the conterminous USA: An integrated assessment: Part 7. Economic analysis of field crops and land use with climate change', *Climatic Change*, 69(1), pp. 127–150. doi: 10.1007/s10584-005-3616-5.

Sarmento, H. et al. (2010) 'Warming effects on marine microbial food web processes: How far can we go when it comes to predictions?' *Philosophical Transactions of the Royal Society B: Biological Sciences*, 365(1549), pp. 2137–2149. doi: 10.1098/rstb.2010.0045.

Savant, N. K. et al. (1999) 'Silicon nutrition and sugarcane production: A review', *Journal of Plant Nutrition*, 22(12), pp. 1853–1903. doi: 10.1080/01904169909365761.

Scholz, K., Eriksson, M. and Strid, I. (2015) 'Carbon footprint of supermarket food waste', *Resources, Conservation and Recycling*, 94, pp. 56–65. doi: 10.1016/j.resconrec.2014.11.016.

Seitzinger, S. P., Kroeze, C. and Styles, R. V. (2000) 'Global distribution of N2O emissions from aquatic systems: Natural emissions and anthropogenic effects', *Chemosphere—Global Change Science*, 2(3–4), pp. 267–279. doi: 10.1016/S1465-9972(00)00015-5.

Seppälä, J. et al. (2006) 'Country-dependent characterisation factors for acidification and terrestrial eutrophication based on accumulated exceedance as an impact category indicator', *International Journal of Life Cycle Assessment*, pp. 403–416. doi: 10.1065/lca2005.06.215.

Sharma, B. et al. (2019) 'Recycling of organic wastes in agriculture: An environmental perspective', *International Journal of Environmental Research*. Springer International Publishing, pp. 409–429. doi: 10.1007/s41742-019-00175-y.

Sharpley, A. N. et al. (2011) 'Critical source area management of agricultural phosphorus: Experiences, challenges and opportunities', *Water Science and Technology*, 64(4), pp. 945–952. doi: 10.2166/wst.2011.712.

Silva, R. C. et al. (2009) 'The effects of enzymatic interesterification on the physical-chemical properties of blends of lard and soybean oil', *LWT—Food Science and Technology*, 42(7), pp. 1275–1282. doi: 10.1016/j.lwt.2009.02.015.

Singh, S. et al. (2021) 'Impact of COVID-19 on logistics systems and disruptions in food supply chain', *International Journal of Production Research*, 59(7), pp. 1993–2008. doi: 10.1080/00207543.2020.1792000.

Sinha, R. K. et al. (2002) 'Vermiculture and waste management: Study of action of earthworms', *The Environmentalist*, 22(3), pp. 261–268.

Solarte-Toro, J. C., Chacón-Pérez, Y. and Cardona-Alzate, C. A. (2018) 'Evaluation of biogas and syngas as energy vectors for heat and power generation using lignocellulosic biomass as raw material', *Electronic Journal of Biotechnology. Pontificia Universidad Catolica de Valparaiso*, pp. 52–62. doi: 10.1016/j.ejbt.2018.03.005.

Spatafora, C. (2012) 'Valorization of vegetable waste: Identification of bioactive compounds and their chemo-enzymatic optimization', *The Open Agriculture Journal*, 6(1), pp. 9–16. doi: 10.2174/1874331501206010009.

Styles, D., Dominguez, E. M. and Chadwick, D. (2016) 'Environmental balance of the of the UK biogas sector: An evaluation by consequential life cycle assessment', *Science of the Total Environment*, 560–561, pp. 241–253. doi: 10.1016/j.scitotenv.2016.03.236.

Sufian, A. et al. (2020) 'Advancements in agriculture strategies and environmental impact: A review', *SSRN Electronic Journal*, pp. 1–15. doi: 10.2139/ssrn.3516438.

Sujatha, E. R. and Sridhar, V. (2018) 'Spatial prediction of erosion risk of a small mountainous watershed using RUSLE: A case-study of the palar sub-watershed in Kodaikanal, South India', *Water (Switzerland)*, 10(11), pp. 1–17. doi: 10.3390/w10111608.

Sun, J. et al. (2018) 'Organic contamination and remediation in the agricultural soils of China: A critical review', *Science of the Total Environment*, 615, pp. 724–740. doi: 10.1016/j.scitotenv.2017.09.271.

Susilawati, A., Maftuah, E. and Fahmi, A. (2020) 'The utilization of agricultural waste as biochar for optimizing swampland: A review', *IOP Conference Series: Materials Science and Engineering*, 980(1). doi: 10.1088/1757-899X/980/1/012065.

Tartowski, S. L. and Howarth, R. W. (2001) 'Nitrogen, nitrogen cycle', in *Encyclopedia of Biodiversity*. Elsevier, pp. 377–388. doi: 10.1016/b0-12-226865-2/00210-8.

Teng, J. et al. (2021) 'Assessing habitat suitability for wintering geese by using Normalized Difference Water Index (NDWI) in a large floodplain wetland, China', *Ecological Indicators*, 122(August 2020), p. 107260. doi: 10.1016/j.ecolind.2020.107260.

Thi, N. B. D., Kumar, G. and Lin, C. Y. (2015) 'An overview of food waste management in developing countries: Current status and future perspective', *Journal of Environmental Management*, 157, pp. 220–229. doi: 10.1016/j.jenvman.2015.04.022.

Tilman, D. et al. (2001) 'Forecasting agriculturally driven global environmental change', *Science*, 292(5515), pp. 281–284. doi: 10.1126/science.1057544.

Tirkey, A. S., Ghosh, M. and Pandey, A. C. (2016) 'Soil erosion assessment for developing suitable sites for artificial recharge of groundwater in drought prone region of Jharkhand state using geospatial techniques', *Arabian Journal of Geosciences*, 9(5). doi: 10.1007/s12517-016-2391-0.

Tonini, D., Albizzati, P. F. and Astrup, T. F. (2018) 'Environmental impacts of food waste: Learnings and challenges from a case study on UK', *Waste Management*, 76, pp. 744–766. doi: 10.1016/j.wasman.2018.03.032.

Tonini, D., Hamelin, L. and Astrup, T. F. (2016) 'Environmental implications of the use of agro-industrial residues for biorefineries: Application of a deterministic model for indirect land-use changes', *GCB Bioenergy*, 8(4), pp. 690–706. doi: 10.1111/gcbb.12290.

Tsai, W. T. (2021) 'Analysis of plastic waste reduction and recycling in Taiwan', *Waste Management and Research*. doi: 10.1177/0734242X21996821.

University of Delhi (2019) 'Study on municipal solid waste management and challenges faced in Indian metropolitan cities', *International Journal of Home Science*, 5(2), pp. 200–205. Available at: www.homesciencejournal.com.

US EPA (2017) 'Climate impacts on agriculture and food supply | climate change impacts | US EPA'. Available at: https://19january2017snapshot.epa.gov/climate-impacts/climate-impacts-agriculture-and-food-supply_.html (Accessed: 4 March 2021).

US EPA (2020) 'Best practices for solid waste management: Best practices for solid waste management: A guide for decision-makers in developing countries', (October), pp. 1–166. Available at: www.iges.or.jp/en/publication_documents/pub/policyreport/en/11066/Best+Practices+for+SWM_Guide+for+Decision+Makers+in+Developing+Countries_Oct+20_0.pdf.

Vaibhav, V., Vijayalakshmi, U. and Roopan, S. M. (2015) 'Agricultural waste as a source for the production of silica nanoparticles', *Spectrochimica Acta—Part A: Molecular and Biomolecular Spectroscopy*, 139, pp. 515–520. doi: 10.1016/j.saa.2014.12.083.

Väisänen, T. et al. (2016) 'Utilization of agricultural and forest industry waste and residues in natural fiber-polymer composites: A review', in *Waste Management*. Elsevier Ltd, pp. 62–73. doi: 10.1016/j.wasman.2016.04.037.

Valenzuela, E. I. and Cervantes, F. J. (2020) 'The role of humic substances in mitigating greenhouse gases emissions: Current knowledge and research gaps', *Science of the Total Environment*, p. 141677. doi: 10.1016/j.scitotenv.2020.141677.

Vermeulen, S. J., Campbell, B. M. and Ingram, J. S. I. (2012) 'Climate change and food systems', *Annual Review of Environment and Resources*, pp. 195–222. doi: 10.1146/annurev-environ-020411-130608.

Vimala Ebenezer, A. et al. (2020) *State of the Art of Food Waste Management in Various Countries, Food Waste to Valuable Resources*. INC. doi: 10.1016/b978-0-12-818353-3.00014-6.

Wang, H. et al. (2015) 'Soil microbial community composition rather than litter quality is linked with soil organic carbon chemical composition in plantations in subtropical China', *Journal of Soils and Sediments*, 15(5), pp. 1094–1103. doi: 10.1007/s11368-015-1118-2.

Wang, H. et al. (2019) 'Effect of irrigation amount and fertilization on agriculture non-point source pollution in the paddy field', *Environmental Science and Pollution Research*, 26(10), pp. 10363–10373. doi: 10.1007/s11356-019-04375-z.

Wang, Y., Liu, G. and Zhao, Z. (2021) 'Spatial heterogeneity of soil fertility in coastal zones: A case study of the Yellow river delta, China', *Journal of Soils and Sediments*. doi: 10.1007/s11368-021-02891-5.

Wansink, B. (2018) 'Household food waste solutions for behavioral economists and marketers', *Journal of Food Products Marketing*, 24(5), pp. 500–521. doi: 10.1080/10454446.2018.1472694.

Wasim Aktar, M. et al. (2010) 'Assessment and occurrence of various heavy metals in surface water of Ganga river around Kolkata: A study for toxicity and ecological impact', *Environmental Monitoring and Assessment*, 160(1–4), pp. 207–213. doi: 10.1007/s10661-008-0688-5.

Wei, J. et al. (2020) 'Research progress of energy utilization of agricultural waste in China: Bibliometric analysis by citespace', *Sustainability (Switzerland)*, 12(3). doi: 10.3390/su12030812.

Woods, J. et al. (2010) 'Energy and the food system', *Philosophical Transactions of the Royal Society B: Biological Sciences. Royal Society*, pp. 2991–3006. doi: 10.1098/rstb.2010.0172.

Wu, Z. et al. (2017) 'Water quality assessment based on the water quality index method in Lake Poyang: The largest freshwater lake in China', *Scientific Reports*, 7(1), pp. 1–10. doi: 10.1038/s41598-017-18285-y.

Yang, Y. B. et al. (2005) 'Study on the transient process of waste fuel incineration in a full-scale moving-bed furnace', *Combustion Science and Technology*, 177(1), pp. 127–150. doi: 10.1080/00102200590883796.

Yoshida, H. et al. (2015) 'A comprehensive substance flow analysis of a municipal wastewater and sludge treatment plant', *Chemosphere*, 138, pp. 874–882. doi: 10.1016/j.chemosphere.2013.09.045.

Yusuf, M. (2017) 'Handbook of ecomaterials', *Handbook of Ecomaterials*, (August). doi: 10.1007/978-3-319-48281-1.

Zheng, Y. et al. (2014) 'Pretreatment of lignocellulosic biomass for enhanced biogas production', in *Progress in Energy and Combustion Science*. Elsevier Ltd, pp. 35–53. doi: 10.1016/j.pecs.2014.01.001.

8 Potential Benefits of Utilization of Kitchen and Agri Wastes

Shalini Dhiman, Kanika Khanna, Jaspreet Kour, Tamanna Bhardwaj, Ravdeep Kaur, Neha Handa and Renu Bhardwaj

CONTENTS

DOI: 10.1201/9781003245773-8

8.1 INTRODUCTION

Kitchen and agri waste generation are expected to increase 40-fold in coming years around the globe. Thus, proper waste management is required; otherwise, these wastes lead to hazardous health issues in all living beings and to the environment. Most of the industries in various countries are working on strategies to convert wastes into wealth. Food and agri wastes have a high numbers of fibres, phenols, essential nutrients, phytochemicals, bioactive compounds and various other beneficial constituents. Thus, from these agri and food wastes residues most of the industries are aiming to produce high-value end products such as bioplastic, biofertilizer, biogas, edible films, probiotics, nanoparticles, carbon dots, microbial media, biochar and biosorbants, which have potential value in the present existing market. Such practices prove that waste is the most valuable resource. The present chapter highlights different kinds of value-added products generated from various foods and agri wastes and their applications in the real world.

8.2 WASTES: AS A VALUABLE RESOURCE

In contemporary times, when society is grappling with challenges such as environmental deterioration and food security for its ever-increasing population, it has become mandatory to change the perception of biomass and food waste from waste to a valuable resource (Fermoso et al., 2018). Sustainable agriculture warrants the replacement of conventional linear economy models based upon the 'take, make and dispose' ideology by sustainable circular bioeconomy (CE) models. The CE intends to maximize natural resource utilization, minimize waste generation and implement waste valorization by adopting innovative technologies and cost-effective production practices (Toop et al., 2017; Sherwood, 2020). The foundation of CE is an industrial symbiosis where waste streams from one industry or sector are the raw material of another (Mirabella et al., 2014). Typically, potential methodologies proposed for the regulation of agriculture waste management into a circular loop include extraction of value-added chemicals, anaerobic digestion and composting (Fermoso et al., 2018). Although first-generation biowaste valourization technologies, anaerobic digestion and composting had limited value, the immense potential of biowaste due to its intrinsic complexity has led to the development of second-generation technologies. These new technologies consider waste as feedstock for diverse industries (Lin et al., 2013).

During the past decade, the biorefinery concept has gained special attention from the scientific fraternity since it focuses on the integration of diverse novel technologies and processes for the synthesis of the broad spectrum of bioproducts (biofuels, bioenergy and other value-added bio-commodity chemicals) from bioresources rather than a single product for waste valourization (Jin et al., 2018). Initially, crops were used as feedstocks in biorefineries; however, these were not sustainable as they not only compromised food security, but the production of crops for these biorefineries also caused the generation of greenhouse gases. The recognition of the drawbacks of these biorefineries prompted the development of second-generation biorefineries that utilize food waste and lignocellulosic biomass as feedstock (De Buck et al.,

2020). Comparatively, biomass biorefineries can potentially create a wider range of chemicals than traditional petroleum-based refineries, as biomass includes a larger percentage of oxygen and a lower amount of carbon and hydrogen than petroleum resources (Cho et al., 2020). Nowadays, the sustainable manufacture of valuable chemicals and biopolymers is completely dependent upon renewable carbon sources. This has not only solved environmental issues, but also reduced reliance on fossil fuels and improved economic efficiency (Cho et al., 2020).

Recently, several studies have proposed and validated the full valourization of agri waste following zero-waste and circular economy concept. For instance, grape pomace can be used both for the extraction of phenolic compounds and the remaining solid residue as a filler to strengthen a biopolymer matrix (Ferri et al., 2020). Similarly, Sen et al. (2020) have proposed a biorefining model for low-grade waste fractions from *Quercus cerris* L. bark which is rich in cork and phloem granules. Since the waste streams are rich in antioxidants, triterpenoids, tannins and flavonoids can be extracted from these fractions. Alternatively, other value-added chemicals could be extracted to produce either plastic additives, biofuels (biochar or bio-oil) or polyurethanes. Finally, the inorganic fractions can be combusted to produce bottom and fly ash, which can be used as a cement additive. Hence, a circular economy offers a promising approach for natural resource and waste management because it promotes maximum resource utilization by preventing the use of fresh resources, manufacturing improvement and encouraging loop closing (Campos et al., 2020).

8.3 WASTES AS A VALUABLE RESOURCE FOR THE PRODUCTION OF ESSENTIAL CHEMICALS

Natural products, including primary and secondary metabolites, enzymes, fibres, pigments, oils and vitamins, are among the most lucrative chemicals considered for mining from biogenic wastes (Zuin and Ramin, 2018). These extracted high value-added chemicals or residues are mostly exploited in food, cosmetic, nutraceutical, pharmaceutical and packaging industries as food additives, colourants, antioxidants, preservatives, medicines, taste enhancers, fragrances, and/or biopesticides (Monteiro et al., 2021). In addition to the reduction in cost for obtaining these chemicals and waste prevention, biorefining of biowaste also provides new functional foods and improved animal feed or fertilizer (Szabo et al., 2018).

8.3.1 Bioactive Compounds

Any substance that can alter or trigger a physiological response in an organism is referred to as a bioactive compound. They are widely known for their health-promoting properties (Leyva-López et al., 2020). Since fruits and vegetables have putative therapeutic effects and their consumption is often linked to lower risk of metabolic disorders, identification, isolation, quantification, characterization and purification of bioactive chemicals from these sources have become a prominent field of research for nutraceutical and pharmaceutical-related studies in recent decades (Coman et al., 2020). In comparison to edible portions, the non-edible parts of fruits, vegetables, cereals and trees such as peels, rinds, seed, pomace, husk and bark contain a higher percentage of bioactive compounds (Ben-Othman et al., 2020; Leyva-López et al., 2020). The bioactivities from plant-based agri-food residues and other waste streams mostly comprise polyphenols, flavonoids, flavanols, anthocyanins, carotenoids, tannins, alkaloids, essential oils, tocopherol, ascorbic acid and dietary fibres, while animal- and dairy-based mainly consist of bioactive peptides, colostrum and whey (Ben-Othman et al., 2020). These bioactive compounds are packed with biological properties such as antioxidative, immunomodulatory, probiotic, anti-tumour, anti-ageing, anti-inflammatory, anti-microbial, anti-mutagenic, and/or anti-anxiety activity (Leyva-López et al., 2020). Because of these bioactive properties and momentous industrial development, bioactive compounds have high demand in a variety of commercial sectors, including food for making edible films, probiotics, functional foods and nutraceuticals (Kumar et al., 2020b), enhancing flavours, taste, fragrance and shelf-life (Yusuf, 2017); cosmetics and pharmaceutical for disease prevention and treatment (Xiong et al., 2019).

Several conventional and non-conventional methods have been used and optimized in isolation or combination for efficient extraction of desired bioactive compounds. Some materials, for instance lignocellulosic material, may have to undergo physical, chemical or biological pretreatment before extraction to avoid any interference (Arun et al., 2020). The conventional extraction methods such as Soxhlet extraction, solvent extraction, hydro distillation and maceration, are not only time-consuming, expensive and less energy-efficient; they also have lower extraction rate, limited selectivity and high solvent consumption. They also require high purity solvents and can cause degradation of thermolabile chemicals (Azmir et al., 2013). To overcome these challenges, ultrasound, microwave heating, supercritical fluids, enzymes, pulsed electric field, accelerated solvent, subcritical water, high pressure and ohmic heating have been explored as green technologies or non-conventional technologies (Azmir et al., 2013). Some of the bioactivities extracted from diverse agri-food waste residues employing different technologies along with their putative commercial applications are summarized in Table 8.1.

8.3.2 Enzymes

As a rich and renewable source of carbohydrates, proteins and lipids, kitchen and agriculture waste have the inherent capacity to be bioconverted to valuable products sustainably. Several chemical modification processes such as hydrolysis, oxidation, esterification, phosphorylation, deamination, glycosylation and hydrogenation can aid biotransformation of these biogenic wastes. For instance, oils, sugars and starches can be esterified to produce biofuels, surfactants, and plastics, respectively. Conventionally, synthetic catalysts were needed to perform these alterations. However, they require more energy, generate by-products and even lack specificity, whereas enzymes/biocatalysts offer a better alternative because of their high specificity, better process efficiency and environmental sustainability (Andler and Goddard, 2018). Thus, factors such as limitations of synthetic catalysts, multifarious industrial usage of enzymes and sustainable development goals have led to the evolution of technologies for the extraction of enzymes from inexpensive substrates such as agriculture and food waste. These residues can serve as an excellent substrate for

TABLE 8.1
Some of the Bioactivities Extracted from Diverse Agri-Food Waste Residues Employing Different Technologies, Along with Their Putative Commercial Applications

Waste	Product	Extraction method	Yield	Bioactivity/ biological properties	Industrial relevance	Reference
Cagnulari *Grape Pomace*	**Polyphenols** (anthocyanins)	Cyclically Pressurized Solid-Liquid Extraction	TPC 4.00 g/L	Antioxidative	Nutraceutical	Posadino et al., 2018
Mango peel	**Carotenoids** (β-Carotene and lutein) **Phenolics** (mangiferin, gallic acid, quercetin, and rutin)	Solvent extraction (acetone/water solution (70:30, *v*/ v)	Carotenoid 3.7–5.7 mg/100 g TPC 2930 to 6624 mg GAE/100 g TFC 502–795 mg CE/100 g	Antioxidant activity	Food, cosmetics and pharmaceutical industries	Marcillo-Parraet et al., 2021
Apple pomace	Phenolics (Triterpenic acids)	Soxhlet extraction (methanol)	TPC (3.48 mg GAE/g DW) Betulinicacid (1.78 mg/100 g dw), Oleanolic acid (3.18 mg/100 g dw), Ursolic acid (6.14 mg/100 g dw, Maslinic acid (0.96 mg/100 g dw), Erythrodiol (0.78 mg/100 g dw), and Uvaol (0.84 mg/100 g dw)	Antioxidant, Anticancer, Anti-inflammatory Anti-urolithic, skin whitening	Functional food, cosmetics and healthcare products development	Nile et al., 2019
Seed coat black soybean (*Glycine max* L.)	Anthocyanins	Microwave assisted extraction (569.46 W, 262.54 s, ethanol to solid ratio of 40:1)	5021.47 mg/l	Antioxidant, anti-inflammatory, anti-cancerous, anti-diabetic effects colourant	Pharmaceutical and nutraceutical industries for functional foods	Kumar et al., 2019
Cotton waste or by-product	Flavonols and anthocyanins	Subcritical water extraction (180 °C, 65% ethanol, solid-liquid ratio 65 mL/g)	Isoquercetrin (110.54 mg/g)			Xu et al., 2021

(*Continued*)

TABLE 8.1
Continued

Waste	Product	Extraction method	Yield	Bioactivity/ biological properties	Industrial relevance	Reference
Camellia sinensis leaves (pruning disposals)	Total phenolics	Sequential combination of Microwave hydrodiffusion and gravity (Pretreatment), ultrasound-assisted (80°C, 80 kHz, from 15 to 60 min) and Pressurized hot water (140 to 200°C)	130 mg GAE/g extract	Antioxidative	Food, cosmetic, and pharmaceutical	Sanz et al., 2020
Onion peels	**Quercetin**	Ethanol	15 mg quercetin g^{-1}	Antioxidative	Functional foods	Črnivec et al., 2021
Spinach and Orange waste	Polyphenols	Ultrasound-assisted extraction (ethanol/ water/HCl 80/19.9/0.1 (v/v/v), 25 °C, 30 min)	TPC- 0.82 mg GAE/ g fw	Antioxidative	Food, pharmaceutical and/or cosmetic industries	Montenegro-Landívar et al 2021
		Pressurized liquid extraction (ethanol/water/HCl 60/39.9/ 0.1 (*v/v/v*), 80 °C, 15 min)	TPC- 3 mg GAE g^{-1} fw			
Peach waste	**Polyphenols** **Flavonoids** **Anthocyanins** **Vitamin C**	Microwave assisted extraction (540 W, 50 s), pretreatment-freezing	309–317 mg GAE 94–120 mgQE 8–9 mg CGE 108 mg/100 g dm			Plazzotta et al., 2020
Olive pomace	**Phenolic compounds**	Ultrasound-assisted enzyme hydrolysis (pH-5.75, t-40 min, tp-55 °C)	4.0%	Antioxidative	Food additive and preservative	Wang et al., 2017
		Gamma radiation (dose rate-5 kGy)	TPC-139 mg/g extract TFC-5.1mg/g extract Hydroxytyrosol (62 mg/g)	Anti-tumour, Anti-ageing, Anti-inflammatory, hepatoxicity therapeutics for cardiovascular disease Taste enhancing		Madureira et al., 2020

Waste	Product	Extraction method	Yield	Bioactivity/ biological properties	Industrial relevance	Reference
Turnip tops Wild cardoon (leaf blade) Radish waste	**Flavonoids** (glycosylated derivatives of kaempferol, quercetin and isorhamnetin) **Phenolic acids** (Hydroxycinnamic acid derivatives) **Phenolic acid** s (mainly 4 - O - caffeoylquinic acid)., **Flavonoids Polyunsaturated fatty acids**, **Tocopherols** **Phenolic compounds** (p -coumaric acid and apigenin)	(ethanol: water, 80: 20, v/v)	TFC 18.8 mg/g Phenolic acid (2.8 mg/g) phenolic acids (23.6 mg/g), TFC (14.7 mg/g) α- linolenic acid (0.843 g/100g dw) TPC (57.6 mg/g)	Antioxidant, Antimicrobial, Anti-inflammatory, Anticancer and Anti-anxiety	food processing, cosmetics, and in the pharmaceutical industry	Chihoub et al., 2019
Waste distilled by-products remaining after steam distillation of *Cupressus lusitanica* and *Cistus ladanifer* waste	**Phenolic compound** (catechins, hydroxycinnamic acid derivatives and quercetin), **Tannins Salicylic acid**	Ultrasound-assisted extraction with 70% acetone	TPC (251.3- 275.6 mg GAE/g extract) **TFC** (6.3-15.2 mg QE/g extract) **Tannins** (82.2- 116.6 mg GAE/g extract)	Anti-inflammatory Antioxidative	Pharmaceutical, food and cosmetic	Tavares, Martins, Faleiro et al., 2020
Citrus peels	p-Coumaric acid trans ferulic acid **Flavonoids** Hesperidin Rutin Narangin	Solid-liquid extraction	0.100 mg/g 0.29-1.38 mg/g 280-673 mg/g 3.3-4.7 mg/g 1.72-115 mg/g	Antioxidative	Pharmaceutical, food and cosmetic	Gómez-Mejía et al., 2019

(Continued)

TABLE 8.1
Continued

Waste	Product	Extraction method	Yield	Bioactivity/ biological properties	Industrial relevance	Reference
Orange peels	**Essential oil**	Microwave Hydro diffusion and Gravity (500 W, 15 Min)	346 g/100 kg citrus	Anti-inflammatory Antibacterial	Food (flavouring), Pharmaceutical, Cosmetics, preparation of soaps, perfumes, and other home care products, green solvent	Boukroufa et al., 2015
	Polyphenols	Ultrasound-assisted extraction (0.956 W/cm2 and 59.83 C)	11.71 g GAE/ 100 kg citrus	Antioxidative	Pharmaceutical, cosmetic and food	
	Pectin	Microwave assisted extraction (500 W, 3 min)	4.83 kg/ 100 kg citrus		Food gelling agent	
Cupressus lusitanica waste products *Cistus ladanifer* waste products	**Essential oils**; EO (α-pinene (13–28 %) and camphene (5–25 %)) and **Hydrolates**; H (2,6,6-trimethyl cyclohexanone (2–12%) and trans-pinocarveol (5–13%)	Steam-distillation (SD) and hydro distillation (HD)	EO yield SD- 0.12-0.26 % (v/dw) HD-0.1 % (v/dw) The EO yields SD- 0.01-0.04 (v/fw) HD- 0.15 % (v/fw)	Anti-inflammatory, Antioxidative	Perfume, Cosmetic, Flavour, and Pharmaceutics industries	Tavares, Martins, Miguel et al., 2020
Sugarcane straw (SS) and coffee husk (CH) xylan substrate	**Xylooligosaccharides**	Enzymatic hydrolysis	SS (10.23 ± 0.56 g/L) CH (8.45 ± 0.65 g/L)	Probiotic Antioxidative	Food and pharmaceutical industry as functional ingredients	Ávila et al., 2020
Fruit peels, groundnut shell, coconut and walnut husk	**Exopolysaccharides**	Submerged fermentation (*Pleurotuspulmonarius*)	5.60 g/L	Antimicrobial, Antioxidant Prebiotic	Food, pharmaceutical, and cosmeceutical industries	Ogidi et al., 2020

Abbreviations: TPC (total polyphenol content), GAE (gallic acid equivalents), TFC (total flavonoid content), QE (quercetin equivalents), dw (dry weight), fw (fresh weight) microbial growth and concomitant synthesis of industrially crucial enzymes (Ahmed et al., 2016). Both submerged and solid-state fermentation techniques are widely used by industrial enzyme manufacturers to produce these biocatalysts (Amin et al., 2019). Although 90% of industrial enzyme extraction is by submerged fermentation techniques, lately, solid-state fermentation has gained more scientific acceptance because of its better suitability for agriculture waste, economic feasibility and ability to provide a higher yield (Kiran, Trzcinski, Ng et al., 2014; Salim et al., 2017). Among various agricultural residues, wheat and rice bran are usually considered as best feedstocks for enzyme extraction (Meng et al., 2019), while mushrooms are regarded as the major player in biomass degradation. They are capable of providing a maximum yield of saccharification enzymes along with suitable substrate. The potential of mushrooms to use lignocellulosic wastes as a growth substrate and simultaneously produce lignocellulolytic enzymes could be studied in detail from the recent review published by Kumla et al. (2020).

Agriculture waste mainly comprises lignocellulosic materials: lignin, cellulose and hemicellulose. These materials are not readily fermentable because of their complex polymeric structures; hence, a pretreatment that converts them to fermentable sugars is necessary before they can be converted to value-added products. The synergistic action of hydrolytic (cellulases and hemicellulases) and oxidative (ligninases) enzymes are required for their saccharification/degradation (Kumla et al., 2020). The saccharification is the second most cost-limiting process in the commercial valourization of biomass after cost incurred due to raw materials. Therefore, the cost-effective production of biomass hydrolyzing enzymes such as cellulases, amylases, and pectinases has been of greater interest of the scientific community (Kiran, Trzcinski, Ng et al., 2014; Saratale et al., 2014). Cellulases comprising endoglucanases, cello bio hydrolases and b-glucosidases are one of the most valuable group industrial enzymes offering innumerable industrial applications. For instance, cellulases are exploited for biofuels, single-cell protein, organic acids and lipids production. They are also used in the laundry, textile, food, feed, brewing, pulp and paper and agriculture industries (Kuhad et al., 2011). A-Amylase catalyzes hydrolyses of starch to oligosaccharides, due to which amylase finds application in the starch, laundry, biofuel production, clarification of beer or fruit juices, baking, desizing, pulp and paper industries (de Souza and de Oliveira Magalhães, 2010) and food waste biodegradation (Msarah et al., 2020). Similarly, pectinases or pectinolytic enzymes that degrade pectin substances are also crucial industrially produced enzymes. Pectinases find their relevance in diverse industrial processes; for example, oil extraction, fruit softening, retting and gumming, tea and coffee fermentation, juice extraction and clarification, brewing, animal feed and industrial wastewater treatment (Patidar et al., 2018; Amin et al., 2019). Proteases, protein-peptide bond degrading enzymes, are involved in leather processing, peptide synthesis, silver recovery from X-ray/photographic films, silk degumming, poultry and leather industry, waste management, cheese production, laundry detergents (Kamal et al., 2017) and valourization of food waste high in proteins (Kumar et al., 2015a). Precisely, proteases contribute about 60% to the total industrial enzyme market (Dadshahi et al., 2016). Table 8.2 summarizes details of various enzymes extracted from agriculture and kitchen waste, along with their industrial significance.

TABLE 8.2
Various Enzymes Extracted from Agriculture and Kitchen Waste, Along with Their Industrial Significance

Enzyme	Waste	Extraction method	Yield	Enzymatic action	Industrial relevance	Reference
Cellulases	Wheat bran	SSF (*Aspergillus niger*)	FPase (17 U/gds) CMCase (310U/gds) β-glucosidase (33 U/gds)	Bioconversion of lignocellulosic biomass to fermentable sugars	Biofuel production, hydrolysis of agro-industrial residues, food processing, textile, laundry detergents, pulp and paper, animal feed and pharmaceuticals.	Bansal et al., 2012
Endoglucanase	Grass powder	SSF (*Phanerochaetechrysosporium*)	188.66 U/gds			Saratale et al., 2014
Exoglucanase			24.22 U/gds			
Cellobiase			244.60 U/gds			
Xylanase			427.0 U/gds			
Xylanase exo-polygalacturonase (exo-PG)	Orange peels	SSF (*Aspergillus awamori*)	Xylanase 31,000 U/kg exo-PG 17,600 U/kg			Marzo et al., 2019
	Sugar beet cossettes		Xylanase 35,000 U/kg exo-PG 28,000 U/kg			
Endoglucanase	Food waste	SSF	17.37U/gds			Tian et al., 2018
Xylanase		*Aspergillus niger*	189.24 U/gds			
Cellulose	By-product of corn and wheat ethanol production	SmF	Cellulose 0.592 IU/ml			Iram et al., 2020
Xylanase		*Aspergillus niger*	Xylanase 34.8 IU/ml			
		Bacillus subtilis	Cellulose 0–0.261 IU/ml Xylanase 1.2–5.2 IU/ml			

Enzyme	Waste	Extraction method	Yield	Enzymatic action	Industrial relevance	Reference
Pectinolytic enzymes						
Pectinases	Carrot peels	LSF (*Bacillus mojavensis*)	63.55 U/ml	Degradation of pectin	Oil extraction	Ghazala et al., 2015
	Wheat bran, orange and lemon peels	SSF (*Aspergillus giganteus*)	20 L/min/kgds			Ortiz et al., 2017
	Citrus peel	SmF (*Aspergillus niger*)	117.1 mM/mL/Min			Ahmed et al., 2016
Amylolytic enzymes						
Amylase	Kitchen food waste	SSF (*Bacillus amyloliquefaciens*)	528 U/gds	Degradation of starch	Biofuel production, clarification of beer or fruit juice, baking, laundry, desizing, waste degradation, and Food, pulp and paper and textile industry	Bhatt et al., 2020
	Wheat bran supplemented with starch	SSF (*Bacillus subtilis*)	640 U/g			Almanaa et al., 2020
	Potato peels	SmF (*Bacillus subtilis*)	1394.145 IU/ml			Mushtaq et al., 2017
Glucoamylase	Grass powder	SSF (*Phanerochaetechrysosporium*)	505.0 U/gds	Partially processed starch/dextrin saccharification to glucose	Glucose and fructose syrups production	Saratale et al., 2014
	Mixed food waste	SSF (*Aspergillus awamori*)	108.47 U/gds			Kiran, Trzcinski, and Liu, 2014
	Rice and cake waste	SSF (*Aspergillus niger*)	180.59 U/g			Meng et al., 2019
Pullulanase	Wheat bran	SSF (*Aspergillus* sp.)	65.33 U/gds	Starch saccharification	Starch-based food industries	Naik et al., 2019

(Continued)

TABLE 8.2
Continued

Enzyme	Waste	Extraction method	Yield	Enzymatic action	Industrial relevance	Reference
Other industrially significant enzymes						
Alkaline protease	Cauliflower leaves	SSF (*Alternaria alternata*)	930 U/gds		Leather, silk, food processing, pharmaceutical, detergent, waste processing and recovery of silver from photographic film	Polley and Ghosh, 2018
Protease	Olive oil cake	SSF (*Candida utilisI*)	110 U/g			Moftah et al., 2012
Lipase	Olive oil cake	SSF (*Candida utilisI*)	25 U/g	Hydrolysis of acylglyceride, fatty acid esters, etc. into glycerol and fatty acids	Fat hydrolysis, transesterification of fats and oils, synthesis of esters and chiral organic compounds, production of biodiesel, cleaning products, cosmetics, drugs, flavouring of dairy products and treatment of lipid-rich wastewater	Moftah et al., 2012
	Jatropha seedcake	SmF (*Pseudomonas aeruginosa*)	0.432 U/ml			Bose and Keharia, 2013
	Palm oil industrial residues	SSF (*Aspergillus niger*)	15.41 IU/mL			Silveira et al., 2016
Fructosyltransferase	Sugar cane bagasse	SSF (*Aspergillus flavus*)	197.10 U/gds	Bioconversion of sucrose to fructooligosaccharides	Nutraceutical, pharmaceutical and food industry	Ganaie et al., 2017

Abbreviations: kilogram of dry substrate (kgds); Unit per gram dry substrate (U/gds); international units per milliliter (IU/mL); carboxymethyl cellulase (CMCase), filter paper activity (FPase); Solid-state fermentation (SSF); Submerged fermentation (SmF); Liquid state fermentation (LSF)

Although the benefits of enzymes outweigh the synthetic catalysts in waste valourization, regulation of their stability and performance under normal food waste processing conditions poses a serious challenge for their industrial implementation. Several enzyme immobilization techniques can improve the overall stability and performance of enzymes in these non-optimal circumstances. In addition, they also enable enzyme reusability, making bio-catalysis more commercially viable (Andler and Goddard, 2018).

8.3.3 Food Flavouring and Preservative Compounds

Agricultural and industrial wastes generated are of low cost and are very sustainable resources due to their application in the production of various value-added food flavouring and preservative compounds. Vanillin is one such flavouring compound, which is of high importance and is produced in the food industry. This compound is produced from ferutic acid, which is recovered as the by-product from agro-food industry. Various by-products from maize, sugar, rice and wheat-producing industries acts as a source of ferulic acid, which is responsible for the production of vanillin at a great level (Zavala et al., 2009).

Various-by products generated from the extraction of citrus species are used as preservatives against microorganisms responsible for spoiling various food products without hampering their functional properties. Another example of the use of wastes into production of preserving compounds is the use of avocado oil producing industry. In the avocado oil producing industry, Permal et al. (2020) observed that with the use of wastewater which is dumped in landfills after the generation of avocado oil (80 kg) from avocado fruit (1000 kg), the wastewater generated contained 500 kg of avocado oil. This wastewater when dried from 110° C to 160° C enhanced the total phenolic content of the dried powder. This also increased the antioxidant capacity of the powder. This powder was further helpful in preventing lipid oxidation and was also used as pork sausages.

Production of aroma compounds is very promising in the Solid-State Fermentation (SSF) field. Flavour compounds are synthesized chemically or can be synthesized from the natural matrix or biotechnological processes (Sales et al., 2018). Currently, the use of enzymes, plant cell cultures and microbial cultures are used in biotechnological processes for aroma production (Longo et al., 2006). A study was conducted by Guneser et al. (2017) in which they produced flavour compound from the waste generated by the olive mill. In this study, they used the microbial fermentation of *Rhizopus oryzae* and *Candida tropicalis*. During this process, $_{D}$-Limonene was produced from these plants, which acts as a flavouring compound.

8.4 WASTES AS A VALUABLE RESOURCE FOR THE PRODUCTION OF ESSENTIAL MATERIAL

A huge amount of waste has been generated from all sources. This waste produced is a burden for the environment because of inappropriate disposal of the waste. It has been observed that production of high-value material from the waste generated is an ideal method for proper management of the waste. This also helps in reducing the volume of the waste. Various methods are used for the conversion of wastes into essential materials like bioplastics, biofertilizers and fermentable organic compounds.

8.4.1 Bioplastics

There is global pollution caused due to plastics which are petroleum-based. These plastics stay in the environment for many hundreds of years due to their non-biodegradable nature. In order to overcome this, bioplastics are considered an ecofriendly alternative to help decrease dependence on fossil resources. Bioplastics are produced from biodegradable and renewable energy sources, and the wastes obtained from foods act as a primary raw material for the production of bioplastics (Acquavia et al., 2021). There are several classifications (shown in Figure 8.1) of bioplastics. According to Reddy et al., 2013, bioplastics are categorized into:

1. Bio-based but non-biodegradable plastics (Drop-in bioplastics)
2. Bio-based non-drop-in bioplastics
3. Fossil-based non-drop-in bioplastics

Biodegradable plastics are further divided into:

i. Blends of polymers and commercial polysters
ii. Polysaccharides and proteins derived from vegetal/animal
iii. Polymers derived from microorganisms
iv. Polymers derived from biotechnology

Pectin, cellulose, starch and vegetable and animal proteins are used as feedstock of bioplastics from agro-polymers (Bayer et al., 2014; Sharma and Luzinov, 2012;

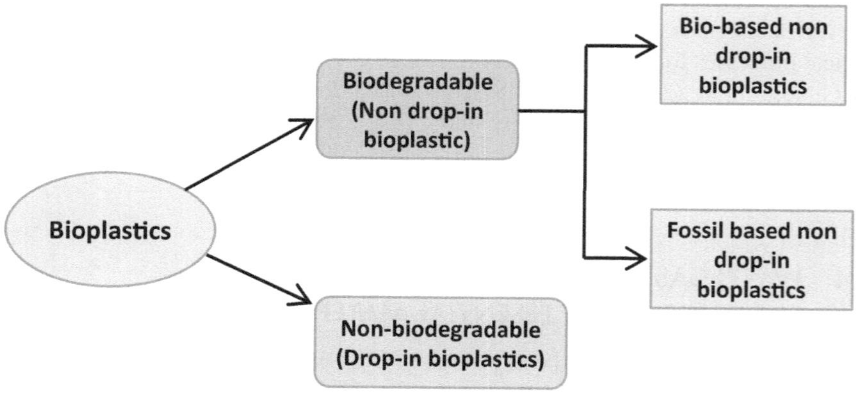

FIGURE 8.1 Classification of bioplastics.

Jimenez-Rosado et al., 2019). According to Emadian et al., 2017, for the production of bioplastics, polymers undergo three steps:

1. Biodeterioration: In this step, there is a decrease in mechanical properties of the matrix. Polymers undergo physical, mechanical and chemical change due to the activity of the microorganisms.
2. Bio fragmentation: In this step, loss in mass and change in the surface features of the matrix occur. The microbes cause the breakdown of polymers into monomers and oligomers.
3. Assimilation: In this step, the fragmented compounds are further used by the microbes and as a result they are converted into end products as biomass, water and carbon dioxide. In this process, there is also loss of mass and change in surface features.

The biological degradation of bioplastics depends on the chemical structure of the material. Polymers which are short chain, have less complex formulas and more amorphous parts can be biodegraded more easily by microorganisms (Ruggero et al., 2019), whereas some additives influence the biodegradability of the matrix (Ciriello et al., 2018). The use of waste generated from foods at all stages (from handling and manufacturing to retail and consumption) for production of bioplastics is a new idea that is emerging (Tsang et al., 2019). Conversion of waste generated from food and kitchen includes several steps, like extraction of biopolymers which is followed by mechanical manufacturing. Mechanical manufacturing includes extrusion, casting, moulding or may also include a combination of these. In a complex process, it involves the bacterial fermentation of the substrate to production of biopolymers. Besides fermentation, extraction of natural components is also done for the production of bioplastics (Tsang et al., 2019) shown in Figure 8.2.

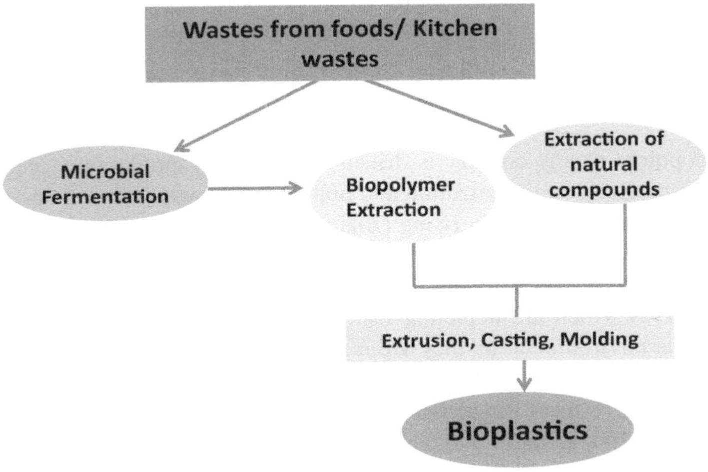

FIGURE 8.2 Steps involved in the manufacturing of bioplastics from food waste.

8.4.2 Biofertilizers

The excess use of chemical fertilizers is affecting living organisms and ecosystems. This also hampers the diversity of microorganisms that is important for the growth and development of plants (Singh et al., 2016). Use of fertilizers of biological origin has developed in order to decrease the risk of environmental degradation, for proper nutrition supply and stimulation and also to reduce environmental pollution (Mahanty et al., 2017). Use of biofertilizers also increases crop production. The term 'biofertilizer' indicates the use of live cells of cellulolytic microorganisms, nitrogen and phosphorus solubilizing bacteria that when applied to soil, improves the soil's properties and also helps in the enhanced growth and development of plants (Singh et al., 2016).

Based on the type of microorganisms contained in the biofertilizers, they are divided into different types, like nitrogen biofertilizer, phosphorus solubilizing biofertilizer and plant growth promoting biofertilizer. *Rhizobium*, blue green algae, *Azolla, Azospirillium and Azotobacter* are the microorganisms that are used in the nitrogen biofertilizer. The microorganisms that are involved in the phosphorus solubilizing biofertilizers include *Pseudomonas* fungi and *Bacillus* (Singh et al., 2016). Biofertilizers are also used for providing potassium to plants and to enhance growth and development. These biofertilizers include potassium solubilizing bacteria like *Bacillus mucilaginosus, Bacillus circulans, Bacillus edaphicus* and *Aspergillus terreus* (Bahadur et al., 2014). The use of biofertilizer in the remediation of heavy metals, reducing nematodes and reducing pesticide contamination is very well-known (Singh et al., 2016).

Composting and anaerobic digestion is considered an effective method for the treatment of waste as well as valourization of biomass (Diacono et al., 2019). During anaerobic digestion, depending on the micro and macro nutrients present in waste, the biofertilizer enhances crop yield and nutrition. It also proves a good method for the mitigation of climatic issues (Alburquerque et al., 2012). It has been shown that the application of biofertilizers decreases the loss of nitrogen by 54% compared to urea, which reduces it by 50% (Sun et al., 2020).

8.5 WASTES AS A VALUABLE RESOURCE FOR THE PRODUCTION OF ENERGY, HEAT AND POWER

The development of any country is driven by its energy production, and with the rise in industrialization, urbanization and population growth, the demand of energy at the global level is constantly rising (Bhatt and Tao, 2020; Nwokolo et al., 2020). Presently, these demands are chiefly met by conventional fossil fuels. The major drawbacks of these energy resources are firstly, they are non-renewable due to limited reserves which are depleting fast, and secondly, these have high negative impact on the environment, as they are a major culprit of global warming and climate change. Hence, dependency on renewable energy resources is fast becoming a favourable option to meet the increasing energy demands. In this regard, resources such as water, wind, sunlight and geothermal heat have been tapped to be used for industrial purposes (Yusuf, 2017). The energy derived from biomass in the form of biofuels has also become a popular alternative source that has high application.

The liquid or gaseous fuels generated from biomass, in particular from kitchen waste and agricultural waste, are called biofuels. This biomass has the ability to produce a variety of fuels such as ethanol, methanol and biodiesel, which are liquid fuels, and hydrogen and methane as gaseous fuels (Demirbas, 2008). The significant difference between petroleum fuels and biofuels is the oxygen levels which are quite high in the latter, thereby making them an environmentally friendly substitute (Demirbas, 2008). In addition, the biomass for the production of biofuels is easily available locally, helps in better waste management, the ecology of the rural area gets benefitted and the end results are sustainable, non-polluting and reliable fuels (Nwokolo et al., 2020). Therefore, both developing and developed nations benefit from biofuel technology because of economic concerns in relation to foreign exchange savings, reduced negative environmental impacts and increased energy security (Demirbas, 2008).

8.5.1 Biogas Generation

Biogas chiefly comprises methane and carbon dioxide along with small quantities of carbon monoxide, hydrogen sulphide, oxygen, ammonia, hydrogen and nitrogen (Chasnyk et al., 2015; Sun et al., 2015). The composition and yield of biogas varies with the types of substrates used for biogas production; hence, in the final product, the quantity of methane varies from 40–75% and carbon dioxide varies from 15–60% (Bhatt and Tao, 2020; Nwokolo et al., 2020). The biodegradation of the waste undergoes a biochemical process known as anaerobic digestion. In this process, the organic substrates are degraded with the help of specific microorganisms in the absence of oxygen, to release the end products as biogas. Apart from biogas, the process also releases a digestate, which is liquid in nature; when treated, it finally gives rise to a combination of solid and liquid effluents (Chen, 2014). The chief microorganisms involved in the process of anaerobic digestion are bacteria, archaebacteria and fungi that are responsible for all the reactions in the process (Wang et al., 2018; Bucker et al., 2020).

Anaerobic digestion involves four steps of sequential breakdown of organic waste materials. These four biochemical reactions are linked, as the product of one step becomes the substrate for the next step. The first step is hydrolysis, in which the complex and insoluble organic matter are hydrolyzed into simpler monomers with the action of hydrolytic microorganisms. These complex insoluble compounds are biopolymers of carbohydrate, lipid or proteinaceous in origin (Bhatt and Tao, 2020; Nwokolo et al., 2020). In this step, the substrate is prepared for the next phase of degradation where fermentative microorganisms act on the products of this phase. The second step is the fermentation phase or acidogenesis, in which the simpler compounds such as glucose, amino acids and lipids are acted upon by bacteria, converting them into organic acids or volatile fatty acids, along with the production of carbon dioxide and hydrogen (Sarkar et al., 2019; Sawyerr et al., 2019). The microorganisms involved in this step are collectively called acidogens and include a complex group of bacteria along with bacteriocides, *Clostridia*, *Bifidobacteria*, *Streptococci* and *Enterobacteriaceae* (Weilant, 2010). The chief organic acid produced at this stage is acetic acid, which becomes a substrate for methanogenic microorganisms

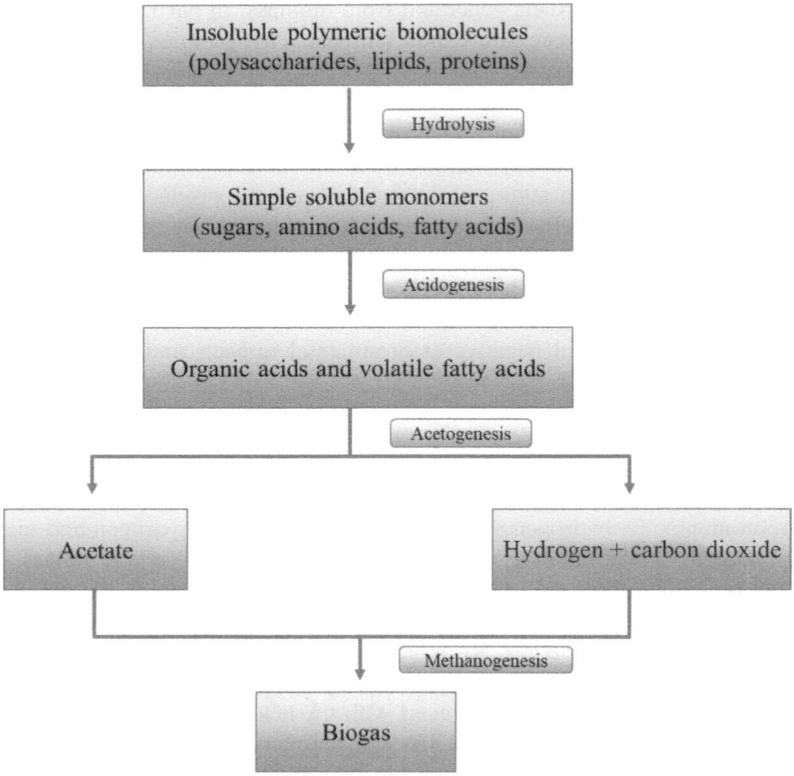

FIGURE 8.3 Schematic outline of the process of biogas generation.

in the next stage (Sarkar et al., 2019). The third stage is known as acetogenesis, in which organic acids, particularly acetic acid and butyric acid, are converted to acetate, carbon dioxide and hydrogen (Nwokolo et al., 2020). The last stage is known as methanogenesis, where acetate is converted into methane gas and carbon dioxide. The decarboxylation of acetate and the formation of methane is brought about by acetophilic methanogenic bacteria. In the same step, carbon dioxide produced also reacts with hydrogen and produces more methane with the help of hydrogenophilic methanogenic bacteria (Anukam et al., 2019). Figure 8.3 represents the outline of the process of biogas generation.

8.6 NOVEL INDUSTRIAL APPLICATION OF KITCHEN AND AGRI WASTES

An abundant amount of agri- and food-based waste is generated at an exasperating rate, and it is likely to rise in coming years at a skyrocketing pace. It has been found by the Food and Agriculture Organization (FAO) that nearly 40% of total food produced in India gets wasted (Plazzotta et al., 2017; Kumar et al., 2020b). Also, the Food Corporation of India estimated approximately 15% of loss of total production. The Ministry of Food Processing Industry of India reported 12 and 21 million ton losses of fruits and vegetables, accounting for about 10.6 billion USD of loss (NAAS, India). More significantly, food and vegetable waste means the indigestible parts that are discarded at the times of collection, handling, harvesting, shipping, processing and storing (Chang et al., 2006). They may also be produced during various stages, from farmers to consumers and both pre- and post-consumer chains of food supplies (Panda et al., 2016). The excessive amounts of phytochemicals and other constituents such as metabolites, phenolic compounds and dietary fibres are present in food and vegetable wastes (Galanakis, 2012). These essential phytochemicals and nutrients are present in peels, shoots, seeds and other organs of the plant (Rudra et al., 2015). For example, the peels of avocadoes, lemons, grape skins and jackfruit and mango seeds comprises more than 15–20% of phenolics compared to fruit pulp (Soong and Barlow, 2004). These useful components can also be extracted in the purified form, and these bioactive metabolites can be further used in the food, textile, agriculture, cosmetic and pharmaceutical industries. This novel technology is mainly focused on the horticultural industries in order to solve environmental issues by acting as a sustainable method through the use of these enriched substances (Sagar et al., 2018). The following sections will discuss the scientific intervention and latest advancements for exploring food and agriculture waste as a valuable tool for the future.

8.6.1 Edible Films

Edible films or coatings comprise a very thin layer applied to food surfaces to maintain their shelf-life, properties, functional abilities and characteristic features at minimal cost (Zambrano-Zaragoza et al., 2018). These films induce their functional abilities through prolonging shelf-life and preventing the spoilage by acting against microbes through acting as a carrier matrix (Prakash et al., 2020). The coating method is highly effective for preserving food products, especially when fruits and vegetables are transported and prone to various microbial pests and insect attacks during pre- and post-harvesting conditions (Raghav et al., 2016). The coating also enables the development of a modified environment to mediate the alterations in minimally processed foods based on sensory qualities, antioxidant properties, ethylene synthesis, firmness and antimicrobial activity. (Ullah et al., 2017). The role of essential oils and their bioactive components have also gained attention owing to the presence of anti-pathogenic properties. The active ingredient of lemongrass oil, citral, has been known to possess antimicrobial properties due to 3,7-dimethyl-2,6-octadienal, which inhibits many food-borne pathogens (Adukwu et al., 2012). Economically, they are safe and impart maximum benefits without any major change in the organoleptic properties of food (Alparslan et al., 2017).

The advancement of edible films has been considered as novel technology that comprises a sum total of nanosystems, nanoemulsions, nanocomposites as well as nanoparticles. They possess antimicrobial properties and act as antioxidants on food surfaces. Due to the presence of various phenolic substances with prevailing antioxidant activities, the veggies and fruit peels are appropriate materials for inclusion within coatings and films. Fish gelatin is another biopolymer that is used for fabricating biofilms due to its biodegradable nature and higher myofibrillar protein levels (Etxabide et al., 2017). Moreover, due to variation in the amino acid sequence, vegetable- and fruit-based films possess lesser permeability towards water compared to mammalian gelatin-based coating or films. However, combining pomegranate peel powder with gelatin films enhances water vapour permeability due to the fact that partial dissolution of pomegranate powder in gelatin results in a heterogeneous structure (Hanani et al., 2019). It has been found that these peels contains both hydrophobic and philic components in order to balance the hygroscopic characteristics without altering the moisture levels. Moreover, the peels also contain cellulose, sugars and starch as well as hemicelluloses, respectively (Borah et al., 2017). The films possessing lower levels of potato peels in contrast to higher levels in biopolymer film lead to more permeability due to larger matrix pore size despite its high density (Borah et al., 2017). The potato peel biopolymer film is best utilized in biodegradable food packaging on a commercial scale. Several fruit- and vegetable-based biofilms with their applications are summarized in Table 8.3.

8.6.2 Probiotics

Since older times, fruits have played the most prominent role for cough remedies and sore throat as medicine. From the past few years, research focus has been oriented towards the demand for the newest functional foods and probiotics worldwide, with the largest consumption rate in food products (Abdel-Hamid et al., 2020). Moreover, the fruits and their peels possess valuable roles with their enriched bioactive compounds. For instance, pomegranate, citrus, fig and mango peels contain functional ingredients in the form of antioxidants, dietary fibres and oligosaccharides, which are also called prebiotics (Coelho et al., 2019). Probiotics and dietary fibres have great potential against diseases like colon cancer and also alleviate constipation (Drago, 2019). In addition, some fibres from fruits show effects on bacterial viability and therefore can be used as an active ingredient in probiotic dairy industries (do Espirito Santo et al., 2012). For decades, several studies have been conducted to enhance the biological potential of probiotics by supplementation of fruit skins or peels. One of the most popular products is probiotic yogurt, which is prepared by using pineapple peel in powder form. It possesses anticancer, antifungal, antibacterial and antioxidant activities, specifically against *Escherichia coli* (Sah et al., 2016). Additionally, supplementing apple, banana and passion fruit peel powder in probiotic yogurt boosted rheological properties with stimulated growth of *Lactobacilli casei, L. acidophilus, L. paracasei* and *Bifidobacterium animalis* respectively (do Espírito Santo et al., 2012). Application of milk with mango peel powder in kefir's microbe growth rate and antioxidant activity has also been determined in fermented foods (Vicenssuto et al., 2020). The peel powder of orange, passion fruit and pineapple was used in 1, 0.5 and 0.7% for forming sugar-free probiotic yogurt (Dias et al., 2020). Anabellied firmness with consumer satisfaction and synergism in lactic acid bacteria was a characteristic feature in yogurt made by incorporating fruit peels.

TABLE 8.3
Fruit- and Vegetable-Derived Edible Films and Their Applications

S. No	Vegetable/ Fruit Source	Matrix type	Food Application/ Treatment	Applications	References
1.	Sweet cherry	Chitosans	Shrimp	Antimicrobial activity with enhanced shelf-life of food product.	Tokatlı and Demirdöven, 2020
2.	Orange	Gelatin	Cupcake	Stimulated peroxide response with reduced microbial growth.	Al-Anbari et al., 2019
3.	Pomegranate	Bean protein	-	Enhanced phenolic compounds, antioxidant capacities, antibacterial activities followed by its usage in bio-functional packaged foods with edible films.	Moghadam et al., 2020
4.	Potato	Essential oil from oregano	Smoked salmon	Antibacterial responses along with reduction in film strength with improved water permeability.	Tammineni et al., 2013
5.	Orange	Chitosan film	Pink shrimp	Prolonged shelf-life of the food products.	Alparslan and Baygar, 2017
6.	Orange	Gelatin	Shrimp	Enhanced shelf-life, preservation, quality extension.	Alparslan et al., 2017
7.	Lemon	Starch and sodium alginate	Tofu and strawberry	Combination of lemon peel and essential oil coating agents reduced the degradation of food.	Alparslan et al., 2017
8.	Orange	Wax and montmorillonite clay	Blood orange	Reduced deformation with excessive acidity and superior quality of fruits.	Nasirifar et al., 2018
9.	Orange	Coating with pectin material	Orange	Edible coating comprising of essential oil prolonged shelf-life without affecting sensory attributes.	Radi et al., 2018
10.	Apple	Carboxymethyl cellulose	Beef	Inhibitory action towards lipid peroxidation and microbial growth on beef.	Shin et al., 2017

8.6.3 Nanoparticles

Fruit and vegetable peels are abundant in various bioactive compounds such as alkaloids, amino acids, enzymes, phenolic compounds, proteins, tannins, terpenoids, saponins, polysaccharides and vitamins. These molecules act as reducing agents during nanoparticle synthesis (Ghosh et al., 2017). Many biomolecules play an essential role during remodelling, thereby directing the growth of particles in a particular direction, whilst many biomolecules also act as capping agents to prevent nanoparticles from being agglomerated (Fawcett et al., 2017). Nanoparticle synthesis by food and vegetable waste has evolved as a novel technology with various characteristics of reliability, sustainability and eco-friendliness. They possess negligible risks towards human and environmental health systems, in contrast to toxic chemicals that are usually solvent-based and manufactured conventionally (Anastas and Warner, 1998).

Interestingly, the significant research in use of nanotechnologies is quite prominent owing to their physicochemical properties and their role in the pharmaceutical and medical industries. Biogenic nanoparticle synthesis is conducted by a bottom-up approach, where atoms and compounds form building blocks followed by self-assembly, resulting in the final synthesis of nanoparticles (Kumar et al., 2020a). There are numerous metal and metal oxide-based nanoparticles that are synthesized from peels of fruits and vegetables, which have been summarized in Table 8.4.

To elucidate, gold nanoparticles are synthesized from onion peel extract that mainly reduces Au^{3+} to onion peel gold nanoparticles in colloidal solution (Patra et al., 2016). The phytoconstituents or active ingredients, specifically cysteine derivatives, present in onions are involved in synthesizing onion peel gold nanoparticles. Zinc oxide nanoparticles were also synthesized using potato peels, through the starch present in peels in order to reduce the metal ion (Bhuvaneswari et al., 2017). Furthermore, banana, citrus, lemon, orange, mango and pomegranate peels have also been shown to reduce silver ions in aqueous media for synthesizing silver nanoparticles (Skiba and Vorobyova, 2019). Similarly, zinc oxide nanoparticles were also reported to be synthesized by grape skin, lemon, citrus, orange and tomato peels (Nava et al., 2017).

8.6.4 Carbon Dots

Food waste is one of the major issues that requires an optimum solution as it is increasing. Food as waste serves as a great source of carbon and hence can be a fitting source for the synthesis of CDs (Fan et al., 2020). Carbon dots (CDs) are small (<10 nm) photoluminescent materials. They are synthesized by top-down and bottom-up synthetic routes (Kumar et al., 2020a). In the top-down method, with the assistance of chemicals accompanied with extreme synthetic conditions, the carbon source is converted into CDs. However, the bottom-up approach requires plants and their by-products as a source. It is considered ideal due to comparatively reduced use of chemicals. They serve as suitable source material since they are rich in components such as dietary fibre, gallic acid, polyphenols, flavonoids and carotenoids (Ajila et al., 2007). The byproducts formed are a source of antibacterial and nutritional dietary fibre (Pérez-Jiménez and Saura-Calixto, 2018). CDs have promising applications in biomedical, environmental

TABLE 8.4
Fruit- and Vegetable-Based Nanoparticles with Their Applications

S. No	Vegetable/Fruit Source	Nanoparticle type	Size	Applications	References
1.	Pomegranate, orange, citrus, apple and banana peels	Silver	48.1 nm	Antibacterial action against bacterial pathogens.	Barros et al., 2018
2.	Orange and banana peels	Silver	-	Antibacterial action against *Staphylococcus aureus*	Shet et al., 2016
3.	Sweet lemon, citrus and orange peels	Silver	-	Antibacterial action against *Pseudomonas aeruginosa* and *E. coli*.	Reenaa and Menon, 2017
4.	Orange peels	Silver	48–54 nm	Photocatalytic activity against methylene blue.	Skiba and Vorobyova, 2019
5.	Pomegranate peels	Silver	50 nm	Antibacterial action against *E. coli, P. aeruginosa* and *S. aureus*.	Shanmugavadivu et al., 2014
6.	Banana peels	Silver	24 nm	Antibacterial action against *E. coli, P. aeruginosa, Bacillus subtilis* and *S. aureus*.	Ibrahim, 2015
7.	Pomegranate peels	Silver	30–40 nm	Anti-pathogenic action against *E. coli, P. aeruginosa, S. aureus* and *Klebsiella pneumonia*.	Devanesan et al., 2018
8.	Apricot peels	Silver	50 nm	Antibacterial action against *E. coli, P. aeruginosa, Bacillus subtilis* and *S. aureus*.	Ajmal et al., 2016
9.	Banana peels	Silver	55 nm	Antibacterial action against *E. coli, P. aeruginosa, B. subtilis, S. aureus* and *Klebsiella pneumonia*.	Kokila et al., 2015
10.	Orange peels	Silver	49 nm	Antibacterial action against *Xanthomonas* sp.	de Barros et al., 2018
11.	Tomato, orange, citrus, lemon and grapefruit peels	Zinc oxide	10–20 nm	Photocatalytic activity against methylene blue.	Nava et al., 2017
12.	Potato peels	Zinc oxide	30–150 nm	Photocatalytic activity against methylene blue and azo dyes.	Bhuvaneswari et al., 2017
13.	Onion peels	Gold	46 nm	Antioxidant activity, antifungal activity against *C. albicans*, and antibacterial activity against *E. coli, S. aureus* and *Listeria monocytogenes*.	Patra et al., 2016
14.	Bottle gourd peels	Silver	5–40 nm	Antibacterial action against *Salmonella typhi* with cytotoxic action against skin and lung carcinoma mutant p53.	Kumar et al., 2015b

TABLE 8.5
Carbon Sources from Different Plants Required to Synthesize CDs

S. No.	Plant	Production conditions	Application	References
1.	*Musa acuminata*	Hydrothermal/ 200 °C/2 h	Selective and sensitive detection of Fe^{3+} ions	Vikneswaran, R., et al., 2014
2.	*Citrus sinensis*	Hydrothermal/ 180 °CC/2 h	Cell imaging	Chatzimitakos, T. et al., 2017
3.	*Garcinia mangostana*	Hydrothermal/ 200 °C/30 min	Cells imaging	Aji and Wiguna, 2017
4.	*Citrullus lanatus*	Hydrothermal/ 220 °C/2 h	Imaging probe	Zhou et al., 2012
5.	*Musa acuminata*	Microwave-assisted/ 500 W/20 min	Determination of colitoxin DNA	Huang et al., 2017
6.	*Citrus maxima*	Hydrothermal/ 200 °C/3 h 0.23	Hg2+ sensing	Lu et al., 2012
7.	*Ananas comosus*	Hydrothermal/ 200 °C/3 h 4.5 nM	Electronic security devices mercury ion (Hg2+) quantification	Vandarkuzhali et al., 2018
8.	*Citrus limon (L.)*	Hydrothermal/200 °C/8 h 73 nM Cr6+ sensing;	Photocatalysis effect	Tyagi et al., 2016
9.	*Citrus limetta*	Hydrothermal/ 180 °C/3 h	Breast cancer detection gene therapy	Ghosh et al., 2019
10.	*Mangifera indica*	Hydrothermal/ 300 °C/2 h 1.2 μM	Cellular abelling ferrous ion (Fe2+) detection	Jiao et al., 2019

studies, detection of heavy metals, determination of pathogens and water purification processes (Huang et al., 2019). Table 8.5 shows carbon sources from different plants required to synthesize CDs.

8.6.5 Microbial Media

In the field of microbiology, there is a need to grow microbes on apt culture media with favourable environments (Jadhav et al., 2018). Commercially available medias are higher on the price list. So, there is a dire need for the production of microbial medias using different substrates (Basu et al., 2015). Many reports stated that fruit and vegetable biowaste can serve as a relevant source for culturing both bacteria and fungi (Deivanayaki and Antony, 2012). Simple and complex sugars from fruit and vegetable waste are easily metabolized by microorganisms, and are thus used for the production of animal feed, bioethanol and biogas (Jamal et al., 2013). Hence, kitchen and agricultural waste can be used for the production of low-cost growth media for microorganisms.

TABLE 8.6
Different types of Agriculture Waste Used for the Production of Low-Cost Growth Medias

S. No.	Plant	Medium Composition	Purpose	References
1.	*Pisum sativum*	Pea peel powder soaked inside urea (0.3 g/L), (NH4)2SO4 (1.4 g/L), KH2PO4 (2.0 g/L), MgSO4 .7H2O (0.3g/L), peptone (1g/L), tween 80 (0.2 g/L), FeSO4 7H2O (0.005 g/L), MnSO4 .7H2O (0.0016g/L), ZnSO4 . 7H2O (0.0014 g/L) CaCl2 .2H2O (0.2 g/L), CoCl2.6H2O (0.2 g/L [102]	Cellulase production using *Trichoderma reesei*	Verma et al., 2011
2.	Orange, Potato, Drum stick *Citrus sinensis, Solanum tuberosum, Moringa oleifera*	Peel powder of orange (0.20 g/100 mL), potato (0.25 g/100 mL), drum stick (1 g/100 mL) and agar (2.5%)	Growth analysis of *Trichoderma*sp., *Aspergillus* sp.	Kumar et al., 2020b
3.	Banana	Banana peel (Autoclave) directly inoculated with fungi	Growth of human fungal pathogens viz. *Lasiodiplodiatheobromae, Nattrassiamangiferae Macrophominaphaseolina, Nigrosporasphaerica, Chaetomium murorum*, and *Schizophyllum commune*	Kindo et al., 2017
4.	*Citrus sinensis*	Peel extract of Orange (19.8 g/L), (NH4)2SO4 (0.6 g/L)	Biodiesel production using oleaginous yeasts	Carota et al., 2020
5.	*Hylocereusundatus*	Peel powder of Dragon fruit (33.3 g/L), peptone (20 mg/mL) and agar (1.5%)]	Viability analysis of *Escherichia coli*	Putri et al., 2017
6.	Orange, Potato, Drum stick *Citrus sinensis, Moringaoleifera, Solanum tuberosum*	Peel powder of orange (0.20 g/100 mL), potato (0.25 g/100 mL), drum stick (1 g/100mL) and agar (2%)	Growth and pigment production analysis of E. coli, *Serratia* sp., *Pseudomonas* sp.	Jadhav et al., 2018
7.	Banana, Melon, *Citrus paradise, Cucumis melo*, Grapefruit	Luria-Bertani medium contained 1% (w/v) starch, banana, grapefruit and melon peel powder	Amylase production from *Bacillus* sp. AY3	Kumar et al., 2020b

8.6.6 Biochar

Biochar is a carbon-rich solid produced post-pyrolysis in which precursor materials are exposed to high temperatures and oxygen-free conditions (Bruno et al., 2009). Conversion of agri waste to biochar is a great move as it helps with clearing waste and also results in the production of functional biochar. Biochar can be further used to produce bioethanol from the waste procured from food processing industries (Yao et al., 2011). Biochar has great application in remediating our natural resources, as it removes harmful pollutants such as heavy metals from contaminated sites (Xu et al., 2013). Table 8.7 enlists sources of plants used for biochar production.

8.6.7 Biosorbants

Biosorbents are the compounds formed when a sorbate (i.e., an atom or ion) reacts with biomass. As a result, sorbate ions glue onto the surface of sorbent, resulting in sequestration of ions from the solution. The final concentration of sorbent ions is reduced in the solution (Niazi, 2016). Multiple biomasses are put to use in developing different kinds of biosorbents. They may include biomasses such as algae, fungi, yeasts and bacteria (Wang and Chen, 2009). Several functional groups are introduced to target the biosorbant. Type of functional group varies from amides, carboxyl, amine, sulfhydryl, hydroxyl, carbonyl, phenolic, sulfonate to phosphate groups (Abdi and Kazemi, 2015). Biosorbents are becoming popular due to their

TABLE 8.7
Sources of Plants Used for Biochar Production

S. No.	Plant	Conditions for biochar production	Applications	References
1.	*Nephelium lappaceum*	Pyrolysis (600 °C) for 3 h	Adsorption for removal of Cu(II) from aqueous solutions of 50 and 100 mg/L at 0.2 and 0.4 g/L adsorbent dosages, respectively	Selvanathan, 2017
2.	*Solanum tuberosum*	Pyrolysis (500 °C) for 5 min	Hydrogen sulfide (H2S) was achieved 53 mg/g at 500 °C, under space velocity (8000 L min-1kg-1)	Sun et al., 2017
3.	*Punica granatum*	Pyrolysis (300 °C) for 2 h	Adsorption of Cu(II) was 52 mg/g	Cao et al., 2019
4.	*Citrus sinensis, Musa*	Pyrolysis (500 °C) for 10 min	Reducing the concentration of chemical oxygen demand (COD), total suspended solid (TSS), biochemical oxygen demand (BOD), and oil and grease of Palm oil Mil effluent (POME) to an acceptable level below the discharge	Lam et al., 2018
5.	*Ananas comosus*	Pyrolysis (200 °C) for 2 h and then heated (650 °C) for 3 h	Sorption of oxytetracycline	Fu et al., 2016
6.	*Litchi chinensis*	Hydrothermal carbonization at 180 °C for 12 h	Adsorption capacity for Congo red (404.4mg/g) and malachite green (2468 mg/g)	Wu et al., 2020
7.	*Banana Musa*	Hydrothermal carbonization at 230 °C for 2 h	Showed excellent lead clarification capability of 359 mg/g and 193 mg/g, respectively	Galanakis (2012).

TABLE 8.8
Biosorbents Production from Agri and Kitchen Wastes and Their Application

S. No.	Plant	Drying Temperature/ Time	Applications	References
1.	*Musa*	RT/4 days	0.3 g adsorbed 81.07% of rhodamine-B	Singh et al., 2018
2.	*Musa paradisiaca*	60 °C/5 h	90% removal of lead (II) and cadmium (II) ions	Ibisi and Asoluka (2018)
3.	*Malus domestica*	60 °C/24 h	Adsorbed 107.52 mg/g of methylene blue	Enniya and Jourani (2017).
4.	*Cucumis sativus*	95 °C/24 h	4 g/L (dosage concentration) adsorbed 81.4% methylene blue	Shakoor and Nasar (2017).
5.	*Luffa acutangular*	60 °C/24 h	8 g/L (dosage concentration) adsorbed 69.64 mg/g of malachite green	Ng et al., 2016
6.	*Hylocereusundatus*	105 °C/24 h	0.06 g (dosage concentration) adsorbed 192.31 mg/g of methylene blue	Jawad et al., 2018
7.	*Solanum tuberosum*/ *Daucus carota* subsp. *Sativus*	60 °C/48 h	3.0 g (dosage concentration) adsorbed 79.3% of nickel	Gill et al. (2013).
8.	*Ananas comosus*	70 °C/48 h	Adsorbed 97.09 mg/g of methylene blue	Krishni et al., 2014

great application in immobilizing heavy metals from contaminated water resources very efficiently (Wang and Chen, 2009). Many reports were used in order to produce biosorbents from agri and kitchen wastes, some of which are referenced in Table 8.8.

8.7 CONCLUSION

Conversion of foods and agri waste to highly valuable compounds is an essentially important step for waste management and safe disposal of pollutants from the environment. This can be achieved with the help of proper technologies and efficient strategies. For attaining such things, we require intense research in the field of waste management, valourization and its conversion strategies. Research in such directions not only provides new opportunities to the researchers to invent new value-added products from wastes residues but also changes the global perception regarding agri and food wastes as a problematic hazardous pollutant to the environment.

8.8 ACKNOWLEDGMENTS

First of all, I would like to thank the Lord who gave me right direction and knowledge to overcome everyday life challenges. My heartfelt gratitude to my supervisor, Prof. Renu Bhardwaj. Without her contribution, I cannot raise the quality of this book chapter. I am also very thankful to all the contributors who helped me finalize the structure of this book chapter. I am also genuinely thankful to Prof. Amit Kumar Tiwari for their time, proper guidance and support for successfully completing and finalizing the book chapter. Last but not least, a special thanks to my family and friends for their constant support, love and care.

8.9 REFERENCES

Abdel-Hamid, M., Romeih, E., Huang, Z., Enomoto, T., Huang, L. and Li, L., (2020). Bioactive properties of probiotic set-yogurt supplemented with Siraitiagrosvenorii fruit extract. *Food Chemistry*, *303*, p. 125400.

Abdi, O. and Kazemi, M., (2015). A review study of biosorption of heavy metals and comparison between different biosorbents. *Journal of Materials and Environmental Science*, *6*(5), pp. 1386–1399.

Acquavia, M.A., Pascale, R., Martelli, G., Bondoni, M. and Bianco, G., (2021). Natural polymeric materials: A solution to plastic pollution from the agro-food sector. *Polymers*, *13*, p. 158.

Adukwu, E.C., Allen, S.C. and Phillips, C.A., (2012). The anti-biofilm activity of lemongrass (C ymbopogon flexuosus) and grapefruit (C itrus paradisi) essential oils against five strains of S taphylococcus aureus. *Journal of Applied Microbiology*, *113*(5), pp. 1217–1227.

Ahmed, I., Zia, M.A., Hussain, M.A., Akram, Z., Naveed, M.T. and Nowrouzi, A., (2016). Bioprocessing of citrus waste peel for induced pectinase production by Aspergillus niger; its purification and characterization. *Journal of Radiation Research and Applied Sciences*, *9*(2), pp. 148–154.

Aji, M.P. and Wiguna, P.A., (2017). Facile synthesis of luminescent carbon dots from mangosteen peel by pyrolysis method. *Journal of Theoretical and Applied Physics*, *11*(2), pp. 119–126.

Ajila, C.M., Bhat, S.G. and Rao, U.P., (2007). Valuable components of raw and ripe peels from two Indian mango varieties. *Food Chemistry*, *102*(4), pp. 1006–1011.

Ajmal, N., Saraswat, K., Sharma, V. and Zafar, M.E., (2016). Synthesis and antibacterial activity of silver nanoparticles from Prunus armeniaca (Apricot) fruit peel extract. *Bulletin of Environment, Pharmacology and Life Sciences*, *5*, pp. 91–94.

Al-Anbari, I.H., Dakhel, A.M. and Adnan, A., (2019). The effect of adding local orange peel powder to microbial inhibition and oxidative reaction within edible film component. *Plant Archives*, *19*, pp. 1006–1012.

Alburquerque, J.A., de la Fuente, C., Ferrer-Costa, A., Carrasco, L., Cegarra, J., Abad, M. and Bernal, M.P., (2012). Assessment of the fertiliser potential of digestates from farm and agroindustrial residues. *Biomass and Bioenergy*, *40*, pp. 181–189.

Almanaa, T.N., Vijayaraghavan, P., Alharbi, N.S., Kadaikunnan, S., Khaled, J.M. and Alyahya, S.A., (2020). Solid state fermentation of amylase production from Bacillus subtilis D19 using agro-residues. *Journal of King Saud University-Science*, *32*(2), pp. 1555–1561.

Alparslan, Y. and Baygar, T., (2017). Effect of chitosan film coating combined with orange peel essential oil on the shelf life of Deepwater pink shrimp. *Food and Bioprocess Technology*, *10*(5), pp. 842–853.

Alparslan, Y., Metin, C., Yapıcı, H.H., Baygar, T., Günlü, A. and Baygar, T., (2017). Combined effect of orange peel essential oil and gelatin coating on the quality and shelf life of shrimps. *Journal of Food Safety and Food Quality*, *68*, pp. 69–78.

Amin, F., Bhatti, H.N. and Bilal, M., (2019). Recent advances in the production strategies of microbial pectinases—A review. *International Journal of Biological Macromolecules*, *122*, pp. 1017–1026.

Anastas, P.T. and Warner, J.C., (1998). Green chemistry. *Frontiers*, *640*, p. 1998.

Andler, S.M. and Goddard, J.M., (2018). Transforming food waste: How immobilized enzymes can valorize waste streams into revenue streams. *NPJ Science of Food*, *2*(1), pp. 1–11.

Anukam, A., Mohammadi, A., Naqvi, M. and Granström, K., (2019). A review of the chemistry of anaerobic digestion: Methods of accelerating and optimizing process efficiency. *Processes*, *7*(8), p. 504.

Arun, K.B., Madhavan, A., Sindhu, R., Binod, P., Pandey, A., Reshmy, R. and Sirohi, R., (2020). Remodeling agro-industrial and food wastes into value-added bio actives and biopolymers. *Industrial Crops and Products*, *154*, p. 112621.

Ávila, P.F., Martins, M., de Almeida Costa, F.A. and Goldbeck, R., (2020). Xylo oligosaccharides production by commercial enzyme mixture from agricultural wastes and their prebiotic and antioxidant potential. *Bioactive Carbohydrates and Dietary Fibre*, *24*, p. 100234.

Ayala-Zavala, J.F., González-Aguilar, G.A. and Del-Toro-Sánchez, L., (2009). Enhancing safety and aroma appealing of fresh-cut fruits and vegetables using the antimicrobial and aromatic power of essential oils. *Journal of Food Science*, *74*(7), pp. R84–R91.

Azmir, J., Zaidul, I.S.M., Rahman, M.M., Sharif, K.M., Mohamed, A., Sahena, F., Jahurul, M.H.A., Ghafoor, K., Norulaini, N.A.N. and Omar, A.K.M., (2013). Techniques for extraction of bioactive compounds from plant materials: A review. *Journal of Food Engineering*, *117*(4), pp. 426–436.

Bahadur, I., Meena, V.S. and Kumar, S., (2014). Importance and application of potassic biofertilizer in Indian agriculture. *Research Journal of Chemical Sciences ISSN*, *2231*, p. 606X.

Bansal, N., Tewari, R., Soni, R. and Soni, S.K., (2012). Production of cellulases from Aspergillus Niger NS-2 in solid state fermentation on agricultural and kitchen waste residues. *Waste Management*, *32*(7), pp. 1341–1346.

Basu, S., Bose, C., Ojha, N., Das, N., Das, J., Pal, M. and Khurana, S., (2015). Evolution of bacterial and fungal growth media. *Bioinformation*, *11*(4), p. 182.

Bayer, I.S., Guzman-Puyol, S., Heredia-Guerrero, J.A., Ceseracciu, L., Pignatelli, F., Ruffilli, R., Cingolani, R. and Athanassiou, A., (2014). Direct transformation of edible vegetable waste into bioplastics. *Macromolecules*, *47*(15), pp. 5135–5143.

Ben-Othman, S., Jõudu, I. and Bhat, R., (2020). Bioactives from agri-food wastes: Present insights and future challenges. *Molecules*, *25*(3), p. 510.

Bhatt, A.H. and Tao, L., (2020). Economic perspectives of biogas production via anaerobic digestion. *Bioengineering*, *7*(3), p. 74.

Bhatt, B., Prajapati, V., Patel, K. and Trivedi, U., (2020). Kitchen waste for economical amylase production using Bacillus amyloliquefaciens KCP2. *Biocatalysis and Agricultural Biotechnology*, *26*, p. 101654.

Bhuvaneswari, S., Subashini, G., Subramaniyam, S., (2017). Green synthesis of zinc oxide nanoparticles using potatopeel and degradation of textile milleuent by photocatalytic activity. *World Journal of Pharmaceutical Research*, 6, 774–785.

Borah, P.P., Das, P. and Badwaik, L.S., (2017). Ultrasound treated potato peel and sweet lime pomace based biopolymer film development. *Ultrasonics Sonochemistry*, *36*, pp. 11–19.

Bose, A, and Haresh, K., (2013). Production, characterization and applications of organic solvent tolerant lipase by *Pseudomonas aeruginosa* Aau2. *Biocatalysis and Agricultural Biotechnology*, *2*(3), pp. 255–266.

Boukroufa, M., Boutekedjiret, C., Petigny, L., Rakotomanomana, N. and Chemat, F., (2015). Bio-refinery of orange peels waste: A new concept based on integrated green and solvent free extraction processes using ultrasound and microwave techniques to obtain essential oil, polyphenols and pectin. *Ultrasonics Sonochemistry*, *24*, pp. 72–79.

Bruno, G., Mike, P., Christelle, B. and Goodspeed, K., (2009). Biochar is carbon negative. *Nature Geosci*, *2*(2).

Bucker, F., Marder, M., Peiter, M.R., Lehn, D.N., Esquerdo, V.M., de Almeida Pinto, L.A. and Konrad, O., (2020). Fish waste: An efficient alternative to biogas and methane production in an anaerobic mono-digestion system. *Renewable Energy*, *147*, pp. 798–805.

Campos, D.A., Gómez-García, R., Vilas-Boas, A.A., Madureira, A.R. and Pintado, M.M., (2020). Management of fruit industrial by-products—a case study on circular economy approach. *Molecules*, *25*(2), p. 320.

Cao, Q., Huang, Z., Liu, S. and Wu, Y., (2019). Potential of Punica granatum biochar to adsorb Cu (II) in soil. *Scientific Reports*, *9*(1), pp. 1–13.

Carota, E., Petruccioli, M., D'Annibale, A., Gallo, A.M. and Crognale, S., (2020). Orange peel waste—based liquid medium for biodiesel production by oleaginous yeasts. *Applied Microbiology and Biotechnology*, *104*(10), 4617–4628.

Chang, J.I., Tsai, J.J. and Wu, K.H., (2006). Composting of vegetable waste. *Waste Management & Research*, *24*(4), pp. 354–362.

Chasnyk, O., Sołowski, G. and Shkarupa, O., (2015). Historical, technical and economic aspects of biogas development: Case of Poland and Ukraine. *Renewable and Sustainable Energy Reviews*, *52*, pp. 227–239.

Chatzimitakos, T., Kasouni, A., Sygellou, L., Avgeropoulos, A., Troganis, A. and Stalikas, C., (2017). Two of a kind but different: Luminescent carbon quantum dots from Citrus peels for iron and tartrazine sensing and cell imaging. *Talanta*, *175*, 305–312.

Chen, L.N.H., (2014). *Anaerobic Digestion Basics*. University of Idaho Extension.

Chihoub, W., Dias, M.I., Barros, L., Calhelha, R.C., Alves, M.J., Harzallah-Skhiri, F. and Ferreira, I.C., (2019). Valorisation of the green waste parts from turnip, radish and wild cardoon: Nutritional value, phenolic profile and bioactivity evaluation. *Food Research International*, *126*, p. 108651.

Cho, E.J., Trinh, L.T.P., Song, Y., Lee, Y.G. and Bae, H.J., (2020). Bioconversion of biomass waste into high value chemicals. *Bioresource Technology*, *298*, p. 122386.

Ciriello, R., Magro, S.L. and Guerrieri, A., (2018). Assay of serum cholinesterase activity by an amperometric biosensor based on a co-crosslinked choline oxidase/overoxidized polypyrrole bilayer. *Analyst*, *143*(4), pp. 920–929.

Coelho, E.M., de Souza, M.E.A.O., Corrêa, L.C., Viana, A.C., de Azevêdo, L.C. and dos Santos Lima, M., (2019). Bioactive compounds and antioxidant activity of mango peel liqueurs (Mangifera indica L.) produced by different methods of maceration. *Antioxidants*, *8*(4), p. 102.

Coman, V., Teleky, B.E., Mitrea, L., Martău, G.A., Szabo, K., Călinoiu, L.F. and Vodnar, D.C., (2020). Bioactive potential of fruit and vegetable wastes. In *Advances in Food and Nutrition Research* (Vol. 91, pp. 157–225). Academic Press.

Črnivec, I.G.O., Skrt, M., Šeremet, D., Sterniša, M., Farčnik, D., Štrumbelj, E., Poljanšek, A., Cebin, N., Pogačnik, L., Možina, S.S. and Humar, M., (2021). Waste streams in onion production: Bioactive compounds, quercetin and use of antimicrobial and antioxidative properties. *Waste Management, 126*, pp. 476–486.

Dadshahi, Z., Homaei, A., Zeinali, F., Sajedi, R.H. and Khajeh, K., (2016). Extraction and purification of a highly thermostable alkaline caseinolytic protease from wastes Penaeus vannamei suitable for food and detergent industries. *Food Chemistry, 202*, pp. 110–115.

de Barros, C.H.N., Cruz, G.C.F., Mayrink, W. and Tasic, L., (2018). Bio-based synthesis of silver nanoparticles from orange waste: Effects of distinct biomolecule coatings on size, morphology, and antimicrobial activity. *Nanotechnology, Science and Applications, 11*, p. 1.

De Buck, V., Polanska, M. and Van Impe, J., (2020). Modeling biowaste biorefineries: A review. *Frontiers in Sustainable Food Systems, 4*, p. 11.

de Souza, M.P., (2010). de Oliveira e and Magalhães. *Application of Microbial-Amylase in Industry-a Review. Brazilian Journal of Microbiology, 41*, pp. 850–861.

Deivanayaki, M. and Antony, I.P., (2012). Alternative vegetable nutrient source for microbial growth. *International Journal of Biosciences (IJB), 2*(5), 47–51.

Demirbas, A., (2008). Biofuels sources, biofuel policy, biofuel economy and global biofuel projections. *Energy Conversion and Management, 49*(8), pp. 2106–2116.

Devanesan, S., AlSalhi, M.S., Balaji, R.V., Ranjitsingh, A.J.A., Ahamed, A., Alfuraydi, A.A., AlQahtani, F.Y., Aleanizy, F.S. and Othman, A.H., (2018). Antimicrobial and cytotoxicity effects of synthesized silver nanoparticles from Punica granatum peel extract. *Nanoscale Research Letters, 13*(1), pp. 1–10.

Diacono, M., Persiani, A., Testani, E., Montemurro, F. and Ciaccia, C., (2019). Recycling agricultural wastes and by-products in organic farming: Biofertilizer production, yield performance and carbon footprint analysis. *Sustainability, 11*(14), p. 3824.

Dias, P.G.I., Sajiwanie, J.W.A. and RMUSK, R., (2020). Formulation and development of composite fruit peel powder incorporated fat and sugar-free probiotic set yogurt. *GSC Biological and Pharmaceutical Sciences, 11*(1), pp. 093–099.

do Espírito Santo, A.P., Cartolano, N.S., Silva, T.F., Soares, F.A., Gioielli, L.A., Perego, P., Converti, A. and Oliveira, M.N., (2012). Fibers from fruit by-products enhance probiotic viability and fatty acid profile and increase CLA content in yoghurts. *International Journal of Food Microbiology, 154*(3), pp. 135–144.

Drago, L., (2019). Probiotics and colon cancer. *Microorganisms, 7*(3), p. 66.

Emadian, S.M., Onay, T.T. and Demirel, B., (2017). Biodegradation of bioplastics in natural environments. *Waste Management, 59*, pp. 526–536.

Enniya, I. and Jourani, A., (2017). Study of Methylene Blue Removal by a biosorbent prepared with Apple peels. *Journal of Materials and Environmental Science, 8*(12), pp. 4573–4581.

Etxabide, A., Urdanpilleta, M., Gómez-Arriaran, I., De La Caba, K. and Guerrero, P., (2017). Effect of pH and lactose on cross-linking extension and structure of fish gelatin films. *Reactive and Functional Polymers, 117*, pp. 140–146.

Fan, H., Zhang, M., Bhandari, B. and Yang, C.H., (2020). Food waste as a carbon source in carbon quantum dots technology and their applications in food safety detection. *Trends in Food Science & Technology, 95*, pp. 86–96.

Fawcett, D., Verduin, J.J., Shah, M., Sharma, S.B. and Poinern, G.E.J., (2017). A review of current research into the biogenic synthesis of metal and metal oxide nanoparticles via marine algae and seagrasses. *Journal of Nanoscience, 2017*.

Fermoso, F.G., Serrano, A., Alonso-Fariñas, B., Fernández-Bolaños, J., Borja, R. and Rodríguez-Gutiérrez, G., (2018). Valuable compound extraction, anaerobic digestion, and composting: A leading biorefinery approach for agricultural wastes. *Journal of Agricultural and Food Chemistry*, *66*(32), pp. 8451–8468.

Ferri, M., Vannini, M., Ehrnell, M., Eliasson, L., Xanthakis, E., Monari, S., Sisti, L., Marchese, P., Celli, A. and Tassoni, A., (2020). From winery waste to bioactive compounds and new polymeric biocomposites: A contribution to the circular economy concept. *Journal of Advanced Research*, *24*, pp. 1–11.

Fu, B., Ge, C., Yue, L., Luo, J., Feng, D., Deng, H. and Yu, H., (2016). Characterization of biochar derived from pineapple peel waste and its application for sorption of oxytetracycline from aqueous solution. *BioResources*, *11*(4), pp. 9017–9035.

Galanakis, C.M., (2012). Recovery of high added-value components from food wastes: Conventional, emerging technologies and commercialized applications. *Trends in Food Science & Technology*, *26*(2), pp. 68–87.

Ganaie, M.A., Soni, H., Naikoo, G.A., Oliveira, L.T.S., Rawat, H.K., Mehta, P.K. and Narain, N., (2017). Screening of low cost agricultural wastes TO maximize the fructosyltransferase production and its applicability in generation of fructooligosaccharides by solid state fermentation. *International Biodeterioration & Biodegradation*, *118*, pp. 19–26.

Ghazala, I., Sayari, N., Romdhane, M.B., Ellouz-Chaabouni, S. and Haddar, A., (2015). Assessment of pectinase production by Bacillus mojavensis I4 using an economical substrate and its potential application in oil sesame extraction. *Journal of Food Science and Technology*, *52*(12), pp. 7710–7722.

Ghosh, P.R., Fawcett, D., Sharma, S.B. and Poinern, G.E., (2017). Production of high-value nanoparticles via biogenic processes using aquacultural and horticultural food waste. *Materials*, *10*(8), p. 852.

Ghosh, S., Ghosal, K., Mohammad, S.A. and Sarkar, K., (2019). Dendrimer functionalized carbon quantum dot for selective detection of breast cancer and gene therapy. *Chemical Engineering Journal*, *373*, pp. 468–484.

Gill, R., Mahmood, A. and Nazir, R., (2013). Biosorption potential and kinetic studies of vegetable waste mixture for the removal of Nickel (II). *Journal of Material Cycles and Waste Management*, *15*(2), pp. 115–121.

Gómez-Mejía, E., Rosales-Conrado, N., León-González, M.E. and Madrid, Y., (2019). Citrus peels waste as a source of value-added compounds: Extraction and quantification of bioactive polyphenols. *Food Chemistry*, *295*, pp. 289–299.

Guneser, O., Demirkol, A., Yuceer, Y.K., Togay, S.O., Hosoglu, M.I. and Elibol, M., (2017). Production of flavor compounds from olive mill waste by *Rhizopus oryzae* and *Candida tropicalis*. *Brazilian Journal of Microbiology*, *48*, pp. 275–285.

Hanani, Z.N., Yee, F.C. and Nor-Khaizura, M.A.R., (2019). Effect of pomegranate (Punica granatum L.) peel powder on the antioxidant and antimicrobial properties of fish gelatin films as active packaging. *Food Hydrocolloids*, *89*, pp. 253–259.

Huang, C.C., Hung, Y.S., Weng, Y.M., Chen, W. and Lai, Y.S., (2019). Sustainable development of carbon nanodots technology: Natural products as a carbon source and applications to food safety. *Trends in Food Science & Technology*, *86*, pp. 144–152.

Huang, Q., Lin, X., Zhu, J.J. and Tong, Q.X., (2017). Pd-Au@ carbon dots nanocomposite: Facile synthesis and application as an ultrasensitive electrochemical biosensor for determination of colitoxin DNA in human serum. *Biosensors and Bioelectronics*, *94*, pp. 507–512.

Ibisi, N.E. and Asoluka, C.A., (2018). Use of agro-waste (Musa paradisiaca peels) as a sustainable biosorbent for toxic metal ions removal from contaminated water. *Chem. International*, *4*(1), 52.

Ibrahim HM., (2015). Green synthesis and characterization of silver nanoparticles using banana peel extract and their antimicrobial activity against representative microorganisms. *Journal of Radiation Research and Applied Sciences*, *18*(3), pp. 265–275.

Iram, A., Cekmecelioglu, D. and Demirci, A., (2020). Screening of bacterial and fungal strains for cellulase and xylanase production using distillers' dried grains with solubles (DDGS) as the main feedstock. *Biomass Conversion and Biorefinery*, pp. 1–10.

Jadhav, P., Sonne, M., Kadam, A., Patil, S., Dahigaonkar, K. and Oberoi, J.K., (2018). Formulation of cost effective alternative bacterial culture media using fruit and vegetables waste. *International Journal of Current Research and Review*, *10*, pp. 6–15.

Jamal, P., Saheed, O.K., Kari, M.I.A., Alam, Z. and Muyibi, S.A., (2013). Cellulolytic fruits wastes: A potential support for enzyme assisted protein production. *Journal of Biological Sciences*, *13*, pp. 379–385.

Jawad, A.H., Kadhum, A.M. and Ngoh, Y.S., (2018). Applicability of dragon fruit (Hylocereuspolyrhizus) peels as low-cost biosorbent for adsorption of methylene blue from aqueous solution: Kinetics, equilibrium and thermodynamics studies. *DESALINATION Water Treat*, *109*, pp. 231–240.

Jiao, X.Y., Li, L.S., Qin, S., Zhang, Y., Huang, K. and Xu, L., (2019). The synthesis of fluorescent carbon dots from mango peel and their multiple applications. *Colloids and Surfaces A: Physicochemical and Engineering Aspects*, *577*, pp. 306–314.

Jiménez-Rosado, M., Zarate-Ramírez, L.S., Romero, A., Bengoechea, C., Partal, P. and Guerrero, A., (2019). Bioplastics based on wheat gluten processed by extrusion. *Journal of Cleaner Production*, *239*, p. 117994.

Jin, Q., Yang, L., Poe, N. and Huang, H., (2018). Integrated processing of plant-derived waste to produce value-added products based on the biorefinery concept. *Trends in Food Science & Technology*, *74*, pp. 119–131.

Kamal, S., Rehman, S. and Iqbal, H.M., (2017). Biotechnological valorization of proteases: From hyperproduction to industrial exploitation—a review. *Environmental Progress & Sustainable Energy*, *36*(2), pp. 511–522.

Kindo, A.J., Tupaki-Sreepurna, A. and Yuvaraj, M., (2017). Banana peel culture as an indigenous medium for easy identification of late-sporulation human fungal pathogens. *Indian Journal of Medical Microbiology*, *34*(4), pp. 457–461.

Kiran, E.U., Trzcinski, A.P. and Liu, Y., (2014). Glucoamylase production from food waste by solid state fermentation and its evaluation in the hydrolysis of domestic food waste. *Biofuel Research Journal*, *1*(3), pp. 98–105.

Kiran, E.U., Trzcinski, A.P., Ng, W.J. and Liu, Y., (2014). Enzyme production from food wastes using a biorefinery concept. *Waste and Biomass Valorization*, *5*(6), pp. 903–917.

Kokila, T., Ramesh, P.S. and Geetha, D., (2015). Biosynthesis of silver nanoparticles from Cavendish banana peel extract and its antibacterial and free radical scavenging assay: A novel biological approach. *Applied Nanoscience*, *5*(8), pp. 911–920.

Krishni, R.R., Foo, K.Y. and Hameed, B.H., (2014). Food cannery effluent, pineapple peel as an effective low-cost biosorbent for removing cationic dye from aqueous solutions. *Desalination and Water Treatment*, *52*(31–33), pp. 6096–6103.

Kuhad, R.C., Gupta, R. and Singh, A., (2011). Microbial cellulases and their industrial applications. *Enzyme Research*, *2011*, p. 280696.

Kumar, H., Bhardwaj, K., Kuča, K., Kalia, A., Nepovimova, E., Verma, R. and Kumar, D., (2020a). Flower-based green synthesis of metallic nanoparticles: Applications beyond fragrance. *Nanomaterials*, *10*(4), p. 766.

Kumar, H., Bhardwaj, K., Sharma, R., Nepovimova, E., Kuča, K., Dhanjal, D.S., Verma, R., Bhardwaj, P., Sharma, S. and Kumar, D., (2020b). Fruit and vegetable peels: Utilization of high value horticultural waste in novel industrial applications. *Molecules*, *25*(12), p. 2812.

Kumar, M., Dahuja, A., Sachdev, A., Kaur, C., Varghese, E., Saha, S. and Sairam, K.V.S.S., (2019). Evaluation of enzyme and microwave-assisted conditions on extraction of anthocyanins and total phenolics from black soybean (Glycine max L.) seed coat. *International Journal of Biological Macromolecules*, *135*, pp. 1070–1081.

Kumar, M.B., Gao, Y., Shen, W. and He, L., (2015a). Valorisation of protein waste: An enzymatic approach to make commodity chemicals. *Frontiers of Chemical Science and Engineering*, *9*(3), pp. 295–307.

Kumar, V., Verma, S., Choudhury, S., Tyagi, M., Chatterjee, S. and Variyar, P.S., (2015b). Biocompatible silver nanoparticles from vegetable waste: Its characterization and bioefficacy. *International Journal of Nanomaterials Science*, *4*, pp. 70–86.

Kumla, J., Suwannarach, N., Sujarit, K., Penkhrue, W., Kakumyan, P., Jatuwong, K., Vadthanarat, S. and Lumyong, S., (2020). Cultivation of mushrooms and their lignocellulolytic enzyme production through the utilization of agro-industrial waste. *Molecules*, *25*(12), pp. 2811.

Lam, S.S., Liew, R.K., Cheng, C.K., Rasit, N., Ooi, C.K., Ma, N.L., Ng, J.H., Lam, W.H., Chong, C.T. and Chase, H.A., (2018). Pyrolysis production of fruit peel biochar for potential use in treatment of palm oil mill effluent. *Journal of Environmental Management*, *213*, pp. 400–408.

Leyva-López, N., Lizárraga-Velázquez, C.E., Hernández, C. and Sánchez-Gutiérrez, E.Y., (2020). Exploitation of agro-industrial waste as potential source of bioactive compounds for aquaculture. *Foods*, *9*(7), p. 843.

Lin, C.S.K., Pfaltzgraff, L.A., Herrero-Davila, L., Mubofu, E.B., Abderrahim, S., Clark, J.H., Koutinas, A.A., Kopsahelis, N., Stamatelatou, K., Dickson, F. and Thankappan, S., (2013). Food waste as a valuable resource for the production of chemicals, materials and fuels. Current situation and global perspective. *Energy & Environmental Science*, *6*(2), pp. 426–464.

Longo, M.A. and Sanromán, M.A., (2006). Production of food aroma compounds: Microbial and enzymatic methodologies. *Food Technology and Biotechnology*, *44*(3), pp. 335–353.

Lu, W., Qin, X., Liu, S., Chang, G., Zhang, Y., Luo, Y., Asiri, A.M., Al-Youbi, A.O. and Sun, X., (2012). Economical, green synthesis of fluorescent carbon nanoparticles and their use as probes for sensitive and selective detection of mercury (II) ions. *Analytical Chemistry*, *84*(12), pp. 5351–5357.

Madureira, J., Dias, M.I., Pinela, J., Calhelha, R.C., Barros, L., Santos-Buelga, C., Margaça, F.M., Ferreira, I.C. and Verde, S.C., (2020). The use of gamma radiation for extractability improvement of bioactive compounds in olive oil wastes. *Science of the Total Environment*, *727*, p. 138706.

Mahanty, T., Bhattacharjee, S., Goswami, M., Bhattacharyya, P., Das, B., Ghosh, A. and Tribedi, P., (2017). Biofertilizers: A potential approach for sustainable agriculture development. *Environmental Science and Pollution Research*, *24*(4), pp. 3315–3335.

Marcillo-Parra, V., Anaguano, M., Molina, M., Tupuna-Yerovi, D.S. and Ruales, J., (2021). Characterization and quantification of bioactive compounds and antioxidant activity in three different varieties of mango (Mangifera indica L.) peel from the Ecuadorian region using HPLC-UV/VIS and UPLC-PDA. *NFS Journal*, *23*, pp. 1–7.

Marzo, C., Díaz, A.B., Caro, I. and Blandino, A., (2019). Valorization of agro-industrial wastes to produce hydrolytic enzymes by fungal solid-state fermentation. *Waste Management & Research*, *37*(2), pp. 149–156.

Meng, S., Yin, Y. and Yu, L., (2019). Exploration of a high-efficiency and low-cost technique for maximizing the glucoamylase production from food waste. *RSC Advances*, *9*(40), pp. 22980–22986.

Mirabella, N., Castellani, V. and Sala, S., (2014). Current options for the valorization of food manufacturing waste: A review. *Journal of Cleaner Production*, *65*, pp. 28–41.

Moftah, O.A.S., Grbavčić, S., Žuža, M., Luković, N., Bezbradica, D. and Knežević-Jugović, Z., (2012). Adding value to the oil cake as a waste from oil processing industry: Production of lipase and protease by Candida utilis in solid state fermentation. *Applied Biochemistry and Biotechnology*, *166*(2), pp. 348–364.

Moghadam, M., Salami, M., Mohammadian, M., Khodadadi, M. and Emam-Djomeh, Z., (2020). Development of antioxidant edible films based on mung bean protein enriched with pomegranate peel. *Food Hydrocolloids*, *104*, p. 105735.

Monteiro, G.C., Minatel, I.O., Junior, A.P., Gomez-Gomez, H.A., de Camargo, J.P.C., Diamante, M.S., Basílio, L.S.P., Tecchio, M.A. and Lima, G.P.P., (2021). Bioactive compounds and antioxidant capacity of grape pomace flours. *LWT*, *135*, p. 110053.

Montenegro-Landívar, M.F., Tapia-Quirós, P., Vecino, X., Reig, M., Valderrama, C., Granados, M., Cortina, J.L. and Saurina, J., (2021). Fruit and vegetable processing wastes as natural sources of antioxidant-rich extracts: Evaluation of advanced extraction technologies by surface response methodology. *Journal of Environmental Chemical Engineering*, *9*(4), p. 105330.

Msarah, M.J., Ibrahim, I., Hamid, A.A. and Aqma, W.S., (2020). Optimisation and production of alpha amylase from thermophilic Bacillus spp. and its application in food waste biodegradation. *Heliyon*, *6*(6), p. e04183.

Mushtaq, Q., Irfan, M., Tabssum, F. and Iqbal Qazi, J., (2017). Potato peels: A potential food waste for amylase production. *Journal of Food Process Engineering*, *40*(4), p. e12512.

Naik, B., Goyal, S.K., Tripathi, A.D. and Kumar, V., (2019). Screening of agro-industrial waste and physical factors for the optimum production of pullulanase in solid-state fermentation from endophytic Aspergillus sp. *Biocatalysis and Agricultural Biotechnology*, *22*, p. 101423.

Nasirifar, S.Z., Maghsoudlou, Y. and Oliyaei, N., (2018). Effect of active lipid-based coating incorporated with nano clay and orange peel essential oil on physicochemical properties of Citrus sinensis. *Food Science & Nutrition*, *6*(6), pp. 1508–1518.

Nava, O.J., Soto-Robles, C.A., Gómez-Gutiérrez, C.M., Vilchis-Nestor, A.R., Castro-Beltrán, A., Olivas, A. and Luque, P.A., (2017). Fruit peel extract mediated green synthesis of zinc oxide nanoparticles. *Journal of Molecular Structure*, *1147*, pp. 1–6.

Ng, H.W., Lee, L.Y., Chan, W.L., Gan, S. and Chemmangattuvalappil, N., (2016). Luffa acutangula peel as an effective natural biosorbent for malachite green removal in aqueous media: Equilibrium, kinetic and thermodynamic investigations. *Desalination and Water Treatment*, *57*(16), pp. 7302–7311.

Niazi, N.K., Murtaza, B., Bibi, I., Shahid, M., White, J.C., Nawaz, M.F., Bashir, S., Shakoor, M.B., Choppala, G., Murtaza, G. and Wang, H., (2016). Removal and recovery of metals by biosorbents and biochars derived from biowastes. In *Environmental Materials and Waste* (pp. 149–177). Academic Press.

Nile, S.H., Nile, A., Liu, J., Kim, D.H. and Kai, G., (2019). Exploitation of apple pomace towards extraction of triterpene acids, antioxidant potential, cytotoxic effects, and inhibition of clinically important enzymes. *Food and Chemical Toxicology*, *131*, p. 110563.

Nwokolo, N., Mukumba, P., Obileke, K. and Enebe, M., (2020). Waste to energy: A focus on the impact of substrate type in biogas production. *Processes*, *8*(10), p. 1224.

Ogidi, C.O., Ubaru, A.M., Ladi-Lawal, T., Thonda, O.A., Aladejana, O.M. and Malomo, O., (2020). Bioactivity assessment of exopolysaccharides produced by Pleurotus pulmonarius in submerged culture with different agro-waste residues. *Heliyon*, *6*(12), p. e05685.

Ortiz, G.E., Ponce-Mora, M.C., Noseda, D.G., Cazabat, G., Saravalli, C., López, M.C., Gil, G.P., Blasco, M. and Albertó, E.O., (2017). Pectinase production by Aspergillus giganteus in solid-state fermentation: Optimization, scale-up, biochemical characterization and its application in olive-oil extraction. *Journal of Industrial Microbiology and Biotechnology*, *44*(2), pp. 197–211.

Panda, S.K., Mishra, S.S., Kayitesi, E. and Ray, R.C., (2016). Microbial-processing of fruit and vegetable wastes for production of vital enzymes and organic acids: Biotechnology and scopes. *Environmental Research*, *146*(April 1), pp. 161–172.

Patidar, M.K., Nighojkar, S., Kumar, A. and Nighojkar, A., (2018). Pectinolytic enzymes-solid state fermentation, assay methods and applications in fruit juice industries: A review. *3 Biotech*, *8*(4), pp. 1–24. https://doi.org/10.1007/s13205-018-1220-4.

Patra, J.K., Kwon, Y. and Baek, K.H., (2016). Green biosynthesis of gold nanoparticles by onion peel extract: Synthesis, characterization and biological activities. *Advanced Powder Technology*, *27*(5) (September 1), pp. 2204–2213.

Pérez-Jiménez, J. and Saura-Calixto, F., (2018). Fruit peels as sources of non-extractable polyphenols or macromolecular antioxidants: Analysis and nutritional implications. *Food Research International*, *111*, pp. 148–152.

Permal, R., Chang, W.L., Chen, T., Seale, B., Hamid, N. and Kam, R., (2020). Optimising the spray drying of avocado wastewater and use of the powder as a food preservative for preventing lipid peroxidation. *Foods*, *9*(9), p. 1187.

Plazzotta, S., Ibarz, R., Manzocco, L. and Martín-Belloso, O., (2020). Optimizing the antioxidant biocompound recovery from peach waste extraction assisted by ultrasounds or microwaves. *Ultrasonics Sonochemistry*, *63*, p. 104954.

Plazzotta, S., Manzocco, L. and Nicoli, M.C., (2017). Fruit and vegetable waste management and the challenge of fresh-cut salad. *Trends in Food Science & Technology*, *63*, pp. 51–59.

Polley, T. and Ghosh, U., (2018). Isolation and identification of potent alkaline protease producing microorganism and optimization of biosynthesis of the enzyme using RSM. *Indian Chemical Engineer*, *60*(3), pp. 285–296.

Posadino, A.M., Biosa, G., Zayed, H., Abou-Saleh, H., Cossu, A., Nasrallah, G.K., Giordo, R., Pagnozzi, D., Porcu, M.C., Pretti, L. and Pintus, G., (2018). Protective effect of cyclically pressurized solid—liquid extraction polyphenols from Cagnulari grape pomace on oxidative endothelial cell death. *Molecules*, *23*(9), p. 2105.

Prakash, A., Baskaran, R. and Vadivel, V., (2020). Citralnanoemulsion incorporated edible coating to extend the shelf life of fresh cut pineapples. *LWT*, *118*, p. 108851.

Putri, C.H., Janica, L., Jannah, M., Ariana, P.P., Tansy, R.V. and Wardhana, Y.R., (2017). Utilization of dragon fruit peel waste as microbial growth media. *Proceedings of the 10th CISAK, Daejeon, Korea*, pp. 91–95.

Radi, M., Akhavan-Darabi, S., Akhavan, H.R. and Amiri, S., (2018). The use of orange peel essential oil microemulsion and nanoemulsion in pectin-based coating to extend the shelf life of fresh-cut orange. *Journal of Food Processing and Preservation*, *42*(2), p. e13441.

Raghav, P.K., Agarwal, N. and Saini, M., (2016). Edible coating of fruits and vegetables: A review. *Education*, *1*, 2455-5630.

Reddy, R.L., Reddy, V.S. and Gupta, G.A., (2013). Study of bio-plastics as green and sustainable alternative to plastics. *International Journal of Emerging Technology and Advanced Engineering*, *3*(5), pp. 76–81.

Reenaa, M. and Menon, A.S., (2017). Synthesis of silver nanoparticles from different citrus fruit peel extracts and a comparative analysis on its antibacterial activity. *International Journal of Current Microbiology and Applied Sciences*, *6*(7), pp. 2358–2365.

Rudra, S.G., Nishad, J., Jakhar, N. and Kaur, C., (2015). Food industry waste: Mine of nutraceuticals. *International Journal of Environmental Science and Technology*, *4*(1), pp. 205–229.

Ruggero, F., Gori, R. and Lubello, C., (2019). Methodologies to assess biodegradation of bioplastics during aerobic composting and anaerobic digestion: A review. *Waste Management & Research*, *37*(10), pp. 959–975.

Sagar, N.A., Pareek, S., Sharma, S., Yahia, E.M. and Lobo, M.G., (2018). Fruit and vegetable waste: Bioactive compounds, their extraction, and possible utilization. *Comprehensive Reviews in Food Science and Food Safety*, *17*(3), pp. 512–531.

Sah, B.N.P., Vasiljevic, T., McKechnie, S. and Donkor, O.N., (2016). Effect of pineapple waste powder on probiotic growth, antioxidant and antimutagenic activities of yogurt. *Journal of Food Science and Technology*, *53*(3), pp. 1698–1708.

Sales, A., Paulino, B.N., Pastore, G.M. and Bicas, J.L., (2018). Biogeneration of aroma compounds. *Current Opinion in Food Science*, *19*, pp. 77–84.

Salim, A.A., Grbavčić, S., Šekuljica, N., Stefanović, A., Tanasković, S.J., Luković, N. and Knežević-Jugović, Z., (2017). Production of enzymes by a newly isolated Bacillus sp. TMF-1 in solid state fermentation on agricultural by-products: The evaluation of substrate pretreatment methods. *Bioresource Technology*, *228*, pp. 193–200.

Sanz, V., Flórez-Fernández, N., Domínguez, H. and Torres, M.D., (2020). Clean technologies applied to the recovery of bioactive extracts from Camellia sinensis leaves agricultural wastes. *Food and Bioproducts Processing*, *122*, pp. 214–221.

Saratale, G.D., Kshirsagar, S.D., Sampange, V.T., Saratale, R.G., Oh, S.E., Govindwar, S.P. and Oh, M.K., (2014). Cellulolytic enzymes production by utilizing agricultural wastes under solid state fermentation and its application for biohydrogen production. *Applied Biochemistry and Biotechnology*, *174*(8), pp. 2801–2817.

Sarker, S., Lamb, J.J., Hjelme, D.R. and Lien, K.M., (2019). A review of the role of critical parameters in the design and operation of biogas production plants. *Applied Sciences*, *9*(9), p. 1915.

Sawyerr, N., Trois, C., Workneh, T. and Okudoh, V.I., (2019). An overview of biogas production: Fundamentals, applications and future research. *International Journal of Energy Economics and Policy*, *9*, 105–116.

Selvanathan, M., Yann, K.T., Chung, C.H., Selvarajoo, A., Arumugasamy, S.K. and Sethu, V., (2017). Adsorption of copper (II) ion from aqueous solution using biochar derived from rambutan (nepheliumlappaceum) peel: Feedforward neural network modelling study. *Water, Air, & Soil Pollution*, *228*(8), 1–19.

Sen, A., Miranda, I., Esteves, B. and Pereira, H., (2020). Chemical characterization, bioactive and fuel properties of waste cork and phloem fractions from Quercus cerris L. bark. *Industrial Crops and Products*, *157*, p. 112909.

Shakoor, S. and Nasar, A., (2017). Adsorptive treatment of hazardous methylene blue dye from artificially contaminated water using cucumis sativus peel waste as a low-cost adsorbent. *Groundwater for Sustainable Development*, *5*, pp. 152–159.

Shanmugavadivu, M., Kuppusamy, S. and Ranjithkumar, R., (2014). Synthesis of pomegranate peel extract mediated silver nanoparticles and its antibacterial activity. *American Journal of Advanced Drug Delivery*, *2*(2), pp. 174–182.

Sharma, S. and Luzinov, I., (2012). Water aided fabrication of whey and albumin plastics. *Journal of Polymers and the Environment*, *20*(3), pp. 681–689.

Sherwood, J., (2020). The significance of biomass in a circular economy. *Bioresource Technology*, *300*, p. 122755.

Shet, A.R., Tantri, S. and Bennal, A., (2016). Economical biosynthesis of silver nanoparticles using fruit waste. *Journal of Chemical and Pharmaceutical Sciences*, *9*(4), pp. 2306–2311.

Shin, S.H., Chang, Y., Lacroix, M. and Han, J., (2017). Control of microbial growth and lipid oxidation on beef product using an apple peel-based edible coating treatment. *LWT*, *84*, pp. 183–188.

Silveira, E.A., Tardioli, P.W. and Farinas, C.S., (2016). Valorization of palm oil industrial waste as feedstock for lipase production. *Applied Biochemistry and Biotechnology*, *179*(4), pp. 558–571.

Singh, M., Dotaniya, M.L., Mishra, A., Dotaniya, C.K., Regar, K.L. and Lata, M., (2016). Role of biofertilizers in conservation agriculture. In *Conservation Agriculture* (pp. 113–134). Springer.

Singh, S., Parveen, N. and Gupta, H., (2018). Adsorptive decontamination of rhodamine-B from water using banana peel powder: A biosorbent. *Environmental Technology & Innovation*, *12*, pp. 189–195.

Skiba, M.I. and Vorobyova, V.I., (2019). Synthesis of silver nanoparticles using orange peel extract prepared by plasmochemical extraction method and degradation of methylene blue under solar irradiation. *Advances in Materials Science and Engineering*, *2019*, 8 pp.

Soong, Y.Y. and Barlow, P.J., (2004). Antioxidant activity and phenolic content of selected fruit seeds. *Food Chemistry*, *88*(3), pp. 411–417.

Sun, B., Gu, L., Bao, L., Zhang, S., Wei, Y., Bai, Z., Zhuang, G. and Zhuang, X., (2020). Application of biofertilizer containing Bacillus subtilis reduced the nitrogen loss in agricultural soil. *Soil Biology and Biochemistry*, *148*, p. 107911.

Sun, Q., Li, H., Yan, J., Liu, L., Yu, Z. and Yu, X., (2015). Selection of appropriate biogas upgrading technology-a review of biogas cleaning, upgrading and utilisation. *Renewable and Sustainable Energy Reviews*, *51*, pp. 521–532.

Sun, Y., Yang, G., Zhang, L. and Sun, Z., (2017). Preparation of high performance H2S removal biochar by direct fluidized bed carbonization using potato peel waste. *Process Safety and Environmental Protection*, *107*, 281–288.

Szabo, K., Cătoi, A.F. and Vodnar, D.C., (2018). Bioactive compounds extracted from tomato processing by-products as a source of valuable nutrients. *Plant Foods for Human Nutrition*, *73*(4), pp. 268–277.

Tammineni, N., Ünlü, G. and Min, S.C., (2013). Development of antimicrobial potato peel waste-based edible films with oregano essential oil to inhibit L isteria monocytogenes on cold-smoked salmon. *International Journal of Food Science & Technology*, *48*(1), pp. 211–214.

Tavares, C.S., Martins, A., Faleiro, M.L., Miguel, M.G., Duarte, L.C., Gameiro, J.A., Roseiro, L.B. and Figueiredo, A.C., (2020). Bioproducts from forest biomass: Essential oils and hydrolates from wastes of Cupressus lusitanica Mill. and Cistus ladanifer L. *Industrial Crops and Products*, *144*, p. 112034.

Tavares, C.S., Martins, A., Miguel, M.G., Carvalheiro, F., Duarte, L.C., Gameiro, J.A., Figueiredo, A.C. and Roseiro, L.B., (2020). Bioproducts from forest biomass II. Bioactive compounds from the steam-distillation by-products of Cupressus lusitanica Mill. and Cistus ladanifer L. wastes. *Industrial Crops and Products*, *158*, p. 112991.

Tian, M., Wai, A., Guha, T.K., Hausner, G. and Yuan, Q., (2018). Production of endoglucanase and xylanase using food waste by solid-state fermentation. *Waste and Biomass Valorization*, *9*(12), pp. 2391–2398.

Tokatlı, K. and Demirdöven, A., (2020). Effects of chitosan edible film coatings on the physicochemical and microbiological qualities of sweet cherry (Prunus avium L.). *Scientia Horticulturae*, *259*, p. 108656.

Toop, T.A., Ward, S., Oldfield, T., Hull, M., Kirby, M.E. and Theodorou, M.K., (2017). AgroCycle—developing a circular economy in agriculture. *Energy Procedia, 123*, pp. 76–80.

Tsang, Y.F., Kumar, V., Samadar, P., Yang, Y., Lee, J., Ok, Y.S., Song, H., Kim, K.H., Kwon, E.E. and Jeon, Y.J., (2019). Production of bioplastic through food waste valorization. *Environment International, 127*, pp. 625–644.

Tyagi, A., Tripathi, K.M., Singh, N., Choudhary, S. and Gupta, R.K., (2016). Green synthesis of carbon quantum dots from lemon peel waste: Applications in sensing and photocatalysis. *RSC Advances, 6*(76), pp. 72423–72432.

Ullah, A., Abbasi, N.A., Shafique, M. and Qureshi, A.A., (2017). Influence of edible coatings on biochemical fruit quality and storage life of bell pepper cv. "Yolo Wonder". *Journal of Food Quality, 2017,* 8 pp.

Vandarkuzhali, S.A.A., Natarajan, S., Jeyabalan, S., Sivaraman, G., Singaravadivel, S., Muthusubramanian, S. and Viswanathan, B., (2018). Pineapple peel-derived carbon dots: Applications as sensor, molecular keypad lock, and memory device. *ACS Omega, 3*(10), pp. 12584–12592.

Verma, N., Bansal, M.C. and Kumar, V., (2011). Pea peel waste: A lignocellulosic waste and its utility in cellulase production by Trichoderma reesei under solid state cultivation. *Bioresources, 6*(2), pp. 1505–1519.

Vicenssuto, G.M. and de Castro, R.J.S., (2020). Development of a novel probiotic milk product with enhanced antioxidant properties using mango peel as a fermentation substrate. *Biocatalysis and Agricultural Biotechnology, 24*, p. 101564.

Vikneswaran, R., Ramesh, S. and Yahya, R.J.M.L., (2014). Green synthesized carbon nanodots as a fluorescent probe for selective and sensitive detection of iron (III) ions. *Materials Letters, 136*, pp. 179–182.

Wang, J. and Chen, C., (2009). Biosorbents for heavy metals removal and their future. *Biotechnology Advances, 27*(2), pp. 195–226.

Wang, P., Wang, H., Qiu, Y., Ren, L. and Jiang, B., (2018). Microbial characteristics in anaerobic digestion process of food waste for methane production—A review. *Bioresource Technology, 248*, pp. 29–36.

Wang, Z., Wang, C., Zhang, C. and Li, W., (2017). Ultrasound-assisted enzyme catalyzed hydrolysis of olive waste and recovery of antioxidant phenolic compounds. *Innovative Food Science & Emerging Technologies, 44*, pp. 224–234.

Weiland, P., (2010). Biogas production: Current state and perspectives. *Applied Microbiology and Biotechnology, 85*(4), pp. 849–860.

Wu, J., Yang, J., Feng, P., Huang, G., Xu, C. and Lin, B., (2020). High-efficiency removal of dyes from wastewater by fully recycling litchi peel biochar. *Chemosphere, 246*, p. 125734.

Xiong, X., Iris, K.M., Tsang, D.C., Bolan, N.S., Ok, Y.S., Igalavithana, A.D., Kirkham, M.B., Kim, K.H. and Vikrant, K., (2019). Value-added chemicals from food supply chain wastes: State-of-the-art review and future prospects. *Chemical Engineering Journal, 375*, p. 121983.

Xu, S., Fang, D., Tian, X., Xu, Y., Zhu, X., Wang, Y., Lei, B., Hu, P. and Ma, L., (2021). Subcritical water extraction of bioactive compounds from waste cotton (Gossypium hirsutum L.) flowers. *Industrial Crops and Products, 164*, p. 113369.

Xu, X., Cao, X. and Zhao, L., (2013). Comparison of rice husk-and dairy manure-derived biochars for simultaneously removing heavy metals from aqueous solutions: Role of mineral components in biochars. *Chemosphere, 92*(8), 955–961.

Yao, Y., Gao, B., Inyang, M., Zimmerman, A.R., Cao, X., Pullammanappallil, P. and Yang, L., (2011). Biochar derived from anaerobically digested sugar beet tailings: Characterization and phosphate removal potential. *Bioresource Technology, 102*(10), 6273–6278.

Yusuf, M., (2017). Agro-industrial waste materials and their recycled value-added applications. *Handbook of Ecomaterials*, pp. 1–11.

Zambrano-Zaragoza, M.L., González-Reza, R., Mendoza-Muñoz, N., Miranda-Linares, V., Bernal-Couoh, T.F., Mendoza-Elvira, S. and Quintanar-Guerrero, D., (2018). Nanosystems in edible coatings: A novel strategy for food preservation. *International Journal of Molecular Sciences*, *19*(3), p. 705.

Zhou, J., Sheng, Z., Han, H., Zou, M. and Li, C., (2012). Facile synthesis of fluorescent carbon dots using watermelon peel as a carbon source. *Materials Letters*, *66*(1), 222–224.

Zuin, V.G. and Ramin, L.Z., (2018). Green and sustainable separation of natural products from agro-industrial waste: Challenges, potentialities, and perspectives on emerging approaches. *Chemistry and Chemical Technologies in Waste Valorization*, pp. 229–282.

9 Utilization of Biomass from Refineries as Additional Source of Energy

Nirupama Prasad, Dan Bahadur Pal and Amit Kumar Tiwari

CONTENTS

9.1 INTRODUCTION

Crude oil currently is utilized for petrochemicals as well as being a source of energy for cheap transportation fuels. The shortfall in energy as well as petrochemicals that could occur at the depletion of crude oil might eventually be made up by nuclear fusion, fuel cells and solar technologies, etc., but currently these are still the subject of ongoing research and have not matured into full technologies (Blazev, 2021). It is also extremely important to find out the substitutes for petrochemicals which are obtained from crude oil, giving gaseous alkanes such as methane, ethane, propane, butane, liquid alkanes such as hexane, diesel, aromatics, oils, semi-solids and solids

DOI: 10.1201/9781003245773-9

like paraffin waxes. The entire chemical industry at this moment uses these as building blocks, and petrochemical plants and refineries are suitably designed to handle these chemicals safely. Since the fuels currently in use are usually alkanes, the solution of the current energy problem is viewed as development of alternate material to be used as fuels. In the following, we present in brief what scientific literature suggests as a solution for the present energy crisis and is based on renewable biomass. The concept of the bio-refinery was proposed around 1990, with the suggestion that the petrochemical refinery (producing fuels and petrochemical) is similar to bio-refineries (producing biofuels and biochemicals) (Huber et al., 2006). The purpose of this chapter is to provide information about the additional energy and range of fine chemicals that can be produced by bio-refineries. Instead of petrochemicals generated from crude oil now, there is a need to generate these building blocks from biomass consisting of carbohydrate, polysaccharides, starches, cellulose, etc (Corma et al., 2007).

Prior to the discovery of fossil fuels as a cheap source of energy and chemicals, our society depended heavily on plants as biomass to meet its energy demands, and there is a need to go back to the old times. The term 'biomasses' refer to agricultural wastes, which include crops residue, weeds, leaf litter, sawdust and livestock waste. These wastes are not being efficiently utilized. These biomasses contain lots of energy if they are being utilized properly (Beltrán-Ramírez et al., 2019). The life cycle of plants consists of utilization of carbon dioxide, water, light and air with nutrients coming from the ground to convert these into biomasses. When we eventually use biomass as a fuel, it combines with the oxygen in air and gives rise to the production of carbon dioxide, moisture and energy, which can be utilized to run transportation vehicles. If one considers the entire process, the total carbon dioxide generated is zero in such arrangements. Biomass is mainly composed of cellulose, hemicellulose and lignin with small amounts of pectin and impurities. Cellulose is crystalline in nature, whereas hemicellulose is amorphous in nature. These components are thermally unstable and combined constitute about 60–90% of the total mass. Cellulose is glucose polymer. Hemicellulose is a sugar polymer with five carbon ring sugars (xylose, arabinose) as well as six carbon ring sugars (glucose, galactose, and mannose). Lignin is thermally more stable and consists about 10–25% of the total mass. Lignin is highly branched and substituted polyaromatic compounds (Amini et al., 2018; Usino et al., 2021). Non-structured biomass can be extracted by hot water; they are mostly starches and glucose polysaccharides, which have 1.4 glucose linkages and are mostly liquid. Some plants convert carbohydrates into terpenes (these are mostly liquids) which are isomeric hydrocarbon of molecular structures $(C_5H_8)_n$, where n can take on values from unity (called monoterpenes) to large values (called polyterpenes). These are high-energy compounds and can be easily used as transportation fuels. In addition to terpenes, other high-energy liquid molecules present in biomass are called triglycerides. Triglycerides are fats and oils which are found in plants as well as the animal kingdom. These fats and oils are hydrophobic materials which form ester bonds when glycerol (1 mole) reacts with fatty acids (3 moles). These fats and oils are mostly utilized for cooking and lubrication. The focus of this chapter is to find a heterogeneous catalyst that can be utilized for the production of biodiesel from these oils (Mathew et al., 2021).

9.2 PROCESSING OF BIOMASS FOR FUEL PURPOSES

The biomass can be utilized for the production of synthesis gas (syngas). Syngas is mainly a mixture of carbon monoxide and hydrogen. Gasification is a process which converts biomass into CO, H_2, CO_2 and N_2. The percentage of these gases depends upon the technology used to produce it. The difference between synthesis gas and producer gas lies in the fact that in the latter there is a higher N_2 content. The gasification technology is similar to the solid coal gasification except for the fact that biomass contains potassium and sodium alkali metals causing slag formation and fouling problems (Mousdale, 2010). The syngas thus produced can further be utilized in standard processes like the Fischer-Tropsch process to produce higher alkanes such as diesels, gasoline, oils and higher molecular waxes (Mok et al., 1992).

The second route to convert biomass is pyrolysis or liquefaction. This is a relatively inefficient conversion method and the higher the amount of water in the biomass, the lower the heating value of the product formed. The thermo-chemical treatment to biomass can yield to oil, and its amount depends upon the reactor residence time, heating rate and the reactor temperature. In this process, two reactions, pyrolysis and liquefaction, occur in tandem. When pyrolysis is the dominant reaction, oil is produced that is water-soluble. A wide range of feedstock can be used such as wood, black liquor, agricultural waste and forest waste, and the product formed has at least 400 compounds including alcohols, acids, aldehyde, esters and aromatics. Bio-oils are commercially used as boiler fuel for the generation of power and heat energy. Bio-oils can also be used for producing fine chemicals. If this is used as transportation fuel, it must be upgraded by mixing with regular feed. This is because of the difficulty of ignition, as it has low heating value and water content. These bio-oils are highly corrosive due to the presence of acids. Coking is another problem that occurs due to the presence of thermally unstable compounds (Matayeva et al., 2019; Osman et al., 2021).

The last and most important step for biomass utilization is hydrolysis, in which cellulose (which is polysaccharide and can be represented by $(C_6H_5O_{10})_n$) can be hydrolyzed by water. This reaction is not total depolymerization, and cellulose forms several degradation products (Cardona et al., 2009). This is because cellulose is solid mass with structure, and the heterogeneous reaction occurs because the acid has to first penetrate into the matrix. This would lead to an initial rapid decrease in the molecular weight, and depending upon the extent of penetration, the degree of polymerization (DP) reaches a lower asymptotic value. If the cellulose is oxidized prior to this reaction using H_2O_2, $NaClO_2$, O_3 or $KBrO_3$, the hydrolysis would give a still lower DP. In addition, this treatment reduces the aldehyde content and increases the carboxylate group concentration in products formed.

9.3 FERMENTATION OF BIOMASS FOR PRODUCTION OF BUILDING BLOCKS OF CHEMICALS

Using the fermentation method, biomasses can be converted into ethanol, sometimes called bioethanol (Demirbas, 2008a). These are blended with gasoline and diesel, and the blends exhibit lower combustion efficiency. In addition, ethanol blended with petrol in lower proportion increases the vapor pressure and leads to smog formation. For higher blends of ethanol, the vapor pressure drops significantly, leading to more difficulty in cold weather. Bioethanol does not mix with diesel, and three methods are used to improve the mixing (Prakash et al., 2018). The first one is to use an emulsifier or additives such as ethyl hexyl nitrate or di-tertiarybutyl peroxide. The second method is to use a dual fuel operation, in which ethanol and diesel are introduced separately. Lastly, an auto ignition method is introduced in the engine design; this makes the engine accept blends of as much as 95% ethanol. In enzymatic hydrolysis, biomass must be pretreated to open up the structure of cellulose. Then, fermentation is carried out using an enzyme (called *trichodermareesei*) and the conversion is limited because the biochemical reaction of cellulose is inhibited by the production of glucose (Mirmohammadsadeghi, 2018). To overcome this problem, people use a mixture of enzymes in which saccharification as well as fermentation both are carried out. In this, cellulose and fermentation enzymes both lead to the formation of ethanol. Hemicelluloses are of two types: slow (typically 65%) and fast hydrolyzing types (typically 35%) (Wyman et al., 2005).

It has already been pointed out that one should not view biomass as only for the production of fuels, but one should also find out if it is possible to produce chemicals from it. On careful examination, it is found that biomasses are a rich source of carbohydrates. These carbohydrates are the basis for generating building blocks for producing chemicals. Biomass has two types of sugars: hexoses (six carbon sugars) and pentoses (five carbon sugars) present in it. In hexoses, glucose is the most common, whereas in pentoses, xylose is the most common. Sugars can be converted in bio-products in two ways: fermentation and chemical transformation.

The use of microbial degradation (the same as fermentation) has several advantages, such as: (1) low energy input, (2) modest chemical plant demand, (3) no environmental burden and (4) less hazardous (Huber et al., 2006). However, it has a few shortcomings, such as lengthy pretreatment times and it requires considerable effort in product purification. A careful selection of organisms or a mixture of organisms can improve the selectivity. In the following, we first examine the production of glucose by fermentation.

Starch, cellulose and lactose present in biomass are the main sources of glucose. Lactic acid is mainly prepared from glucose fermentation. In the fermentation process, a major problem is the recovery of the product from the broth (John et al., 2007). It requires the precipitation of calcium lactate and then recovering the lactic acid from its salt. The side products such as succinic acid, 3-hydroxyl propionic acid and glutamic acid cannot be subjected to distillation; because of the difficulty of separation, purification accounts for 50% of the production cost. Lactic acid is extensively utilized in the food, pharmaceutical and cosmetics industries. These lactic acids exist in optically active isomers from L (+) and D (-) (Escamilla-Alvarado et al., 2017). It has acid as well as hydroxyl functional groups and is amenable to various chemical reactions. The primary product of maleate hydrogenation is succinic acid (anhydride and esters). In recent times, these succinic acids were produced from butane through maleic anhydride or through fermentation of glucose. Since most of the organisms require a basic or neutral medium, only the salts of succinic acid are produced. Succinic acid is a versatile compound and can serve as important raw material for several industrially important chemicals. Some of the chemicals are 1,4-butanediol, γ-butyrolactose, N-methyl pyrrolilone and linear aliphatic esters (Saxena et al., 2017). Itaconic acid is an unsaturated dicarboxylic acid produced by fermentation of xylose. In the past, most of the research has been focused on the fermentation of glucose, and only recently has micro-organisms working on xylose as well as pentose been discovered. The property that makes itaconic acid an important chemical is the presence of a methylene group along with two carboxylic acid groups. Double bonds present in these groups can be treated as an α-substituted acrylic acid in addition to polymerization. The polymerized products from the methyl, ethyl and vinyl esters can be utilized as plastics, elastomers and coatings (Gopaliya et al., 2021).

The biomass must undergo physical and chemical pretreatment before fermentation. Under physical pretreatment, one carries out milling and irradiation, which affects it only marginally. Chemical pretreatment implies hydrolysis reaction techniques such as acids fermentation (acids at room temperature or steam, steam expulsion, sulfur dioxide), basic pretreatment (sodium hydroxide, NaOH+ peroxide, NaOH+ steam expulsion, aqueous ammonia) and/or solvent (methanol, ethanol or acetone) pretreatment (Taherzadeh and Karimi, 2008). An example is the chemical pretreatment of wheat straw, in which the complex molecules break down into simpler molecules which can be taken up for fermentation. The chemical degradation of hemicelluloses produces xylose and furfural commercially. Similarly, the dilute acid hydrolysis of lignocellulosic biomass has been carried out with following reaction steps (Lee et al., 1999; Prasad and Shih, 2016):

1. Batch reactor operates at 220°C.
2. Plug flow reactors where solid and liquid are present at 230°C.
3. Percolation reactors have two stage reverse flow. Unfortunately, fermentation is limited by the growth of inhibitory product, peroxypyrole. These compounds are known to produce formic acid. They also produce various inhibitors for ethanol producing organisms. Enzymology of cellulose degradation has been studied considerably in connection with biofuel fermentation.

The organism, cellulase, has four enzymes carrying out different reactions (Munir and Levin, 2016):

1. Endoglucanases, which decrease the degree of polymerization of cellulose.
2. Cellodextrinases attack the chain ends of the cellulose, liberating glucose.
3. Cellobiohydrolases attack the chain length of cellulose, liberating disaccharides called cellobiose.
4. β-glucosidases hydrolyze soluble cellodextrins and cellobiose to glucose.

Cellulases are widely distributed in the world, where bacteria and fungi produce this set of enzymes. From the biological aspects, the fungal and bacterial cellulases are different and have been the focus of attention. More than 60 fungi have been reported, giving different cellulases and some problems in the use of cellulase (Sharma et al., 2016).

Cellulases have often been described as catalytically inferior and depending upon the pretreatment of cellulose. This is because, in the crystalline cellulose, enzymes penetrate with greater difficulty and it is further observed that (a) after hydrolysis of cellulose, the catalytic efficiency improves, and (b) celluloses have poor stability for temperatures higher than 50°C. For example, the fungal celluloses have a half-life of 10 minutes at 65°C temperature. (c) In cellulase-mediated saccharification, the action is inhibited by cellobiose, which is the major intermediate product. As a result of the third observation, there is a high initial rate of degradation, which slows down with the passage of time in a batch reactor. In flow reactors, flow rates can be increased by reducing the concentration of cellobiose; this way, higher rates of degradation can be achieved.

Fats and oil are acquired from vegetable and animal sources, and they are a mixture of triglycerides having fatty acid moieties (Singhvi et al., 2021).

9.4 CATALYSTS FOR BIODIESEL PRODUCTION

Catalysis is being used to improve yield and selectivity of the product. Also, a developed catalyst must be easily separated from the product, can be reused and offers minor waste formation. It may be observed that there are three phases present in the reaction: the first is the oil phase; the second phase consists of water and methanol; the third phase is the catalyst. The solubility of methanol at low temperatures in the oil phase is limited and the reaction between the oil and methanol occurs in the third phase, which is the heterogeneous catalyst. Since the oil always contains free fatty acid, the most desirable catalyst is one which can do esterification as well as transesterification. A second important observation is that the products of this reaction have an emulsifying property, as a result of which they have a tendency to leach various catalyst metals due to entrainment. An ideal heterogeneous catalyst would be one that could resist this leaching (Xu et al., 2008; Atadashi et al., 2013).

Among various catalysts for biodiesel production reported are unsupported Hetero Poly Acids (HPA) and HPA supported on Ta_2O_5 (Furuta et al., 2004), on ZrO_2 (Li et al., 2009), HPA doped with Zinc (Leng, Wang, Zhu, Ren et al., 2009), ionic liquid (Leng, Wang, Zhu, Wu et al., 2009) and HPA supported on clay (Alsalme et al., 2008; Bokade and Yadav, 2007; Narasimharao et al., 2007), which operate at low temperatures giving high conversions. The alkali metals CaO/MgO, Si/MgO (Kiss et al., 2008), different alkali metals loaded on CaO and Zeolite (Singh and Fernando, 2007) have also been reported. The acid catalysts have also been used (Pena et al., 2009; Alsalme et al., 2008; Kiss et al., 2008; Yan et al., 2008; Li et al., 2008; Narasimharao et al., 2007; Singh and Fernando, 2007; Dalai et al., 2006; Kima et al., 2004) for biodiesel production, in which supports such as montmorrillonite clays and other clays have been studied (Kawashima et al., 2009; Cho et al., 2009). In these, acid clays, alumino silicates and MCM-41 (Umdu et al., 2009) have been used. There are several studies where the basic metals and basic functional groups are bonded on the surfaces; this way, one gets base functionalized catalysts (Marciniuk et al., 2009; Umdu et al., 2009; Vyas et al., 2009). All these references show that the major issue is the leaching of the catalyst. In our laboratory, we have synthesized several of these catalysts and have carried out experiments with them. We observed that all clay catalyst is leached out and there is about a 10% loss of material at every run. Complex catalysts have also been reported in which lead complex, zinc complex, and NiCu catalyst have diphosphate functional groups (Vyas et al., 2009) and have been bonded on surfaces. Support materials such as zirconia, sulfated silica zirconia (Suppes et al., 2004) and mesoporous sulfated silica zirconia (Jaenicke et al., 2000) have been used for biodiesel production.

These studies reveal that generally, catalysts such as MgO, CeO and La_2O_3 alkali/MgO (Li/MgO and Na/MgO) are highly active. The initial activity of various metal oxides was found to be in the order of La_2O_3>MgO=CeO_2>>ZnO. In addition to this, the uses of supports such as Mg-Al-MCM-41 have been reported for the production of different methyl ester by transesterification (Miao et al., 2009; Demirbas, 2008b).

9.5 SYNTHESIS OF ETHANOL AND FUELS FROM BIOMASS

The process of converting CO and H_2 mixture into liquid hydrocarbon is known as Fischer Tropsch synthesis (FTS). The first FTS plant began in Germany in 1938 and was shut down for safety reasons. In 1955, in Sasolburg, South Africa, Sasol started production of liquid fuels and chemicals commercially using coal. For the conversion of carbon-based feedstock to diesel (liquid fuels), FTS is an essential step. Advantages of FTS include: (i) feed stocks flexibility (natural gas, coal, biomass), (ii) the large availability of raw materials, (iii) low sulfur content products and (iv) its suitability for converting difficult-to-process resources. However, FTS has some major drawbacks, such as the polymerization-like nature of the process, and it yields product with a wide range of spectrum. FTS can be used for both low (such as methane) and high molecular weight (waxes) products (Demirbas, 2007, 2010).

Syngas generation, syngas conversion, and hydro processing are the main processing steps in FTS. In this synthesis, a wide range of products can be produced. These products include olefins, paraffins and oxygenated products (alcohols, aldehydes, acids and ketones). Process parameters that influence product quality are feed gas composition, process temperature and pressure, catalyst type and promoters. The high temperature fluidized bed reactor uses iron-based catalysts for the generation of high amounts of linear olefins. A ratio of Al_2O_3 and SiO_2 in iron-based catalysts has a significant influence on its activity and selectivity (Demirbas, 2007).

The bio-syngas is obtained from a biomass, and it consists of mainly H_2O, CO, CO_2 and CH_4. Bio-based syngas composition can be changed by CH_4 water-based syngas and CO_2 removal. The basic reactions in FTS are as follows:

$$nCO+2nH_2O \rightarrow (CH_2) + nH_2O-- \quad \text{(i)}$$

$$nCO+(2n+1)H_2 \rightarrow C_2H_{2n+1} + nH_2O \quad \text{(ii)}$$

$$nCO+(2n+m/2)H_2 \rightarrow C_nH_{2n} + nH_2O \quad \text{(iii)}$$

In this, n is hydrocarbon chain's average length and M is the number of hydrogen atom/carbon atoms. All reactions are exothermic and products produced are a mixture of paraffin and olefin. Tar, hydrogen sulfide, carbonyl sulfide, ammonia, hydrogen cyanide, alkalis and dust particles lead to catalyst poisoning. Thus, these components must be removed thoroughly to avoid catalyst poisoning. The process of diesel from bio-syn-gas can be represented by the following scheme (Figure 9.1).

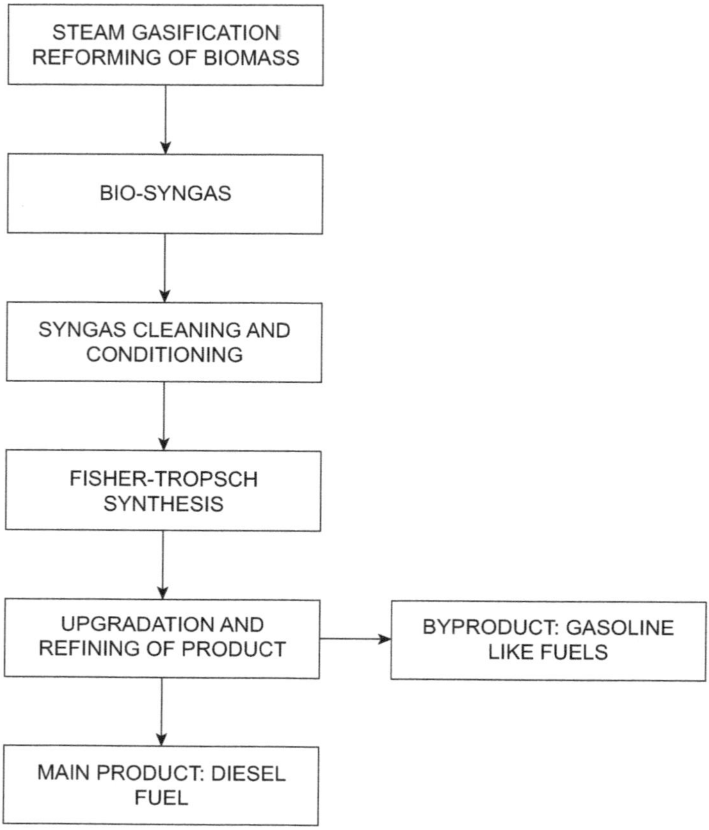

FIGURE 9.1 Fischer-Tropsch synthesis for diesel fuel production from bio-syngas.

9.6 PROCESSES FOR CONVERSION OF BIOMASS

Thermo-chemical biomass conversion involves biomass heating at high temperature. There are two basic processes: The first one is the gasification conversion to hydrocarbon, while in the second approach, it is liquefied by high temperature and pressure pyrolysis (Gallezot, 2012; Damartzis and Zabaniotou, 2011; Zhang et al., 2010).

9.6.1 Gasification

In this process, biomass is partially oxidized in the temperature range of 800–900°C in order to convert it into a combustible gas mixture. The reactions occurring are as follows:

$$C - O_2 \rightarrow CO_2$$
$$C + 1/2O_2 \rightarrow CO$$

$$CO + 1/2O_2 \rightarrow CO_2$$
$$CO_2 + 1/2O_2 \rightarrow 2CO_2$$

CH_4 and H_2 are formed simultaneously by splitting the organic material, and the hydrogen present could reduce carbon monoxide to methane as follows:

$$CO_2 + 4H_2 \rightarrow CH_4 + 2H_2O$$

The resulting gas is a mixture of methane, water, and nitrogen along with carbon oxide.

9.6.2 Pyrolysis

In this process, biomass is heated in an inert atmosphere in the temperature range of 300–700°C. Pyrolysis process parameters such as heating rate, reaction temperature and pressure and reaction time have a significant influence on the quantity and quality of the products formed. Products obtained in this process are char, liquid fuels, gaseous fuels and water vapor. The basic phenomenon of the pyrolysis process follows the following steps: (a) heat transfer to the biomass, (b) initiation of the pyrolysis releasing of volatiles and chars, (c) outflow of volatiles, (d) condensation of volatiles, (e) autocatalytic pyrolysis reactions. Depending on the reacting system temperature, the pyrolysis process can be endothermic or exothermic.

9.6.2.1 Slow and Fast Pyrolysis

When heating rate is kept slow, around 5–7°C/minute, it is called slow pyrolysis and leads to a greater amount of char produced. Fast pyrolysis is a considerably better process where the heating rate is high (300–500°C/min). Fluidized-bed reactors are best suited for the process, which requires high heating rates and rapid devolatilization. A fluidized-bed reactor is relatively easy to operate.

9.6.2.2 Flash Pyrolysis

It is reached within a few seconds of carrying out at atmospheric pressure. There is a need for rapid heating of the biomass, and it is supplied in small particles after grinding. They can further be categorized as flash hydro-pyrolysis, solar flash pyrolysis, rapid thermal process, vacuum flash pyrolysis and catalytic biomass pyrolysis. The use of the catalyst improves the quality of the product, and catalysts used are usually zeolites and basic materials.

9.6.3 Chemistry of Pyrolysis of Cellulose

The low-temperature pyrolysis process consists of degradation, decomposition, and charring due to rapid volatilization accompanying the formation of levoglucosan. The chemistry involves the breakage of the glucose chain, followed by splitting of one mole of water to give glucosan. The degradation reaction involves the depolymerization, hydrolysis, oxidation, dehydration, and decarboxylation as follows:

$$\underset{\text{Cellulose}}{\left(C_6H_{10}O_5\right)_x} \overset{535K}{\longleftrightarrow} \underset{\text{Levoglucosa}}{xC_6H_{10}O_5}$$

$$\underset{\text{Levoglucosa}}{\left(C_6H_{10}O_5\right)_5} \longleftrightarrow H_2O + \underset{\text{(Methyl glycoxal)}}{2CH_3-CO-CHO}$$

$$2CH_3-CO-CHO + 2H_2 \longleftrightarrow \underset{\text{Acetal}}{2CH_3-CO-CH_2OH}$$

$$\underset{\text{(Propyleneglycol)}}{2CH_3-CO-CH_2OH} + 2H_2 \longleftrightarrow 2CH_3-CHOH-CH_2OH$$

$$\underset{\text{(Isopropyl alcohol)}}{2CH_3-CHOH-CH_2OH} + H_2 + H_2 \longleftrightarrow 2CH_3-CHOH-CH_2OH + H_2O$$

9.7 CONCLUSION

Biomass can be utilized for the production of syngas, liquid fuel, gaseous fuel and fine chemicals. Syngas can be obtained by the gasification or pyrolysis of biomass. Pyrolysis or liquefaction is a relatively inefficient conversion method. The moisture content present in biomass has a significant influence on the heating value of the product. The higher the moisture content of the biomass, the lower the heating value of the product. The syngas is used in Fischer-Tropsch synthesis to produce liquid fuels such as diesel, gasoline, oils and higher molecular waxes. Using the fermentation method, biomasses can be converted into ethanol, sometimes called bioethanol. These bioethanols are blended with gasoline and diesel for utilization. Biomasses are a rich source of carbohydrates, which are the basis of generating building blocks for producing chemicals. Sugars present in the biomass can be converted in bio-products in two ways: fermentation (microbial degradation) and chemical transformation. Fermentation has several advantages, such as low energy input, modest chemical plant demand, no environmental burden and is less hazardous. However, this fermentation process has a few shortcomings, such as lengthy pretreatment times and requiring considerable effort in product purification.

Biomass has the potential to turn out to be a sustainable feed stock for the bulk production of value-added products. However, full-fledged production from biomass at a commercial scale is still challenging. Various researchers are searching for cost-effective conversion techniques. The major issue with this renewable feedstock is its availability, which remains uncertain and is the subject of political risks. Thus, several things still to be properly assembled and optimized before it acquires efficient industrial configuration.

9.8 ACKNOWLEDGEMENTS

All the authors acknowledge their parent institutes, e.g., Birla Institute of Technology, Mesra, Ranchi, Jharkhand and the Indian Institute of Technology (BHU), Varanasi for providing space, facility for characterization and raw materials respectively, and NPIU (TEQIP-III), Govt. of India for the financial support.

9.9 CONFLICT OF INTEREST

The authors declare that there is no conflict of interest regarding the publication of this article.

9.10 REFERENCES

Alsalme, A., Kozhevnikova, E.F. and Kozhevnikov, I.V., 2008. Heteropoly acids as catalysts for liquid-phase esterification and transesterification. *Applied Catalysis A: General*, *349*(1–2), pp. 170–176.

Amini, N., Haritos, V.S. and Tanksale, A., 2018. Microwave assisted pretreatment of eucalyptus sawdust enhances enzymatic saccharification and maximizes fermentable sugar yield. *Renewable Energy*, *127*, pp. 653–660.

Atadashi, I.M., Aroua, M.K., Aziz, A.A. and Sulaiman, N.M.N., 2013. The effects of catalysts in biodiesel production: A review. *Journal of Industrial and Engineering Chemistry*, *19*(1), pp. 14–26.

Beltrán-Ramírez, F., Orona-Tamayo, D., Cornejo-Corona, I., González-Cervantes, J.L.N., de Jesús Esparza-Claudio, J. and Quintana-Rodríguez, E., 2019. Agro-industrial waste revalorization: The growing biorefinery. In *Biomass for bioenergy-Recent trends and future challenges, IntechOpen*. (pp. 83–102).

Blazev, A.S., 2021. *Energy security for the 21st century*. CRC Press.

Bokade, V.V. and Yadav, G.D., 2007. Synthesis of bio-diesel and bio-lubricant by transesterification of vegetable oil with lower and higher alcohols over heteropolyacids supported by clay (K-10). *Process Safety and Environmental Protection*, *85*(5), pp. 372–377.

Cardona, C.A., Sanchez, O.J. and Gutierrez, L.F., 2009. *Process synthesis for fuel ethanol production*. CRC Press.

Cho, Y.B., Seo, G. and Chang, D.R., 2009. Transesterification of tributyrin with methanol over calcium oxide catalysts prepared from various precursors. *Fuel Processing Technology*, *90*(10), pp. 1252–1258.

Corma, A., Iborra, S. and Velty, A., 2007. Chemical routes for the transformation of biomass into chemicals. *Chemical reviews*, *107*(6), pp. 2411–2502.

Dalai, A.K., Kulkarni, M.G. and Meher, L.C., 2006, May. Biodiesel productions from vegetable oils using heterogeneous catalysts and their applications as lubricity additives. In *2006 IEEE EIC climate change conference* (pp. 1–8). IEEE.

Damartzis, T. and Zabaniotou, A., 2011. Thermochemical conversion of biomass to second generation biofuels through integrated process design—A review. *Renewable and Sustainable Energy Reviews*, *15*(1), pp. 366–378.

Demirbas, A., 2007. Progress and recent trends in biofuels. *Progress in Energy and Combustion Science*, *33*(1), pp. 1–18.

Demirbas, A., 2008a. Biodiesel from triglycerides via transesterification. *Biodiesel: A Realistic Fuel Alternative for Diesel Engines*, pp. 121–140.
Demirbas, A., 2008b. Comparison of transesterification methods for production of biodiesel from vegetable oils and fats. *Energy Conversion and Management*, *49*(1), pp. 125–130.
Demirbas, A., 2010. Fuels from biomass. *Biorefineries: For Biomass Upgrading Facilities*, pp. 33–73.
Escamilla-Alvarado, C., Pérez-Pimienta, J.A., Ponce-Noyola, T. and Poggi-Varaldo, H.M., 2017. An overview of the enzyme potential in bioenergy-producing biorefineries. *Journal of Chemical Technology & Biotechnology*, *92*(5), pp. 906–924.
Furuta, S., Matsuhashi, H. and Arata, K., 2004. Biodiesel fuel production with solid superacid catalysis in fixed bed reactor under atmospheric pressure. *Catalysis Communications*, *5*(12), pp. 721–723.
Gallezot, P., 2012. Conversion of biomass to selected chemical products. *Chemical Society Reviews*, *41*(4), pp. 1538–1558.
Gopaliya, D., Kumar, V. and Khare, S.K., 2021. Recent advances in itaconic acid production from microbial cell factories. *Biocatalysis and Agricultural Biotechnology*, p. 102130.
Huber, G.W., Iborra, S. and Corma, A., 2006. Synthesis of transportation fuels from biomass: Chemistry, catalysts, and engineering. *Chemical Reviews*, *106*(9), pp. 4044–4098.
Jaenicke, S., Chuah, G.K., Lin, X.H. and Hu, X.C., 2000. Organic—inorganic hybrid catalysts for acid-and base-catalyzed reactions. *Microporous and Mesoporous Materials*, *35*, pp. 143–153.
John, R.P., Nampoothiri, K.M. and Pandey, A., 2007. Fermentative production of lactic acid from biomass: An overview on process developments and future perspectives. *Applied Microbiology and Biotechnology*, *74*(3), pp. 524–534.
Kawashima, A., Matsubara, K. and Honda, K., 2009. Acceleration of catalytic activity of calcium oxide for biodiesel production. *Bioresource Technology*, *100*(2), pp. 696–700.
Kim, H.J., Kang, B.S., Kim, M.J., Park, Y.M., Kim, D.K., Lee, J.S. and Lee, K.Y., 2004. Transesterification of vegetable oil to biodiesel using heterogeneous base catalyst. *Catalysis Today*, *93*, pp. 315–320.
Kiss, A.A., Dimian, A.C. and Rothenberg, G., 2008. Biodiesel by catalytic reactive distillation powered by metal oxides. *Energy & Fuels*, *22*(1), pp. 598–604.
Lee, Y.Y., Iyer, P. and Torget, R.W., 1999. Dilute-acid hydrolysis of lignocellulosic biomass. In *Recent progress in bioconversion of lignocellulosics* (pp. 93–115). Springer.
Leng, Y., Wang, J., Zhu, D., Ren, X., Ge, H. and Shen, L., 2009. Heteropolyanion-based ionic liquids: Reaction-induced self-separation catalysts for esterification. *Angewandte Chemie*, *121*(1), pp. 174–177.
Leng, Y., Wang, J., Zhu, D., Wu, Y. and Zhao, P., 2009. Sulfonated organic heteropolyacid salts: Recyclable green solid catalysts for esterifications. *Journal of Molecular Catalysis A: Chemical*, *313*(1–2), pp. 1–6.
Li, E. and Rudolph, V., 2008. Transesterification of vegetable oil to biodiesel over MgO-functionalized mesoporous catalysts. *Energy & Fuels*, *22*(1), pp. 145–149.
Li, J., Wang, X., Zhu, W. and Cao, F., 2009. Zn1.2H0.6PW12O40 nanotubes with double acid sites as heterogeneous catalysts for the production of biodiesel from waste cooking oil. *ChemSusChem: Chemistry & Sustainability Energy & Materials*, *2*(2), pp. 177–183.
Marciniuk, L.L., Garcia, C.M., Muterle, R.B. and Schuchardt, U., 2009. Bioenergy II: Acid diphosphates as catalysts for the production of methyl and ethyl esters from vegetable oils. *International Journal of Chemical Reactor Engineering*, *7*(1).
Matayeva, A., Basile, F., Cavani, F., Bianchi, D. and Chiaberge, S., 2019. Development of upgraded bio-oil via liquefaction and pyrolysis. In *Studies in surface science and catalysis* (Vol. 178, pp. 231–256). Elsevier.

Mathew, G.M., Raina, D., Narisetty, V., Kumar, V., Saran, S., Pugazhendi, A., Sindhu, R., Pandey, A. and Binod, P., 2021. Recent advances in biodiesel production: Challenges and solutions. *Science of The Total Environment*, p. 148751.

Miao, X., Li, R. and Yao, H., 2009. Effective acid-catalyzed transesterification for biodiesel production. *Energy Conversion and Management*, *50*(10), pp. 2680–2684.

Mirmohammadsadeghi, M., 2018. *Investigation of diesel-ethanol and diesel-gasoline dual fuel combustion in a single cylinder optical diesel engine* (Doctoral dissertation, Brunel University London).

Mok, W.S., Antal Jr, M.J. and Varhegyi, G., 1992. Productive and parasitic pathways in dilute acid-catalyzed hydrolysis of cellulose. *Industrial & Engineering Chemistry Research*, *31*(1), pp. 94–100.

Mousdale, D.M., 2010. *Introduction to biofuels*. CRC Press.

Munir, R. and Levin, D.B., 2016. Enzyme systems of anaerobes for biomass conversion. *Anaerobes in Biotechnology*, pp. 113–138.

Narasimharao, K., Brown, D.R., Lee, A.F., Newman, A.D., Siril, P.F., Tavener, S.J. and Wilson, K., 2007. Structure—activity relations in Cs-doped heteropolyacid catalysts for biodiesel production. *Journal of Catalysis*, *248*(2), pp. 226–234.

Osman, A.I., Mehta, N., Elgarahy, A.M., Al-Hinai, A., Ala'a, H. and Rooney, D.W., 2021. Conversion of biomass to biofuels and life cycle assessment: A review. *Environmental Chemistry Letters*, pp. 1–44.

Pena, R., Romero, R., Martinez, S.L., Ramos, M.J., Martinez, A. and Natividad, R., 2009. Transesterification of castor oil: Effect of catalyst and co-solvent. *Industrial & Engineering Chemistry Research*, *48*(3), pp. 1186–1189.

Prakash, T., Geo, V.E., Martin, L.J. and Nagalingam, B., 2018. Effect of ternary blends of bio-ethanol, diesel and castor oil on performance, emission and combustion in a CI engine. *Renewable Energy*, *122*, pp. 301–309.

Prasad, M.N.V. and Shih, K. eds., 2016. *Environmental materials and waste: Resource recovery and pollution prevention*. Academic Press.

Saxena, R.K., Saran, S., Isar, J. and Kaushik, R., 2017. Production and applications of succinic acid. In *Current developments in biotechnology and bioengineering* (pp. 601–630). Elsevier.

Sharma, A., Tewari, R., Rana, S.S., Soni, R. and Soni, S.K., 2016. Cellulases: Classification, methods of determination and industrial applications. *Applied Biochemistry and Biotechnology*, *179*(8), pp. 1346–1380.

Singh, A.K. and Fernando, S.D., 2007. Reaction kinetics of soybean oil transesterification using heterogeneous metal oxide catalysts. *Chemical Engineering & Technology: Industrial Chemistry-Plant Equipment-Process Engineering-Biotechnology*, *30*(12), pp. 1716–1720.

Singhvi, M.S., Deshmukh, A.R. and Kim, B.S., 2021. Cellulase mimicking nanomaterial-assisted cellulose hydrolysis for enhanced bioethanol fermentation: An emerging sustainable approach. *Green Chemistry*, *23*(14), pp. 5064–5081.

Suppes, G.J., Dasari, M.A., Doskocil, E.J., Mankidy, P.J. and Goff, M.J., 2004. Transesterification of soybean oil with zeolite and metal catalysts. *Applied Catalysis A: General*, *257*(2), pp. 213–223.

Taherzadeh, M.J. and Karimi, K., 2008. Pretreatment of lignocellulosic wastes to improve ethanol and biogas production: A review. *International Journal of Molecular Sciences*, *9*(9), pp. 1621–1651.

Umdu, E.S., Tuncer, M. and Seker, E., 2009. Transesterification of Nannochloropsisoculata microalga's lipid to biodiesel on Al2O3 supported CaO and MgO catalysts. *Bioresource Technology*, *100*(11), pp. 2828–2831.

Usino, D.O., Ylitervo, P., Moreno, A., Sipponen, M.H. and Richards, T., 2021. Primary interactions of biomass components during fast pyrolysis. *Journal of Analytical and Applied Pyrolysis*, *159*, p. 105297.

Vyas, A.P., Subrahmanyam, N. and Patel, P.A., 2009. Production of biodiesel through transesterification of Jatropha oil using KNO3/Al2O3 solid catalyst. *Fuel*, *88*(4), pp. 625–628.

Wyman, C.E., Decker, S.R., Himmel, M.E., Brady, J.W., Skopec, C.E. and Viikari, L., 2005. Hydrolysis of cellulose and hemicellulose. *Polysaccharides: Structural Diversity and Functional Versatility*, *1*, pp. 1023–1062.

Xu, L., Yang, X., Yu, X. and Guo, Y., 2008. Preparation of mesoporous polyoxometalate—tantalum pentoxide composite catalyst for efficient esterification of fatty acid. *Catalysis Communications*, *9*(7), pp. 1607–1611.

Yan, S., Lu, H. and Liang, B., 2008. Supported CaO catalysts used in the transesterification of rapeseed oil for the purpose of biodiesel production. *Energy & Fuels*, *22*(1), pp. 646–651.

Zhang, L., Xu, C.C. and Champagne, P., 2010. Overview of recent advances in thermo-chemical conversion of biomass. *Energy Conversion and Management*, *51*(5), pp. 969–982.

10 Effects of Agricultural Wastes on Environment and Its Control Measures

Arun Dev Singh, Palak Bakshi, Pardeep Kumar, Jaspreet Kour, Shalini Dhiman, Mohd. Ibrahim, Isha Madaan, Dhriti Kapoor, Bilal Ahmad Mir and Renu Bhardwaj

CONTENTS

10.1 INTRODUCTION

Agricultural wastes are the residues or by-products of crop cultivation and processing activities. These raw materials include fruits, vegetables, cultivated crops, fodder and dairy products. These non-product deposits from the agricultural raw and processed activities are of low economic value to mankind with respect to the cost of their management. These agro-wastes mostly contain waste such as sugarcane bagasses, cornstalks, pruning, vegetables, drop fruits and hazardous wastes from modern agricultural practices such as pesticides, inorganic fertilizers and other animal wastes (manures) (Nagendran, 2011; Xue et al., 2016).

DOI: 10.1201/9781003245773-10

Population explosion negatively impacts crop production and yields. Thus, industrialization of agriculture has led to a detrimental impact on the ecological and environmental backlashes worldwide. Modern agricultural practices are contributing significantly to environmental pollution and waste generation (Sabiiti, 2011). These agro-production activities include excessive use of inorganic fertilizers and chemical pesticides, which put pressure on groundwater pools and further leads to environmental chaos like water stress, low soil nutrients and toxic trace elements. However, burning of agricultural wastes also contributes a minimal amount of air pollution, but the use of inorganic fertilizers and chemicals like pesticides contributes in large part; they pollute the groundwater as well as surface water resources, change the soil texture and produce trace elements that are toxic to human populations. Runoff from fertilizers, mainly phosphorus, causes eutrophication and further alters the smell and taste of surface water resources, whereas nitrate leachates to groundwater are harmful to human health. Pesticide runoff leads to ecological imbalance by hampering the reproductive cycles of top predators (Obi et al., 2016).

Salinity is among the common problems associated with excessive irrigation, cropping and land cleaning processes. Salinization hampers crop productivity as well as changes the texture of the soil. It also restricts the normal functioning of the plant's nutrient and water absorption processes through the roots. China and India, with irrigated areas of 44.88 and 42.10 (Mha), are troubled with salinization in irrigation area with 6.70 and 7.00 (Mha) of lands (Ghassemi et al., 1995; Nagendran, 2011).

Climate change is the most challenging development in the world's ecosystem. It is contributed by both natural and anthropogenic activities and further leads to global warming. Agricultural activities contribute to these global climate changes through GHG production, irrigation practices and during the processing of raw materials. Smith and Olesen (2010) found that agricultural N_2O (nitrous oxide) and CH4 (methane) emission have risen by 17 percent globally from 1990 to 2005, and it contributes 5.1 to 6.1×10^9 tons of CO_2 per year. Thus, there is an urgent need for the introduction of global agro-waste management strategies and alternative fuels to safeguard biological ecosystems and to ensure the availability of fossil fuels for future generations.

Through vermicomposting, the available agro-wastes are treated with earthworms as they are organic-rich compounds. The energy-rich complex organic materials are broken down and are finally decomposed into humus or vermicompost with high nutrient concentrations (Sabiiti, 2011; Gupta et al., 2019). These organic manures are sustainable replacements for inorganic fertilizers (Karmakar et al., 2015).

Agricultural wastes are also used as biological adsorbents against different pollutants such as heavy metals, dyes, pesticide residues, aromatic compounds etc. (Dai et al., 2018). These adsorbents are characterized by loose porous structures and are provided with the functional groups containing hydroxyl and carboxyl, which are characteristics of effective adsorbents. For example, banana wastes act as the best adsorbents for the removal of zinc (Zn) ions from the aqueous medium (Ogundipe and Babarinde, 2017). These bio-adsorbents work effectively for the removal of phosphorus and nitrogen ions e.g., NO^{3-} and $PO4^{3-}$ etc. (Park et al., 2020). Furthermore, the activated carbon or biochar, a product of pyrolysis of many of the agricultural wastes, is effectively being used as a bio-adsorbent for the removal of pollutants in aqueous and gaseous phases e.g., dyes, heavy metals, CO_2 and formaldehyde etc. (Inyang et al., 2016; Luka et al., 2018).

However, the sustainable energy recovery from the agricultural waste biomass is a global initiative in minimizing the exploitation of fossil fuels for energy. This initiative requires technologies like thermochemical processes including gasification and combustion, anaerobic digestion (a biochemical process) and Fischer Tropsch synthesis, a diesel fuel-yielding process (Antonelli et al., 2015). These energy recovery exercises are confirmed to be important for lowering the environmental pollution, CO2 levels and in the use of fossil fuels for energy demands. Çelik and Demirer (2015) examined pistachio hull waste for biogas generation through biochemical anaerobic digestion process.

10.2 AGRICULTURAL WASTE AND THEIR RELEVANT SOURCES

Agricultural waste is a vast reservoir of untapped biomass resources, its by-products and co-products that are typically characterized as plant or animal wastes. These wastes that aren't (or aren't further processed into) food or feed can cause significant environmental and economic costs in the agricultural sector. This includes harvest waste like rice husk, wheat straw, maize husk and cobs, millet stovers, pineapple peels, groundnut shell, fruits leftovers and sugarcane bagasses, molasse, pesticides, fertilizers and other wastes like poultry, slaughterhouse waste, etc (Vaibhav et al., 2015; Dai et al., 2018). Agricultural wastes are lignocellulosic materials with lignin, cellulose, and hemicellulose as the main structural components (Salleh et al., 2011). The technological contribution of the green revolution, which influenced productivity, and the fast rise of people have all contributed to an increase in agricultural production of more than three times in the last 50 years (FAO and OECD, 2019). This increase in global output has put more strain on the environment, resulting in severe effects on soil, air, and water resources. To differentiate and process the agricultural waste in a more productive manner, it is categorized in 4 sections: crop waste, animal waste, processing waste and hazardous waste (Figure 10.1). The rationalization of agricultural wastes is crucial for efficiently reducing environmental contamination. The efficient management of agricultural wastes is critical for reducing pollution and improving the ecological environment.

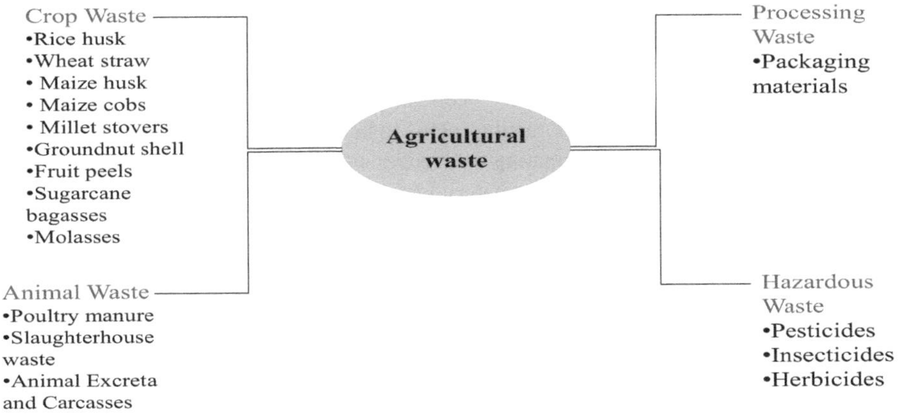

FIGURE 10.1 Different types of agricultural wastes.

10.3 AGRICULTURAL WASTE AND ENVIRONMENTAL ISSUES

Agricultural progress seems intensely affected by effluents resulting from the excessive usage of agricultural production as well as the usage of chemicals employed in agriculture, both of which have a significant impact on the affected ecosystems as well as the global environment. The amount of waste produced has been determined by the sort of farming production initiated. Although the tropical weather seems suitable for growing crop production, it also allows the production or spread of insects as well as weeds (Obi et al., 2016). As a result of this circumstance, there is a growing trend for pesticides to control pests as well as minimize the risk of infectious disease; this requirement frequently leads to chemical usage through growers. The majority of pesticide vials, as well as packaging, are discarded into crops and irrigation canals after they've been applied. According to the Plant Protection Department (PPD), around 1.8 percent of chemicals are still in their vial (Dien and Vong, 2006). Due to its remarkably long-lasting as well as harmful compounds, such residues tend to create unforeseen environmental implications like dietary contamination, inadequate nutrient sanitation, as well as polluted cropland.

Furthermore, current stagnant and wasted pesticides, as well as pesticide packages with waste material from the original contents, can have negative effects on the environment whether they are preserved or dumped inappropriately, causing them to pollute the environment via osmosis and causing damage to the environment (Nagendran, 2011). Fertilizers, for instance, are critical in crop output for sustaining crop quantity and profitability. Inorganic fertilizers are low-cost as well as high-yielding. Several growers, on the other hand, add more fertilizer to their fields than the plants require (Hai and Tuyet, 2010). The severe outcome of these massive fertilizer applications is that it has been utilized to enhance overall crop output. The rate of absorption of nitrogen, phosphorous or potassium fertilizer components varies based on land features, plant kinds, as well as fertilization strategy (Thao, 2003). A component of the fertilizer's massive quantities has been preserved inside the soil, while another portion enters reservoirs, lakes, and/or waterways as a consequence from either surface runoff and the rainwater harvesting used, tending to result in surface water pollution; a fraction reaches groundwater, while another portion evaporates and becomes denigrated, leading to polluted air.

Waste materials, like manure as well as organic matter in the abattoir; affluent, like urine, cage wash water, wastewater from animal bathing as well as hygiene in abattoirs; contaminants, like H_2S and CH_4; and odors are all examples of waste from agricultural production. Because most of them have been situated around suburban buildings, contamination produced by livestock raising seems to be a severe issue (Agamuthu, 2009). Air quality comprises odors originating from cages as a result of livestock waste decomposition, the rotting flesh of organic material in compost, animal urination, as well as repetitive food products. The quantity of the odor is based on the number of animals involved, movement, heat and moisture (Dien and Vong, 2006). The percentage of NH_3, H_2S, and CH_4 fluctuates with the phases of processing and has also been influenced by organic compounds, key nutrients, bacteria and the condition of the livestock (Obi et al., 2016). This contaminated or non-recyclable debris material can produce greenhouse gases even while depleting soil quality and polluting the water. Water covers 75–95 percent of the total volume in livestock waste, with the remaining consisting of organic matter, inorganic material, as well as a variety of microbes or parasite eggs (Hai and Tuyet, 2010). These microorganisms as well as chemicals have the potential to transfer infections to humans or have a variety of detrimental consequences on the atmosphere.

10.3.1 Biocides

Rodenticides, insecticides, nematicides, fungicides and herbicides are all considered biocide in farming. Herbicides or insecticides are frequently seen as agro-saviors. Most of these are prey predator-specific chemical compositions. Their utilization during the prior several centuries has revealed a lot of information on the perceived outcomes and the obvious harmful repercussions. There is a plethora of information on pesticides as well as agrochemicals (Green, 2005). Despite fertilizers, pesticides' effects are enhanced by the separate and co-action of additional frequently employed substances like wetting agents, emulsifiers, solvents, and so forth. In some instances, chemicals formed during the decomposition of the primary active substance unite as well as enhance the effect. Various biocides are used at various phases of production in rainfed land, resulting in emission as well as accumulation of a variety of harmful compounds. These are recognized as having adverse environmental implications. The methods for assessing these environmental influences as well as their health consequences have been extensively analyzed (Nagendran, 2011). Analysts from across the globe are familiar with persistent toxic substances (PTS). Ding et al. (2005) on PAH, Mai et al. (2005) on PCB, Cao et al. (2007) on DDT and Hu et al. (2009) on HCHs, DDT, PCBs, and PAHs have all significantly contributed. These studies emphasized the important connections which have been obtained between various PTS. Pesticides from certain categories are being linked to genetic as well as hormonal changes or alterations in both humans and animals. In the case of developing countries, this issue includes procedures, as agricultural pollutants in food and water can have major consequences for human health as well as fertility.

10.3.2 Salinization

In many nations, salinization is seen as one of the biggest potential hazards to ecological resources and human health, affecting around 1 billion hectares globally, and approximately 7 percent of the planet's geographical area (Metternicht and Zinck, 2003; Yensen, 2008). In all parts of the globe, salinization has been the most noticeable effect of farming. It is caused by both natural and human activity, such as agricultural removal and inadequate irrigation. Throughout these processes, increasing wastewater absorbs salts that are accumulated at or near the surface (Omami, 2007). Saline soils affect efficiency while also rendering the soil worthless for other applications (production). It is well recognized that high concentrations in the soil inhibit crop absorption of water as well as micronutrients. Construction vehicles, sewage or gasoline pipelines, factories and dwellings are among the other prominent effects. The environmental impact of soil salinity has been seen in limited disposable rates and increased productivity losses. For mitigating arid fields as well as irrigation saline environments, physiological, biological and mechanical approaches are being produced, which are evaluated in the analysis by Nagendran (2011).

Although most crop plants are sensitive to salinity generated by excess amounts of salts in the soil, salinization is among the most hazardous climatic variables affecting agricultural crop production, and the farmland influenced by it is expanding day by day. Salinity affects not only the productivity of most plants but also the physicochemical aspects of the soil as well as the region's natural ecosystems. Reduced crop yields, limited economic yields and soil depletion are all consequences of salinization (Hu and Schmidhalter, 2004). Seed germination, plant growth, as well as nutrient and water intake are all affected by salinization activities, which are the result of extensive interactions between morphological, physiological, as well as biological changes (Akbarimoghaddam et al., 2011). Salinization has an impact on essentially all aspects of plant development, particularly germination, seed quality, or reproductive growth. Crops are susceptible to ion toxicity, osmotic stress, nutritional (N, Ca, K, P, Fe, Zn) deficiencies, as well as oxidative stress as a result of soil salinization, which inhibits moisture intake from the soil. Although phosphate ions accumulate with calcium ions in salinized soils, crop phosphorus (P) absorption is greatly reduced (Bano and Fatima, 2009).

10.3.3 Fertilizers

With the rapid increase in population, the demand for food is increased. Then the chemical fertilizers are used to enhance agricultural productivity, crop growth and enriching the soil nutrients (Omidire et al., 2015). However, the unscientific or overuse of fertilizers causes serious environmental problems such as leaching, global warming, soil degradation and water pollution (Herrero et al., 2010; Eickhout et al., 2006). Some of the fertilizers are shown and their environmental issues are seen in Table 10.1.

TABLE 10.1
Different Fertilizers and the Associated Environmental Issues

S. No.	Chemical fertilizers	Environmental issues	References
1.	Nitrogen	Accumulation of nitrate or nitrite in plants Water contamination of water by nitrate	Bhattacharyya et al., 2016
2.	Nitrogen	Leaching and contaminate groundwater	Signor and Cerri, 2013
3.	Nitrogen	Eutrophication	Tayefeh et al., 2018
4.	Nitrogen fertilizer	Nitrous oxide (N2O) leads global warming and ozone destruction.	Rütting et al., 2018
5.	NH_3	Cause Acid rain which damage to aquatic and terrestrial life.	Ameeta and Ronak, 2017
6.	Long-term overuse of P fertilizers	Toxicity in soil	Cheraghi et al., 2012
7.	Phosphorus (P)	Eutrophication of fresh waters	Tilman et al., 2001
8.	NO_x	Atmospheric smog and acidification of soil Terrestrial eutrophication	Tilman et al., 2001 Nikkhah et al., 2015
9.	NH4-based N fertilizers	Soil acidity	Mahler and Harder, 1984
10.	N_2O	Global warming	Snyder et al., 2009
11.	K fertilizer	Environmental Pollution	Zhang et al., 2014
12.	Phosphorus (P)	Water quality degrade	Conley et al., 2009

TABLE 10.2
Greenhouse Gases and Their Percentage (%) of Emission at Different Stages

S. No.	Percentage of greenhouse gas emission	Stages of GHG emission	Major gas emission	References
1.	3.8–5.2%	Chemical fertilizer and pesticides	CO_2	Vermeulen et al., 2012.
2.	40.8–55.7%	Direct emission in productivity stage	N_2O and CH_4	
3.	23.9–43.8%	Indirect emission stage	CO_2	
4.	10.2–16.7%	Post-production processing stage	CO_2, N_2O, and CH_4	

10.3.4 Climate Change

As the population growth increases the waste generation is also increasing, which leads to climate change (Dhar et al., 2017). Agriculture waste is one of the main sources of environmental pollution (Wang et al., 2016). Burning of these agriculture wastes is involved in the emission of CO_2, N_2O and hydrocarbons (Yevich et al., 2003). Some other gases like methane gas are emitted from peatland (Agus et al. 2013), rice cultivation and crop residues (Smith et al., 2014). These harmful gases increase the risk of environmental pollution (Hosseinzadeh-Bandbafha et al., 2018). The aquatic environment is contaminated by agriculture waste, which is discharged into the water body (Wang et al., 2016). GHGs releasing and production in the atmosphere occurs through improper manure storage and decomposition (Niles, 2008). Some of the greenhouse gases and their percentage of emission at different stages are shown in Table 10.2.

10.4 AGRICULTURAL WASTE MANAGEMENT

Agricultural residues and wastes that are disposed of onto the land act as natural fertilizer for the soil. The excess disposal of these wastes is responsible for the emission of toxic gases, releasing greenhouse gases, affecting the environmental climate and also harming the health of the public (Ye et al., 2013). In order to overcome these difficulties generated by agro-wastes, there are several eco-friendly methods that help in the proper management of the agricultural wastes. These methods include the conversion of bio-waste into biogas, biochar, biofertilizers, bioadsorbants, biofuel and vermicompost.

10.4.1 Biogas

Biogas is a much convinced alternate for gaseous biofuel. It is produced from various organic wastes and resources. The most utilized alternative method for the production of biogas is anaerobic digestion of organic waste. This method generates valuable fuel and helps to subsequently reduce the volume of the waste which is disposed. Biogas helps in fighting against climate change because it is not a global warming gas. The carbon produced during the burning of biogas is consumed by plants during photosynthesis, and therefore less carbon is added into the atmosphere (Environmental Benefits of Biomethane, 2012). The composition of biogas listed here is given by Renato et al., 2013:

Gas	Percentage
Methane	40–75%
Carbon dioxide	15–60%
Water as moisture	1–5%
Nitrogen	0–5%
Hydrogen	
Hydrogen sulfide	Traces
Oxygen	
Trace gases	
Ammonia	

There are many uses of biogas, like cooking, heating, lighting, heat and power generation in industries and transportation (Kristensson et al., 2007). Among all these uses, biogas is primarily used for cooking, followed closely by lighting. Production of biogas from anaerobic digestion by the use of organic solids is a very conventional method that helps in the reduction of waste and also helps in the utilization of renewable resources (Sunitkunaporn et al., 2014). During the production of biogas, the organic waste undergoes four major bioreactions. The first reaction includes hydrolysis which is followed by acidogenesis, acetogenesis and finally methanogenesis (Thamsiriroj and Murphy, 2011). The schematic representation of biogas production is shown in Figure 10.2.

The studies conducted by Mussoline et al., 2012 and Ye et al., 2013 have found that when agricultural material is added to organic waste, it enhances the production of biogas. For the production of biogas, there are various steps required to be checked before the process starts, and these steps are represented through the flowchart shown in Figure 10.3.

10.4.2 Vermicompost

In the present era, the production of agricultural wastes is increasing day by day. The most common reasons for this are food processing industries and ungraded agricultural produce coming to market. Improper management of agricultural wastes not only disturbs the ecosystem's stability but also creates unhygienic conditions which may cause several harmful health disorders among people (Mane and Raskar, 2012). The process of vermicomposting is an eco-friendly solution to this problem of inadequate waste management. This process involves joint activity of epigenic earthworms and microorganisms, which results in the conversion of harmful wastes into a useful finished product known as vermicompost (Bhat et al., 2013). Earthworms are commonly called "unheralded soldiers of mankind," "farmers' friends" and "the

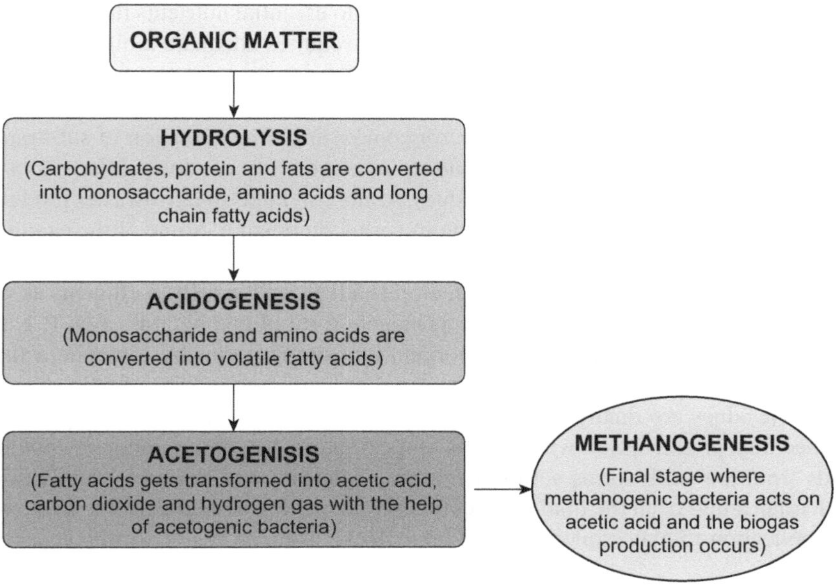

FIGURE 10.2 Representation of production of biogas.

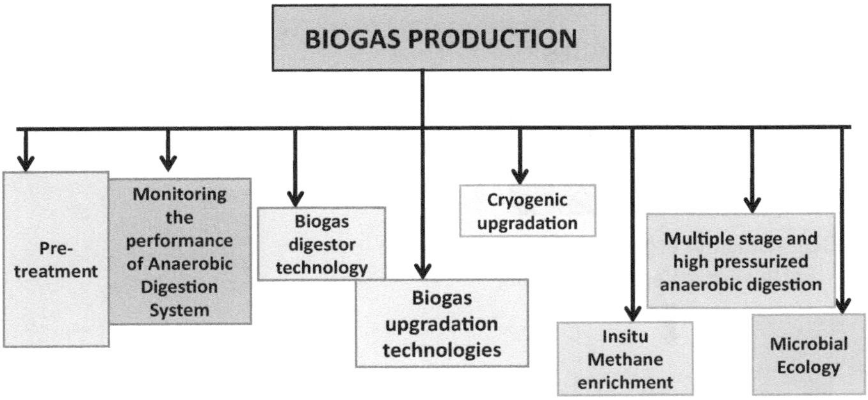

FIGURE 10.3 Flowchart depicting the different strategies for biogas production.

intestine of earth" due to their incredible potential to digest different kinds of organic residues and convert them into useful nutrient-rich vermicompost in a relatively short period of time without utilizing a large proportion of the residues (Jambhekar, 1992; Sanchez-Monedero et al., 2001). *Eisenia foetida* is one of the most common species of earthworm used for vermicomposting. This is a non-thermophilic multi-step process during which organic wastes are broken down into essential nutrients that are beneficial for soil as well as plant health. The process requires adequate moisture content, plant-based wastes and not animal wastes, as well as an environment free of rodents, birds and termites as they may consume earthworms as feed, thereby hindering the process (Gupta et al., 2019). During vermicomposting, fragmentation of substrate is accomplished by the earthworms, resulting in availability of more surface area for the microbial action on the organic residues. Almost all kinds of agro-wastes produced by different sectors may act as substrate in vermicomposting. Some of the examples of substrate used are residues from food preparation, such as fruit and vegetable peels, leftover food, scrap paper, animal dung, crop residues, and industrial effluents as well as yard trimmings. The complex and unavailable form of compounds of N, P, K, Ca etc present in these residues are transformed into simple forms by catabolic activity of enzymes secreted by the alimentary canal of earthworms. The simple digested organic residues are finally shed out as "casts" from the earthworm's body and the finished final product is known as vermicompost. Moreover, the excretion of coelomic fluids from the earthworms act on the waste's parasites and bacteria, removing the unpleasant smell from the final vermicompost and making it pathogen-free to be used for application to plant-soil systems (Figure 10.4). The process of vermicomposting

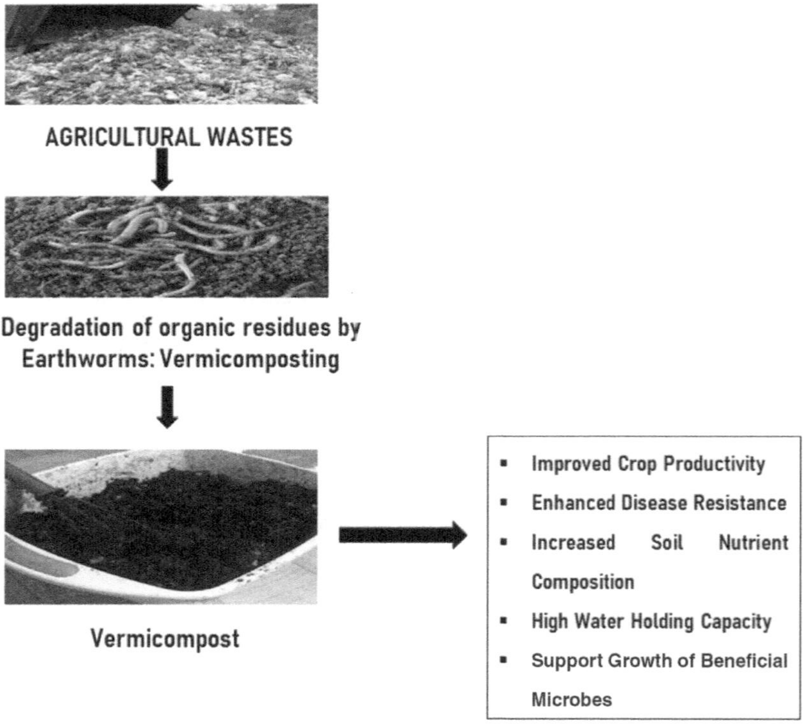

FIGURE 10.4 Process of vermicomposting and its benefits.

alters the physical as well as chemical properties of the soil, making it more porous, thereby enhancing water-holding capacity and availability to the crop plants (Bhat et al., 2018). Also, a number of hormones and enzymes are added to soil during this process, which promotes the growth of helpful microorganisms to further promote plant growth (Pathma and Sakthive, 2012).

Vermicompost is an organic manure rich in essential macronutrients such as N, P, K and Ca. These nutrients are definitely required by every plant for proper growth and development. Vermicompost also adds to the soil structure through cation exchange capacity of the soil, preventing soil erosion and salinity (Joshi et al., 2015). It can be utilized for all different kinds of agricultural crops including cereals, oilseeds, legumes and horticultural crops. It has been proven in research studies that biomass of crops grown in vermicompost medium is significantly higher compared to those cultivated in garden soil (Nithya and Lekeshmanaswamy, 2010). Therefore, vermicompost may act as an effective substitute for the harmful inorganic chemicals used for enhancing the yield of agricultural crops (Atiyeh et al., 2000). It also lessens the probability of attack by pathogens and pests on the crops, thereby acting as disease control agents. Also, vermicompost application results in reduction of toxic heavy metals such as cadmium (Cd) and lead (Pb) in soil. Use of vermicompost in crop cultivation will not only be beneficial to the crops but will also serve the purpose of detoxification of tons of industrial and agro-wastes being produced on a large scale on a daily basis in an environmentally-friendly manner (Alwaneen, 2016). Vermicompost maintenance systems can be easily acquired even by small farmers, as it has been found that cow manure of approximately 450 kg provides profit of more than 200 percent (Lalander et al., 2015). The material subjected to vermicomposting is neither landfilled nor burned; rather, it is recycled. Moreover, vermicompost can also be applied to greenhouses in establishing new plants such as root development in vineyards.

Hence, vermicompost is a cost-effective, eco-friendly solution to the problem of the bulk of agro-residues, which positively maintains the nutritional composition of the soil and improves agricultural productivity. It is an attractive alternative to endorse sustainable agriculture and innocuous management of agricultural wastes.

10.4.3 Biochar

Production of biochar from agro-residues is another efficient method for the eco-safe management of the latter. Usually, the wastes from agricultural produce are burned openly in uncontrolled conditions, which leads to the emission of greenhouse gases, global warming and tremendous pollution of basic components of the environment, i.e., air, water and soil. Therefore, it is necessary to cease these activities and promote eco-friendly management of agricultural wastes. “Biochar” is a combination of the words “bio,” meaning “living,” and “char,” meaning “charcoal.” Basically, charcoal is produced from burning of wood in environments of limited oxygen (pyrolysis) and is used as a fuel. Biochar is the charcoal-like product that is obtained by the pyrolysis of agricultural and forest wastes like living plant biomass. Burning of agro-residues in inert atmospheres at low to high temperatures (350–7000 ^{o}C) is called pyrolysis, and it converts large macromolecules of residues into smaller fragments

(Wang et al., 2015b). The final product formed is black in color, carbon-rich, porous with high surface area, and is called biochar. Production of biochar does not cause any kind of ecosystem degradation; instead, it is a highly useful product itself (Wu et al., 2015). As a source of recalcitrant organic carbon, it alleviates the increasing global warming. There are several areas of application of biochar, such as waste disposal, soil amelioration, energy restoration and climate change mitigation (Ahmad et al., 2014). Biochar is utilized as an adsorbent to remove toxic contaminants being released from different industries including textiles, pharmaceutical, dye and kraft bleaching. Recently, a more useful form of biochar called Magnetic Biochar (MBC) has been produced by the attachment of different metal ions such as Fe, Ca, Mg, and Ni on the surface of the biochar, which enhances its magnetic properties and adds to its beneficial values for application in different fields (Das et al., 2015). MBC overcomes the drawbacks of conventional biochar, which has to be activated by additional centrifugation for separation from liquid or by oxygen plasma method (Mukherjee et al., 2011; Goda et al., 2014). Methods employed for the production of MBC are conventional heating in electrical furnace, calcination, microwave heating and co-precipitation (Ma et al., 2015). There are many useful applications of biochar, listed in Table 10.3.

Biochar can be stored in terrestrial ecosystems for longer periods of time as it is not susceptible to biological decay. Owing to its long-term stability, it is crucial in providing ongoing environmental benefits. Use of biochar must be promoted to obtain economic benefits in the agricultural as well as industrial sectors.

10.4.4 Biofertilizers

Most farmers traditionally used synthetic fertilizers as they enhanced crop yield to significant levels. But with the course of time, it was found that these chemical fertilizers not only decreased soil fertility tremendously, but also killed the beneficial microorganisms present in soil that promoted plant growth. Moreover, serious ailments were found in human beings who consumed the crops grown under application of these synthetic fertilizers. Pollution is another major threat. Therefore, use of these synthetic chemicals needed to be replaced by safe substances that could provide the crops with more benefits. As a result, production and application of biofertilizers was promoted (Alam and Seth, 2014).

The word "biofertilizer" is a combination of two words: "bio," which means "living," and "fertilizer," which means "substance that enhances soil fertility." So, the living substances which improve soil fertility and act as a supplement to enhance crop produce may be considered biofertilizer. There are many kinds of biofertilizers utilized commercially on the basis of their nature and function, such as N-fixing biofertilizer, P-solubilizing biofertilizer, Phosphate-mobilizing biofertilizer and plant growth promoting biofertilizer.

Processed agro-wastes can also be potent biofertilizers as they carry the potential to promote crop yield by recuperating soil health. Biofertilizers are produced from carriers with specific properties, as shown in Table 10.4.

Various kinds of agro-wastes such as fruits, vegetables, cow dung and weeds may act as biofertilizer carrier and are utilized for the production of biofertilizer through

TABLE 10.3
Applications of Biochar in Different Sectors

S. No.	Source of Biochar	Benefits	Reference
1.	Rice-husk	Enhanced soil N and P content. Elevated yield and nutritional composition of *Dolichos lablab* (Rongai legume). Decreased soil salinity	Ouna et al., 2020
2.	Bagasse	Substitute for concrete in cement industry	Zeidabadi et al., 2018
3.	Coconut shells	Used as Green Concrete, thereby reducing global warming due to cement production from synthetic concrete	Mo et al., 2016
4.	Rhizomes, Stems and Corncobs of Cassava	Promoted the colonization of beneficial microbes in soil Reduced leaching of nutrients from rhizosphere High moisture content and porosity in soil	Wijitkosum and Jiwnok, 2019
5.	Hollow fruit bunches	Improved soil fertility by enhancing fertilizer retention time in soil	Chan et al., 2008
6.	Corn Straw	Adsorption of Copper and Zinc contaminants from polluted water	Chen et al., 2011
7.	Pinewood MBC	Removal of Pb, Cd, As from wastewater	Reddy et al., 2014; Wang et al., 2015a
8.	Oak wood and bark MBC	Eradication of Cr, Pb & Cd from industrial effluents	Zhu et al., 2014
9.	Cotton stalk MBC	Act as dopant to enhance capacitance value of supercapacitor	Chen et al., 2015
10.	Date Pits	Immobilization of heavy metals in soil, making them unavailable to the plants	Li et al., 2017
11.	Oil Palm Shells	Enhanced cation exchange capacity, acid neutralization tendency and mineral content of soil	Millan et al., 2020
12.	Alfa-alfa hays	Removal of tetracycline from water	Jang and Kan, 2019
13.	Wheat straws	Adsorption of methylene blue and acid orange dyes from industrial effluents	Pirbazari et al., 2014

TABLE 10.4
Properties of Biofertilizer Carriers

S. No.	Property	Characteristic for biofertilizer carrier
1.	Physical form	Powdered or Granular.
2.	Biological activity	Enable to support microbial growth and its release into the soil.
3.	Physiological feature	Good aeration with high moisture holding and pH buffering capacity.
4.	Ecological property	Environmentally safe and non-toxic in nature.
5.	Availability	Cost-effective and easy to store and use.

different methods. The most common and cost-effective method for the conversion of agricultural residues into biofertilizers is Solid-State Fermentation (SSF) (Lim and Matu, 2015). This is the process of cultivation of microbes using solid substrates under controlled conditions of limited availability or no availability of water (Yasin et al., 2012). When different kinds of agro-wastes, such as guava peels, watermelon peels or papaya peels are subjected to SSF, a soluble fermented solution is formed that supports the growth of beneficial microbes that may be bacteria, yeast and/or other fungi (Devi and Sumathy, 2017). The process of SSF is extremely sensitive to the amount of moisture content. Low levels of moisture may decrease the solubility of nutrients present in substrates, whereas excess moisture may reduce the porosity of the solid matrix, resulting in steric hindrance to growth of microbial strains and hence less enzyme yield (Lim and Matu, 2015). However, this is a simple and cheap process and can be followed by farmers with low income generation. The biofertilizers have been found to enhance the crop yield, improve soil characteristics by maintaining accurate nutrient composition and develop resistance to pathogen attacks in plants.

Since they are eco-friendly and non-toxic, biofertilizers can be used as substitutes for harmful chemical fertilizers for improved soil and crop health. This will also support the concept of sustainable agriculture by maintaining soil richness as well as human health, as those consuming crop products obtained from the crops grown with biofertilizer application will no longer be at risk (Singh et al., 2013).

10.4.5 Bioadsorbents

The abundantly produced agricultural biomass like straw, husk, droves and crumbs have eco-friendly, stable, low-cost and green energy characteristics that provide an efficient way to use them in wastewater treatment. Anastopoulos et al. (2017) demonstrated the application of agro-wastes in adsorption of heavy metals due to their high surface area and porosity. Cellulose and lignin were the main components of agricultural waste found to sequester heavy metal ions in wastewater (Sud et al., 2008). Use of agriculture wastes as bioadsorbents seem to be a promising technology for pollution treatment (Basu et al., 2017). Adsorption of cadmium ion through the application of lentil husks has been reported by Basu et al. (2017). Similarly, adsorption of cadmium by using husk functionalized with xanthates and black rice husks was reported (Saeed et al., 2022). The most important by-product consists of agricultural straw, referring to the remaining rice, wheat, beans, corn, oil, cotton and sugarcane, which are found to be crucial as bio-adsorbent to treat heavy metal-contaminated water (Ekman et al., 2013). Cadmium ion adsorption in water through alfa-alfa straw has been investigated by Gardea-Torresdey et al. (1988). They performed the task at room temperature and reported the adsorption of 7.1 mg g^{-1} cadmium by 5 mgL^{-1} adsorbents. Similar observations were reported by using modified and unmodified rice straw (Ding et al., 2012). Thus, using agriculture waste as a bio-adsorbent seems to be a promising green technology in treating heavy metal-polluted water.

10.4.6 Biofuel

Cellulose, lignin and hemicellulose are the main components of agricultural waste. The content of these chemicals varies from crop to crop. For instance, Tumagyan et al. (2016) demonstrated 35–50 percent of cellulose, 20–30 percent lignin and 15–30 percent hemicellulose. Lignocellulose accounts for most of the biomass in plants (Anwar et al., 2014). Through biological pathways, cellulose and hemicellulose can be converted into biofuels such as bioethanol and biobutanol (Ren et al., 2016).

10.5 CONCLUSION

Agriculture and associated wastes generated through processing and industrialization of the raw materials is a concern to the world's ecosystem. Agricultural by-products, like inorganic fertilizers and other toxic pesticides, are hampering the biological activities of various organisms as well as leading to environmental pollution. The burning of agricultural wastes leads to air pollution and contributes to global temperature fluctuations. Thus, these wastes must be managed in an eco-friendly manner. Recently, the world has used some new technology to increase the economic values of these by-products in the production of various useful products, such as biochar, biofuels, bio-adsorbents, organic fertilizers and energy production. However, more research and advanced technology are required to deal with these agricultural wastes.

10.6 ACKNOWLEDGEMENT

The coordinator is grateful and acknowledges the generous support from Editors, Corresponding Authors and all the Contributors who spared their valuable time in writing the chapter for this book, and sincerely thanks the publisher for their tireless efforts in making this book. We express our special thanks to Prof. (**Dr.**) **Renu Bhardwaj** for their kind support and insights for the manuscript completion and for providing the opportunity to accelerate in the research field. We are really grateful to **Dr. Amit Kumar Tiwari** for their valuable comments and suggestions regarding the manuscript's development.

10.7 REFERENCES

Agamuthu, P. (2009, November). Challenges and opportunities in agro-waste management: An Asian perspective. In *Inaugural meeting of first regional 3R forum in Asia*, University of Malasiya (Vol. 11, No. 12, pp. 4153–4158).

Agus, F., Henson, I. E., Sahardjo, B. H., Harris, N., Van Noordwijk, M., & Killeen, T. J. (2013). Review of emission factors for assessment of CO2 emission from land use change to oil palm in Southeast Asia. Reports from the technical panels of the 2nd Greenhouse Gas Working Group of the Roundtable on Sustainable Palm Oil (RSPO).

Ahmad, M., Rajapaksha, A. U., Lim, J. E., Zhang, M., Bolan, N., Mohan, D., & Ok, Y. S. (2014). Biochar as a sorbent for contaminant management in soil and water: A review. *Chemosphere*, *99*, 19–33.

Akbarimoghaddam, H., Galavi, M., Ghanbari, A., & Panjehkeh, N. (2011). Salinity effects on seed germination and seedling growth of bread wheat cultivars. *Trakia Journal of Sciences*, *9*(1), 43–50.

Alam, S., & Seth, R. K. (2014). Comparative study on effect of chemical and bio-fertilizer on growth, development and yield production of paddy crop (Oryza sativa). *International Journal of Science and Research*, *3*(9), 411–414.

Alwaneen, W. S. (2016). Cow manure composting by microbial treatment for using as potting material: An overview. *Pakistan Journal of Biological Sciences: PJBS*, *19*(1), 1–10.

Ameeta, S., & Ronak, C. (2017). A review on the effect of organic and chemical fertilizers on plants. *International Journal for Research in Applied Science and Engineering Technology*, *5*(2), 677–680.

Anastopoulos, I., Bhatnagar, A., Hameed, B. H., Yong, S. O., & Omirou, M. (2017). A review on waste-derived adsorbents from sugar industry for pollutant removal in water and wastewater. *Journal of Molecular Liquids*, *240*, 179–188.

Antonelli, M., Baccioli, A., Francesconi, M., Psaroudakis, P., & Martorano, L. (2015). Technologies for energy recovery from waste biomasses: A study about Tuscan potentialities. *Energy Procedia*, *81*, 450–460.

Anwar, Z., Gulfraz, M., & Irshad, M. (2014). Agro-industrial lignocellulosic biomass a key to unlock the future bio-energy: A brief review. *Journal of Radiation Research and Applied Sciences*. http://dx.doi.org/10.1016/j.jrras.2014.02.003.

Atiyeh, R. M., Subler, S., Edwards, C. A., Bachman, G., Metzger, J. D., & Shuster, W. (2000). Effects of vermicomposts and composts on plant growth in horticultural container media and soil. *Pedobiologia*, *44*(5), 579–590.

Bano, A., & Fatima, M. (2009). Salt tolerance in Zea mays (L). following inoculation with rhizobium and pseudomonas. *Biology and Fertility of Soils*, *45*(4), 405–413.

Basu, M., Guha, A. K., & Ray, L. (2017). Adsorption behavior of cadmium on husk of lentil. *Process Safety and Environmental*, *106*, 11–22.

Bhat, S. A., Singh, J., & Vig, A. P. (2013). Vermiremediation of dyeing sludge from textile mill with the help of exotic earthworm Eisenia fetida Savigny. *Environmental Science and Pollution Research*, *20*(9), 5975–5982.

Bhat, S. A., Singh, S., Singh, J., Kumar, S., & Vig, A. P. (2018). Bioremediation and detoxification of industrial wastes by earthworms: Vermicompost as powerful crop nutrient in sustainable agriculture. *Bioresource Technology*, *252*, 172–179.

Bhattacharyya, R., Ghosh, B. N., Dogra, P., Mishra, P. K., Santra, P., Kumar, S., . . . Parmar, B. (2016). Soil conservation issues in India. *Sustainability*, *8*(6), 565.

Cao, H. Y., Liang, T., Tao, S., & Zhang, C. S. (2007). Simulating the temporal changes of OCP pollution in Hangzhou, China. *Chemosphere*, *67*(7), 1335–1345.

Çelik, İ., & Demirer, G. N. (2015). Biogas production from pistachio (Pistacia vera L.) processing waste. *Biocatalysis and Agricultural Biotechnology*, *4*(4), 767–772.

Chan, K. Y., Van Zwieten, L., Meszaros, I., Downie, A., & Joseph, S. (2008). Agronomic values of greenwaste biochar as a soil amendment. *Soil Research*, *45*(8), 629–634.

Chen, M. D., Wumaie, T., Li, W. L., Song, H. H., & Song, R. R. (2015). Electrochemical performance of cotton stalk based activated carbon electrodes modified by MnO2 for supercapacitor. *Materials Technology*, *30*(Supl), A2–A7.

Chen, X., Chen, G., Chen, L., Chen, Y., Lehmann, J., McBride, M. B., & Hay, A. G. (2011). Adsorption of copper and zinc by biochars produced from pyrolysis of hardwood and corn straw in aqueous solution. *Bioresource Technology*, *102*(19), 8877–8884.

Cheraghi, M., Lorestani, B., & Merrikhpour, H. (2012). Investigation of the effects of phosphate fertilizer application on the heavy metal content in agricultural soils with different cultivation patterns. *Biological Trace Element Research*, *145*(1), 87–92.

Conley, D. J., Paerl, H. W., Howarth, R. W., Boesch, D. F., Seitzinger, S. P., Havens, K. E., . . . Likens, G. E. (2009). Controlling eutrophication: Nitrogen and phosphorus. *Science, 323*(5917), 1014–1015.

Dai, Y., Sun, Q., Wang, W., Lu, L., Liu, M., Li, J., . . . Zhang, Y. (2018). Utilizations of agricultural waste as adsorbent for the removal of contaminants: A review. *Chemosphere, 211*, 235–253.

Das, O., Sarmah, A. K., & Bhattacharyya, D. (2015). A novel approach in organic waste utilization through biochar addition in wood/polypropylene composites. *Waste Management, 38*, 132–140.

Devi, V., & Sumathy, V. J. H. (2017). Production of biofertilizer from fruit waste. *European Journal of Pharmaceutical and Medical Research, 4*(9), 436–443.

Dhar, H., Kumar, S., & Kumar, R. (2017). A review on organic waste to energy systems in India. *Bioresource Technology, 245*, 1229–1237.

Dien, B. V., & Vong, V. D. (2006). Analysis of pesticide compound residues in some water sources in the province of Gia Lai and DakLak. *Vietnam Food Administrator.*

Ding, A. F., Pan, G. X., & Zhang, X. H. (2005). Contents and origin analysis of PAHs in paddy soils of Wujiang County. *Journal of Agro-Environment Science, 24*(6), 1166–1170.

Ding, Y., Jing, D. B., Gong, H. L., Zhou, L. B., & Yang, X. S. (2012). Biosorption of aquatic Cadmium(II) by unmodified rice straw. *Bioresource Technology, 114*, 20–25.

Eickhout, B., Bouwman, A. V., & Van Zeijts, H. (2006). The role of nitrogen in world food production and environmental sustainability. *Agriculture, Ecosystems & Environment, 116*(1–2), 4–14.

Ekman, A., Wallberg, O., Joelsson, E., & Borjesson, P. (2013). Possibilities for sustainablebio-refineries based on agricultural residues—a case study of potential straw-basedethanol production in Sweden. *Applied Energy, 102*, 299–308.

Environmental Benefits of Biomethane. (2012). www.biomethane.org.uk/ environmental-benefits-of-biomethane.htmlm.

Food and Agriculture Organization of the United Nations (FAO) & Organization for Economic Co-operation and Development (OECD). (2019). Background notes on sustainable, productive and resilient agro-food systems: Value chains, human capital, and the 2030 agenda. A Report to the G20 Agriculture Deputies July 2019.

Gardea-Torresdey, J. L., Gonzalez, J. H., Tiemann, K. J., Rodriguez, O., & Gamez, G. (1998). Phytofiltration of hazardous cadmium, chromium, lead and zinc ions by biomass of Medicago sativa (Alfalfa). *Journal of Hazardous Materials, 57*, 29–39.

Ghassemi, F., Jakeman, A. J., & Nix, H. A. (1995). *Salinisation of land and water resources: Human causes, extent, management and case studies.* CAB International.

Goda, K., Sreekala, M. S., Malhotra, S. K., Joseph, K., & Thomas, S. (2014). *Advances in polymer composites: Biocomposites–state of the art, new challenges, and opportunities.* Edited by Sabu Thomas. Kuruvilla Joseph, SK Malhotra.

Green, S. A. (2005). *Sittiq's handbook of pesticides and agricultural chemicals.* William Andrew. Inc., Norwich, NY.

Gupta, C., Prakash, D., Gupta, S., & Nazareno, M. A. (2019). Role of vermicomposting in agricultural waste management. In *Sustainable green technologies for environmental management* (pp. 283–295). Springer, Singapore.

Hai, H. T., & Tuyet, N. T. A. (2010). Benefits of the 3R approach for agricultural waste management (AWM) in Vietnam: Under the framework of joint project on Asia resource circulation research working paper series. Institute for Global Environmental Strategies supported by the Ministry of Environment, Japan.

Herrero, M., Thornton, P. K., Notenbaert, A. M., Wood, S., Msangi, S., Freeman, H. A., . . . Rosegrant, M. (2010). Smart investments in sustainable food production: Revisiting mixed crop-livestock systems. *Science, 327*(5967), 822–825.

Hosseinzadeh-Bandbafha, H., Nabavi-Pelesaraei, A., Khanali, M., Ghahderijani, M., & Chau, K. W. (2018). Application of data envelopment analysis approach for optimization of energy use and reduction of greenhouse gas emission in peanut production of Iran. *Journal of Cleaner Production, 172*, 1327–1335. www.oecd-ilibrary.org/docserver/dca82200-en.pdf?expires ¼1563959111&id¼id&accname¼guest&checksum¼5BD0A7A51327DB165936B4AE57A0E CE. (Accessed 15 June 2021).

Hu, G. J., Chen, S. L., Zhao, Y. G., Sun, C., Li, J., & Wang, H. (2009). Persistent toxic substances in agricultural soils of Lishui County, Jiangsu Province, China. *Bulletin of Environmental Contamination and Toxicology, 82*(1), 48–54.

Hu, Y., & Schmidhalter, U. (2004). Limitation of salt stress to plant growth. In *Plant toxicology* (pp. 205–238). CRC Press.

Inyang, M. I., Gao, B., Yao, Y., Xue, Y., Zimmerman, A., Mosa, A., . . . Cao, X. (2016). A review of biochar as a low-cost adsorbent for aqueous heavy metal removal. *Critical Reviews in Environmental Science and Technology, 46*(4), 406–433.

Jambhekar, H. A. (1992). Use of earthworm as a potential source of decompose organic wastes. In *Proceeding of the national seminar on organic farming, Mahatama Phule Krishi Vidyapeeth, Pune* (pp. 52–53).

Jang, H. M., & Kan, E. (2019). Engineered biochar from agricultural waste for removal of tetracycline in water. *Bioresource Technology, 284*, 437–447.

Joshi, R., Singh, J., & Vig, A. P. (2015). Vermicompost as an effective organic fertilizer and biocontrol agent: Effect on growth, yield and quality of plants. *Reviews in Environmental Science and Bio/Technology, 14*(1), 137–159.

Karmakar, S., Adhikary, M., Gangopadhyay, A., & Brahmachari, K. (2015). Impact of vermicomposting in agricultural waste management vis-à-vis soil health care. *Journal of Environmental Science and Natural Resources, 8*(1), 99–104.

Kristensson I, et al. (2007). Biogas 2007 på gasnätet utan propantillsats. Rapport SGC 176, 1102–7371, ISRN SGC-R-176-SE, pp. 6–18.

Lalander, C. H., Komakech, A. J., & Vinnerås, B. (2015). Vermicomposting as manure management strategy for urban small-holder animal farms–Kampala case study. *Waste Management, 39*, 96–103.

Li, H., Dong, X., da Silva, E. B., de Oliveira, L. M., Chen, Y., & Ma, L. Q. (2017). Mechanisms of metal sorption by biochars: Biochar characteristics and modifications. *Chemosphere, 178*, 466–478.

Lim, S. F., & Matu, S. U. (2015). Utilization of agro-wastes to produce biofertilizer. *International Journal of Energy and Environmental Engineering, 6*(1), 31–35.

Luka, Y., Highina, B. K., & Zubairu, A. (2018). The promising precursors for development of activated carbon: Agricultural waste materials-A review. *Journal Impact Factor, 3*, 46.

Ma, H., Li, J. B., Liu, W. W., Miao, M., Cheng, B. J., & Zhu, S. W. (2015). Novel synthesis of a versatile magnetic adsorbent derived from corncob for dye removal. *Bioresource Technology, 190*, 13–20.

Mahler, R. L., & Harder, R. W. (1984). The influence of tillage methods, cropping sequence, and N rates on the acidification of a northern Idaho soil. *Soil Science, 137*(1), 52–60.

Mai, B., Zeng, E. Y., Luo, X., Yang, Q., Zhang, G., Li, X., . . . Fu, J. (2005). Abundances, depositional fluxes, and homologue patterns of polychlorinated biphenyls in dated sediment cores from the Pearl River Delta, China. *Environmental Science & Technology, 39*(1), 49–56.

Mane, T. T., & Raskar Smita, S. (2012). Management of agriculture waste from market yard through vermicomposting. *Research Journal of Recent Sciences, 1*(ISC-2011), 289–296.

Metternicht, G. I., Zinck, J. A. (2003). Remote sensing of soil salinity: Potentials and constraints. *Remote Sensensing Environment, 85*, 1–20.

Millan, L. M. R., Vargas, F. E. S., & Nzihou, A. (2020). Characterization of steam gasification biochars from lignocellulosic agrowaste towards soil applications. *Waste and Biomass Valorization*, 1–15.

Mo, K. H., Alengaram, U. J., Jumaat, M. Z., Yap, S. P., & Lee, S. C. (2016). Green concrete partially comprised of farming waste residues: A review. *Journal of Cleaner Production*, *117*, 122–138.

Mukherjee, A., Zimmerman, A. R., & Harris, W. (2011). Surface chemistry variations among a series of laboratory-produced biochars. *Geoderma*, *163*(3–4), 247–255.

Mussoline, W., Esposito, G., Lens, P., Garuti, G., & Giordano, A. (2012). Design considerations for a farm-scale biogas plant based on pilot-scale anaerobic digesters loaded with rice straw and piggery wastewater. *Biomass and Bioenergy*, *46*, 469–478.

Nagendran, R. (2011). Agricultural waste and pollution. In *Waste* (pp. 341–355). Academic Press.

Nikkhah, A., Khojastehpour, M., Emadi, B., Taheri-Rad, A., & Khorramdel, S. (2015). Environmental impacts of peanut production system using life cycle assessment methodology. *Journal of Cleaner Production*, *92*, 84–90.

Niles, M. (2008). Sustainable soils: Reducing, mitigating, and adapting to climate change with organic agriculture. *Sustainable Development Law & Policy*, *9*, 19.

Nithya, G., & Lekeshmanaswamy, M. (2010). Production and utilization of vermicompost in agro industry. *Journal of Phytology*, *2*(2), 68–72.

Obi, F. O., Ugwuishiwu, B. O., & Nwakaire, J. N. (2016). Agricultural waste concept, generation, utilization and management. *Nigerian Journal of Technology*, *35*(4), 957–964.

Ogundipe, K. D., & Babarinde, A. (2017). Comparative study on batch equilibrium biosorption of Cd (II), Pb (II) and Zn (II) using plantain (Musa paradisiaca) flower: Kinetics, isotherm, and thermodynamics. *Chem International*, *3*(2), 135–149.

Omami, E. N. (2007). *Responses of Amaranth to salinity stress* (Doctoral dissertation, University of Pretoria).

Omidire, N. S., Shange, R., Khan, V., Bean, R., & Bean, J. (2015). Assessing the impacts of inorganic and organic fertilizer on crop performance under a microirrigation-plastic mulch regime. *Professional Agricultural Workers Journal (PAWJ)*, *3*(174–2016–2179).

Ouna, E., Njoka, J., Keya, S., & Wanjogu, R. (2020). Converting agro-waste into biochar: Improving soil fertility and productivity in ASAL ecosystems. *Kenya Policy Briefs*, *1*(1), 7–8.

Park, S., Lee, C., Lee, J., Jung, S., & Choi, K. Y. (2020). Applications of natural and synthetic melanins as biosorbents and adhesive coatings. *Biotechnology and Bioprocess Engineering*, *25*(5), 646–654.

Pathma, J., & Sakthivel, N. (2012). Microbial diversity of vermicompost bacteria that exhibit useful agricultural traits and waste management potential. *SpringerPlus*, *1*(1), 1–19.

Pirbazari, A. E., Saberikhah, E., & Kozani, S. H. (2014). Fe3O4–wheat straw: Preparation, characterization and its application for methylene blue adsorption. *Water Resources and Industry*, *7*, 23–37.

Reddy, D. H. K., & Lee, S. M. (2014). Magnetic biochar composite: Facile synthesis, characterization, and application for heavy metal removal. *Colloids and Surfaces A: Physicochemical and Engineering Aspects*, *454*, 96–103.

Ren, N. Q., Zhao, L., Chen, C., Guo, W. Q., & Cao, G. L. (2016). A review on bioconversion of lignocellulosic biomass to H2: Key challenges and new insights. *Bioresource Technology*, *215*, 92–99.

Renato, B., Ennio, C., Giulia, C., Renato, G., Lidia, L., Tommaso, O., et al. (2013). Performance of a biogas upgrading process based on alkali absorption with regeneration using air pollution control residues. *Waste Manag*, 2694–2705.

Rütting, T., Aronsson, H., & Delin, S. (2018). Efficient use of nitrogen in agriculture. *Nutrient Cycling in Agroecosystems*, *110*(1), 1–5.

Mathew, S., Soans, J. C., Rachitha, R., Shilpalekha, M. S., Gowda, S. G. S., Juvvi, P., & Chakka, A. K. (2022). Green technology approach for heavy metal adsorption by agricultural and food industry solid wastes as bio-adsorbents: A review. *Journal of Food Science and Technology*, 1–10.

Sabiiti, E. N. (2011). Utilising agricultural waste to enhance food security and conserve the environment. *African Journal of Food, Agriculture, Nutrition and Development*, *11*(6).

Salleh, M. A. M., Mahmoud, D. K., Karim, W. A. W. A., & Idris, A. (2011). Cationic and anionic dye adsorption by agricultural solid wastes: A comprehensive review. *Desalination*, *280*(1–3), 1–13.

Sanchez-Monedero, M. A., Roig, A., Paredes, C., & Bernal, M. P. (2001). Nitrogen transformation during organic waste composting by the Rutgers system and its effects on pH, EC and maturity of the composting mixtures. *Bioresource Technology*, *78*(3), 301–308.

Signor, D., & Cerri, C. E. P. (2013). Nitrous oxide emissions in agricultural soils: A review. *Pesquisa Agropecuária Tropical*, *43*(3), 322–338.

Singh, A. K., Masih, H., Nidhi, P., Kumar, Y., Peter, J. K., & Mishra, S. K. (2013). Production of Biofertilizer from agro-waste by using Thermotolerant Phosphate Solubilising Bacteria. *International Journal of Bioinformatics and Biological Science*, *1*(2), 227–244.

Smith, P., Clark, H., Dong, H., Elsiddig, E. A., Haberl, H., Harper, R., . . . Tubiello, F. (2014). Agriculture, forestry and other land use (AFOLU).

Smith, P., & Olesen, J. E. (2010). Synergies between the mitigation of, and adaptation to, climate change in agriculture. *The Journal of Agricultural Science*, *148*(5), 543–552.

Snyder, C. S., Bruulsema, T. W., Jensen, T. L., & Fixen, P. E. (2009). Review of greenhouse gas emissions from crop production systems and fertilizer management effects. *Agriculture, Ecosystems & Environment*, *133*(3–4), 247–266.

Sud, D., Mahajan, G., & Kaur, M. P. (2008). Agricultural waste material as potential adsorbent for sequestering heavy metal ions from aqueous solutions—a review. *Bioresource Technology*, 99, 6017–6027.

Suntikunaporn, M., Echaroj, S., Rimpikul, W., Kaewprapha, P., Puttarak, N., Geethalakshmi, B., & Yiarayong, P. (2014). Evaluation of agricultural wastes for biogas production. *Thammasat International Journal of Science and Technology*, *19*(1), 1–8.

Tayefeh, M., Sadeghi, S. M., Noorhosseini, S. A., Bacenetti, J., & Damalas, C. A. (2018). Environmental impact of rice production based on nitrogen fertilizer use. *Environmental Science and Pollution Research*, *25*(16), 15885–15895.

Thamsiriroj, T., & Murphy, J. (2011). Modelling mono-digestion of grass silage in a 2-stage CSTR anaerobic digester, bio. *Technology*, *102*(2). 948–959.

Thao, L. T. H. (2003). Nitrogen and phosphorus in the environment. *Journal of Survey Research*, *15*(3), 56–62.

Tilman, D., Fargione, J., Wolff, B., D'antonio, C., Dobson, A., Howarth, R., . . . Swackhamer, D. (2001). Forecasting agriculturally driven global environmental change. *Science*, *292*(5515), 281–284.

Tumagyan, A., Sargsyan, H., Madoyan, R., 2016. Device and method for processing crop unlock the future bio-energy: A brief review. *Journal of Radiation Research and Applied Sciences*, *7*, 163–173.

Vaibhav, V., Vijayalakshmi, U., & Roopan, S. M. (2015). Agricultural waste as a source for the production of silica nanoparticles. *Spectrochimica Acta Part A: Molecular and Biomolecular Spectroscopy*, *139*, 515–520.

Vermeulen, S. J., Campbell, B. M., & Ingram, J. S. (2012). Climate change and food systems. *Annual Review of Environment and Resources*, *37*, 195–222.

Wang, B., Dong, F., Chen, M., Zhu, J., Tan, J., Fu, X., . . . Chen, S. (2016). Advances in recycling and utilization of agricultural wastes in China: Based on environmental risk, crucial pathways, influencing factors, policy mechanism. *Procedia Environmental Sciences*, *31*, 12–17.

Wang, H., Gao, B., Wang, S., Fang, J., Xue, Y., & Yang, K. (2015a). Removal of Pb (II), Cu (II), and Cd (II) from aqueous solutions by biochar derived from KMnO4 treated hickory wood. *Bioresource Technology*, *197*, 356–362.

Wang, S., Gao, B., Li, Y., Mosa, A., Zimmerman, A. R., Ma, L. Q., . . . Migliaccio, K. W. (2015b). Manganese oxide-modified biochars: Preparation, characterization, and sorption of arsenate and lead. *Bioresource Technology*, *181*, 13–17.

Wijitkosum, S., & Jiwnok, P. (2019). Elemental composition of biochar obtained from agricultural waste for soil amendment and carbon sequestration. *Applied Sciences*, *9*(19), 3980.

Wu, C. H., Chang, S. H., & Lin, C. W. (2015). Improvement of oxygen release from calcium peroxide-polyvinyl alcohol beads by adding low-cost bamboo biochar and its application in bioremediation. *CLEAN–Soil, Air, Water*, *43*(2), 287–295.

Xue, L., Gao, B., Wan, Y., Fang, J., Wang, S., Li, Y., . . . Yang, L. (2016). High efficiency and selectivity of MgFe-LDH modified wheat-straw biochar in the removal of nitrate from aqueous solutions. *Journal of the Taiwan Institute of Chemical Engineers*, *63*, 312–317.

Yasin, M., Ahmad, K., Mussarat, W., & Tanveer, A. (2012). Bio-fertilizers, substitution of synthetic fertilizers in cereals for leveraging agriculture. *Crop and Environment*, *3*(1–2), 62–66.

Ye, J., Li, D., Sun, Y., Wang, G., Yuan, Z., Zhen, F., & Wang, Y. (2013). Improved biogas production from rice straw by co-digestion with kitchen waste and pig manure. *Waste Management*, *33*(12), 1–6.

Yensen, N. P. (2008). Halophyte uses for the twenty-first century. In *Ecophysiology of high salinity tolerant plants* (pp. 367–396). Springer, Dordrecht.

Yevich, R., & Logan, J. A. (2003). An assessment of biofuel use and burning of agricultural waste in the developing world. *Global Biogeochemical Cycles*, *17*(4).

Zeidabadi, Z. A., Bakhtiari, S., Abbaslou, H., & Ghanizadeh, A. R. (2018). Synthesis, characterization and evaluation of biochar from agricultural waste biomass for use in building materials. *Construction and Building Materials*, *181*, 301–308.

Zhang, L., Jiang, J., Holm, N., & Chen, F. (2014). Mini-chunk biochar supercapacitors. *Journal of Applied Electrochemistry*, *44*(10), 1145–1151.

Zhu, J., Gu, H., Guo, J., Chen, M., Wei, H., Luo, Z., . . . Wei, S. (2014). Mesoporous magnetic carbon nanocomposite fabrics for highly efficient Cr (VI) removal. *Journal of Materials Chemistry A*, *2*(7), 2256–2265.

11 Kitchen and Agri Waste as Renewable, Clean and Alternative Bioenergy Resource

Neerja Sharma, Nitika Kapoor, Sukhmeen Kaur Kohli, Pooja Sharma, Jaspreet Kour, Kamini Devi, Ashutosh Sharma, Rupinder Kaur, Amrit Pal Singh and Renu Bhardwaj

CONTENTS

11.1 INTRODUCTION

Expansion of the agricultural sector due to increasing global population and food demand has led to the generation of huge amount of vegetable and agri waste, whose proper disposal is necessary to protect the ecological environment from degradation and pollution. In this century, one of the most noteworthy challenges is to meet the increasing energy demands. It is reported that about 50 million tonnes of vegetable

DOI: 10.1201/9781003245773-11

waste is produced in India, which is 30% of its total production (Verma et al. 2011). At landfill sites, the wastes produce greenhouse gases (GHGs), i.e., methane and carbon dioxide, which cause global warming and climate change. Moreover, the rapidly rising population is putting continuous pressure on natural resources to meet its demands, deteriorating the resources day by day. So all these anthropogenic activities have driven the development and focus on bioenergy or sustainable fuel production. Bioethanol, biohydrogen and biodiesel are some biofuels which are renewable energy sources, and by the use of these alternative fuels the emission of GHGs is reduced, which further decreases environmental pollution (Fivga et al. 2019).

Lignocellulosic agri waste like corncobs, wheat straws, rice husks, sugarcane bagasse, cassava peels and wood chips are good sources of renewable energy because of their low cost and abundance (Sivamani et al. 2018). Economically and technologically important chemical butyric acid used in textile, cosmetics, food and pharmaceutical industries is produced from fossil fuels, but due to environmental concerns it is generated from lignocellulosic waste (wheat straw) through microbial fermentation (Câmara-Salim et al. 2021).

Presently, bioethanol is treated as the most eco-friendly biofuel for transportation, and it is produced by corn and sugarcane crops on a large scale for global supply (Zabed et al. 2016). Fruit peels are rich in sugars and vegetable wastes are rich in cellulose, hemicellulose and lignin, considered significant feedstock for the production of bioethanol. Vegetable peels from kitchen and food processing industries or market-rejected vegetables that are spotted and rotting cause heavy loss and are left in the open for composting/decomposing, creating environmental issues (Bhuvaneswari and Sivakumar 2019). So these huge amounts of wastes, which are rich in bioenergy molecules, are converted into bioethanol through fermentation, followed by acid and enzyme pre-treatments (Singh et al. 2012). On the other side, an important fuel like hydrogen is directly used in combustion engines to generate electricity and is also used in automobile fuels (Alves et al. 2013). A pathway for the production of hydrogen is not eco-friendly; therefore, its sustainable form, i.e., biohydrogen, is produced from cellulosic waste through the process of dark fermentation.

Waste management by transformation into biogas through the process of anaerobic digestion is the most eco-friendly method, as it does not require more energy inputs or difficult processing (Xu et al. 2018). The waste at landfill sites or thrown in the open for burning transforms most of the hydrogen and methane into useful bioenergy resources using digesters through anaerobic digestion (Prabakara et al. 2018).

11.2 BIOCHEMICAL TRANSFORMATION OF KITCHEN AND AGRI WASTE INTO BIOFUELS

Basically, biowaste comprises three main constituents: cellulose, hemicellulose and lignin. Lignin provides more energy content in comparison to cellulosic biomass for biofuel synthesis. Biochemical transformation includes the utilization of the fungi or particular bacteria which convert biomass into bioenergy. The main processes for the production of biofuels are anaerobic digestion and alcoholic fermentation.

11.2.1 Anaerobic Digestion of Biological Waste

The primary solution for management of biological waste is production of methane employing anaerobic digestion (Paritosh et al. 2017). The benefits of using the anaerobic digestion approach is that it is cost-effective and has minimal residual waste generation, and results in usage of biological waste for production of renewable sources of energy (Nasir et al. 2012; Morita and Sasaki 2012). Biological digestion by employing anaerobic digestion is generally categorized into three stages: i) enzymatic hydrolysis, ii) acidogenesis (acid formation), and iii) methane production (Figure 11.1). In the enzymatic stage, the larger-sized polymers molecules which are not transferred through the membranes are hydrolyzed by hydrolases enzymes, produced by facultative or obligatory bacteria which are resistant to an anaerobic environment. The process of hydrolysis breaks complex polymers into simpler oligomers, and monomer units such as oligosaccharides are converted to simpler monosaccharide. Breakdowns of starch molecules lead to formation of glucose and peptides; amino acids are formed by hydrolysis of proteins, and lipids breakdown into glycerol and fatty acids (Paritosh et al. 2017).

After the hydrolysis through enzymatic cleavage, the generated products are taken by acidogenic microorganisms via membrane diffusion (Lier et al. 2008). These microorganisms undergo fermentation to form volatile fatty acids; for example, acetate is obutyrate, butyrate, propionate and valerate accompanied by generation of hydrogen, ammonia and carbon dioxide (Bryant 1979). During this stage, the facultative bacteria exploit the available oxygen and carbon and creates anaerobic condition for initiation of methanogenesis. Products like carbon, oxygen and acetates

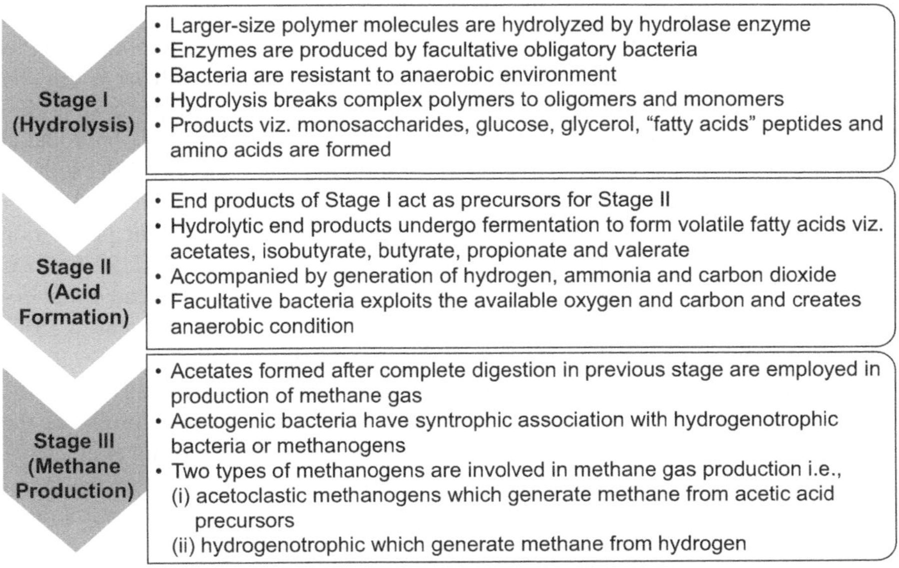

FIGURE 11.1 Various stages involved in anaerobic digestion of biological waste.

act as direct constitutes required for the production of methane. The acetates formed after complete digestion are employed in the production of methane gas in the next stage.

The third and last phase of anaerobic biological digestion of waste is methanogenesis, in which methanogenic microorganisms produce methane by using available intermediates (Ferry 2010). Methanogenic microorganisms belong to the obligate anaerobic archaea group, which is very sensitive to oxygen. Environmental factors such as higher pH and low redox potential are required for methanogenesis (Wolfe 2011). Furthermore, two types of methanogens are involved in methane gas production: i) acetoclastic methanogens, which generate methane from acetic acid precursors, and ii) hydrogenotrophics, which generate methane from hydrogen. Figure 11.1 shows the stages of anaerobic digestion by hydrolysis, acid formation and methane production.

11.2.2 Factors Affecting Anaerobic Digestion

Optimal environmental cues are prerequisites for anaerobes to work with elevated metabolic activity. Specifically, the methanogens are comparatively sensitive to alteration in endurance conditions. Hence it is pivotal to regulate controlled environmental conditions, so that the process of methanation occurs efficiently (Paritosh et al. 2017). One of the imperative factors that affect the methanogenesis is seeding, which may increase the steadiness of the digestion process. Generally, digested sludge is used for seeding, which is acquired from sewage plants, cow dung slurry or landfill sites. One of the earlier reports suggests that use of goat rumen fluid as inoculums was quite efficient in generation of biogas (Paritosh et al. 2017).

The microorganisms involved in methanogenesis are metabolically most active at 35ºC (mesophilic bacteria) and 55ºC (thermophilic bacteria), respectively. One of the most significant concerns related to activity of anaerobic microorganisms is environmental temperature, as the conversion of acetic acid to methane is directly dependent on temperature (Van Haandel and Lettinga 1994). Bouallagui et al. (2004) affirmed that at 4% total solid, the methane production was 58%, 62%, 65% and at temperatures 20°C, 55°C, 35°C respectively. Similarly, at 8% total solid, methane production was observed to be 59% and 57% at 55°C and 35°C respectively. Another report given by Kim et al. (2006) revealed similar results where methane production was found to be 58.9%, 67.4%, 66.25% and 65.6% at temperatures 55°C, 50°C, 45°C and 40°C respectively. Experimental results by Gou et al. (2014) indicated that co-digestion of waste-activated charcoal showed maximum production of gas at 55°C, which is 1.3 times greater than production at 45°C and 35°C.

Another imperative factor that influences the rate of methanogenesis is C/N ratio. Simultaneous digestion of chicken, dairy and wheat straw manure produces the highest methane when the C/N ratio is 27.2 and the pH is stable (Wang et al. 2012). A study conducted by Karthikeyan and Visvanathan (2012) revealed that the maximum rate of anaerobic digestion was found at C/N ratio of 27. Similar to C/N ratio, pH is another major factor that influences the microbial activity in the bioreactors and furthermore influences its efficacy. Wang et al. (2012) suggested the optimal pH range for maximum microbial activity ranges from 6.3–7.8. At first, due to a large extent of carbon dioxide production, the pH is low, i.e., 6.2, but after 10–12 days the pH starts increasing and gets stable around 7 and 8. Another observation made by Lee et al. (2009) affirmed that an ideal range of pH is 6.5–8.2 for methanogenesis to occur in food waste leachate. One of the main rationales for alteration in pH is changes in Volatile Fatty Acids (VFAs) and bicarbonate levels that further influence the alkalinity of the system. The synthesis of VFAs and their accumulation showed suppressive and detrimental impact on the digestion process, which consequently might result in lowered gas production (Siegert and Banks 2005; Labatut et al. 2011; Vijayaraghavan et al. 2012). The activity of the acid-sensitive enzymes is lost in response to a decline in pH and enhancement in levels of VFAs, hence the biological activity of methanogenic bacteria is also lowered (Bouallagui et al. 2005).

Organic Loading Rate (ORL) is the amount of fodder treated per unit volume of reactor per day. In a study conducted by Nagao et al. (2012), it was observed that the volumetric biogas production rate was elevated by 6.6, 5.8, 4.2 and 2.7 L/L/d as OLR was augmented to 9.2, 7.4, 5.5 and 3.7 kg-VSm^3/d respectively. Further elevation of OLR to 12.9 kg-VSm^3/d resulted in reduction in gas production rate. Experimental evidence was provided by Tampio et al. (2014), where the study was conducted for comparison of digestion of autoclaved and unprocessed food remains. The results revealed that the highest methane production was observed at OLR of 3kgVS/m^{3}d in unprocessed food remains and at 4 kg VS/m3d for autoclaved food waste respectively. Agyeman and Tao (2014) revealed that digestion of food waste anaerobically at varied ORLs led to an elevation in biogas production by 101–116% when ORL was augmented from 1 to 2 g VS/L/d, and further elevation from 2 to 3 g VS/L/d led to only 25–38% increase in gas production. Moreover, co-digestion of food employing activated sludge under the influence of mesophilic and thermophilic anaerobic systems at varied OLR revealed that the thermophilic system showed comparatively more load-bearing capacity at high OLR of 7 g VS/L/d, while the mesophilic system showed the best load-bearing capacity at <5gVS/L/d OLRs (Carlsson et al. 2008).

The biodigestion of proteins and other nutrient-rich substrates results in generation of ammonia and exits the system in the form of ammonium ions and amines (Whelan et al. 2010; Yenigun and Demirel 2013). Augmented ammonia production has both favourable and injurious impact on microbial growth and activity (Whelan et al. 2010; Walkers et al. 2011). Ammonia has been affirmed to play a pivotal role in maintaining C/N ratio, which is essential for regulating the efficacy of the digestion process (Wang et al. 2012). Microbes require different nutrients to fulfil their energy requisite (carbon) and for ultra-structure maintenance (nitrogen). In addition to this, few nutrients are required in small quantities, such as Ca, Mg, K, Na and Cl (Raposo et al. 2011). In enzyme biosynthesis and regulation of enzyme activity, various metal ions viz. Cr, Ni and Zn are necessary for methanogenesis (Schattauer et al. 2011; Facchin et al. 2013).

11.2.3 Bioethanol Production via Alcoholic Fermentation

A number of transportation fuels can be produced from kitchen or agriculture wastes by employing various methods. The biofuels produced can reduce the level of greenhouse gases in the environment, thereby reducing pollution levels. Bioethanol produced from corn and sugarcane and biodiesel produced from soy and rapeseed dominates the current market of biofuels. They are the most widely available source of clean and renewable transportation energy (Ghosh 2016).

The United States, Brazil, Europe, China, Argentina and India are the major bioethanol-producing countries. But the United States contributes about 58% of the total world ethanol production (RFA 2017). More oxygen content and higher octane numbers in bioethanol helps in early ignition of engines and total combustion of fuel, which causes less release of greenhouse gases and particulate matter (80%) (Krylova et al. 2008). So, amalgamation of bioethanol with petrol is a better and more ecologically sustainable option (Yao et al. 2009).

The major feedstocks used for production of bioethanol are corn, wheat, cassava, barley, potatoes, sugarcane, sorghum, sugar beet, whey, barley, bagasse, lignocellulosic and cellulosic biomass, switchgrass, giant miscanthus, etc. (Ghosh 2016). The main three crop residues which are recognized for having higher bioethanol production potential are wheat straw, rice straw and corn stover (Saini et al. 2015). The bioethanol generation capacity of the feedstock is dependent upon the composition of cellulose and hemicellulose of the concerned crop. Lignin, lignocellulosic content, greater ash and silica content in crop residues deterred the hydrolysis and the conversion of biological biomass into bioethanol (Binod et al. 2010). Hence, bioresidues are subjected to various pre-treatments including alkali treatment, hydrolysis and fermentation using cellulose enzyme, which enhances the yield of bioethanol and decreases the production of hazardous compounds like acetic acid, furfural, etc. (Yuan et al. 2018).

Various other agro-industrial processed wastes, for example, de-oiled seed cakes and fruit wastes, have also been explored for bioethanol production. Tayeh et al. (2014) reported that dry olive de-oiled seed cake can yield up to 3% of bioethanol. Other non-oiled seed cakes such as sesame, sunflower, canola, soy and peanut are used to produce bioethanol. These non-oiled seed cakes are good source of protein and nutrients, most importantly fibres. The cellulose, hemicellulose, and lignin containing fibres can be transformed into bioethanol by using appropriate pretreatments, agitation, and hydrolysis processes (Balan et al. 2009). Several fruit wastes like pineapple peel (8.34%), banana peel (7.45%), a mixture of apple and banana (38%) and palm oil empty fruit bunch (14.5%) have been assessed for the generation of bioethanol (Gupta and Verma 2015). Fruit and vegetable wastes, being rich in polysaccharides and available in abundance, can be subjected to solid state fermentation for the production of bioethanol (Laufenberg et al. 2003; Tang et al. 2008; Singh et al. 2012).

Several steps are involved in production of biofuels from agricultural/kitchen wastes; these include pretreatment (physicochemical and biological); detoxification, hydrolysis, alcoholic fermentation, anaerobic digestion and transesterification (Pattanaik et al. 2019). Pretreatment techniques are helpful in processing of lignocellulosic biomass (lignin, cellulose and hemicellulose), which hinders the accessibility of microorganisms/enzymes for degradation of agri/kitchen wastes, thereby enhancing bioethanol production potential (Sarkar et al. 2012).

Sugars released during the hydrolysis/saccharification process of residues can be converted into bioethanol via alcoholic fermentation by using various microorganisms, including bacteria and yeast. Hydrolysis is carried out by using acid/alkali and enzymes. Acid treatment is cheap, but it may alter the sugar into undesirable forms. On the other hand, enzymatic treatments are more effective, but they are expensive and slow (Günerken et al. 2015; Lee et al. 2019).

Bacteria, yeasts and fungi are often utilized for raising ethanol production by the process of fermentation. Out of the various microorganisms studied for ethanol production from wheat straw, *S. cerevisiae* is found to be the most effective strain. Heat-resistant bacteria such as *Thermoanaerobacter* sp. and *Clostridium* sp. are found to be appropriate for the fermentation of ethanol anaerobically. But these thermophilic bacteria are very much less tolerant to ethanol, i.e., less than 30 g/L (Talebnia et al. 2010; Pattanaik et al. 2019). From 180.73 million tons of sugarcane bagasse biomass, about 51.3 GL bioethanol is produced (Saini et al. 2015), whereas H_2SO_4 pretreated biomass (1.25%) enhanced its potential to 59.1 GL (Cardona et al. 2010).

According to Singh et al. (2012), *Zymomonas mobilis, Thermocellum, Saccharomyces cerevisiae* and recombinant *Escherichia coli* are prospective microorganisms for ethanol production by means of fermentation. Out of these microorganisms, *S. cerevisiae*, because of its higher efficiency, is best suited for generation of ethanol from monosaccharide with higher resistance to ethanol levels and other toxic complexes generated during the delignification process (Lin and Tanaka 2006). *S. cerevisiae* can only ferment hexose sugar, whereas other microorganisms like *Candida shehatae, Pachysolan tannophilus, Kluyveromyces marixianus* and *Pichia stipitis* can ferment both pentose and hexose sugars simultaneously (Pattanaik et al. 2019). Major challenges faced during ethanol production from agri/kitchen wastes include a reduction in higher lignin content of plant biomass, development of standard techniques for immediate consumption of pentoses as well as hexoses from same container, and development of ethanol-tolerant strains of microorganisms for enhanced yields of ethanol (Chantanta et al. 2008; Chen and Dixon 2007; Jørgensen et al. 2007). Studies are also in progress for development of techniques where saccharification and fermentation processes can be completed in the same container so that further growth of microorganisms by end-products can be minimized (Singh et al. 2012).

Batch method and semi-continuous or continuous methods of fermentation can be used to generate bioalcohols. Fermentation methods can be combined with saccharification to improve yield and instantaneously drop the production cost of ethanol. Saccharification can be integrated with fermentation by various methods, including simultaneous saccharification and fermentation (SSF), sequential hydrolysis and fermentation (SHF), and consolidated bioprocessing (CBP), simultaneous saccharification and co-fermentation (SSCF) and direct microbial conversion (DMC). Due to final product inhibition and contamination, the yield of bioalcohols is lower in the SHF method. Saccharification and fermentation of hexoses are carried out in the same vessel. This process is cost-effective as it produces a higher yield and can eliminate the phenomenon of product inhibition when compared to the SHF method. The SSCF method includes co-fermentation of hexoses and pentoses sugars. So this technique requires co-culture of microorganisms like *C. shehatae* and *S. cerevisiae*, which show growth at the same optimum temperature and pH (das Neves et al. 2007; Pattanaik et al. 2019). The CBP/DMC method of fermentation is cost-effective as it involves hydrolysis followed by fermentation of sugars by microorganisms. But the yield of this method is very low, and it takes a longer time. Microorganisms involved in CBP/DMC methods are bacteria *Clostridium thermocellum* and some fungi, including *Fusarium oxysporum, Neurospora crassa* and *Paecilomyces sp.* (Balat 2007). The process of formation of bioethanol from kitchen/agri waste is schematically represented in Figure 11.2.

11.2.4 Manufacture of Biohydrogen

Biohydrogen is an environmentally friendly and high-energy biofuel with zero carbon dioxide emission (Gomez et al. 2011). Biohydrogen can be produced via light-dependent fermentative methods or dark fermentative methods. Out of these two methods, the dark fermentative method is cost-effective and takes place under anaerobic conditions (Singh et al. 2012). Breakdown of carbohydrates releases molecular hydrogen (H_2) by facultative/obligate microorganisms in the dark fermentation process. Organic substrates are oxidized by microbes, and released electrons then react with hydrogen ions to form molecular hydrogen (H_2) under anaerobic conditions (Das and Veziroglu 2008, Lee et al. 2019).

Substrate required for biohydrogen production should be carbohydrate-rich and nitrogen-deficient. Vegetable wastes are considered to be the most potential substrate for biohydrogen production because of their high organic content and high degradation potential (Kapdan and Kargi 2006; Show et al. 2011). Several

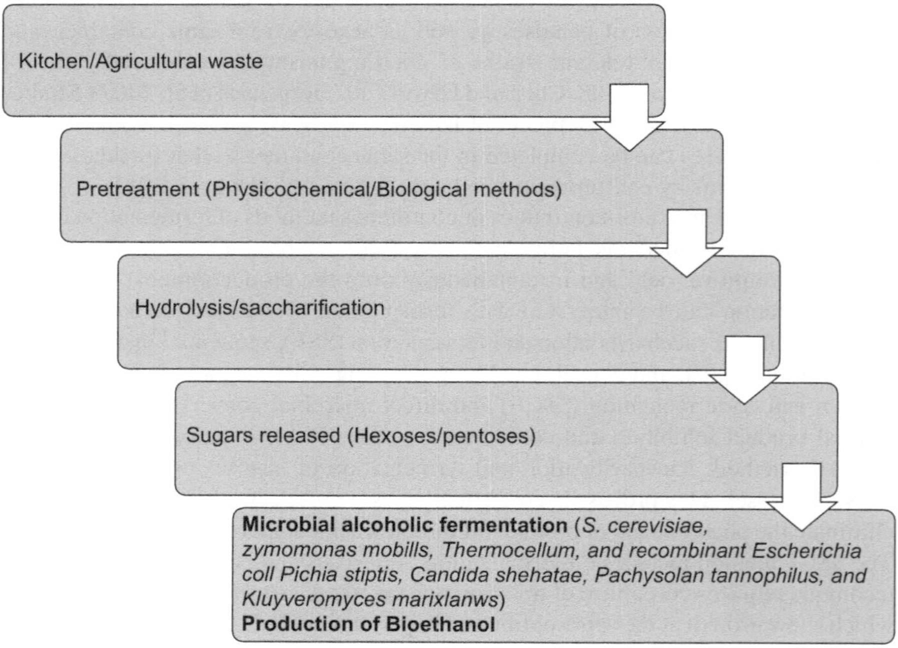

FIGURE 11.2 Production of bioethanol from kitchen/agri waste.

crop leftovers such as wheat bran, paddy straws and barley straws which are rich in cellulose and hemicellusose also produce biohydrogen. These biomasses are needed to undergo physical/chemical pre-treatments to enhance the yield of biohydrogen. The pre-treatment enhances the hydrolytic activity and hydrogen production efficiency of microorganisms (Pattanaik et al. 2019). It has been reported that pre-treated corn stalk yields higher amounts of biohydrogen compared to non-treated ones. NaOH (0.5%) treatment enhanced the biohydrogen production by 19 times, whereas HCl (0.2%) treatment along with thermal treatment enhanced the biohydrogen generation by approximately 50 times compared to non-treated corn stalks. Steam and microwave treatments (thermal pre-treatment) are also found to be effective for enhancing the production of biohydrogen (Ivanova et al. 2009). Organic manure-derived biohydrogen is relatively less efficient than that of crop leftovers because the latter is nitrogen-rich and produces ammonia during biohydrogen production (Cavinato et al. 2012). Methanogenic activity is evaded during biohydrogen production by physicochemical pre-treatment of manure, controlling thermophilic conditions and codigestion of manure with carbon-rich sources (Wu et al. 2009).

Sugar beet and potato processing effluent, starch industrial unit leftovers, cheese whey, winery and microbrewery leftovers are good sources of carbohydrates which provide suitable substrates for the generation of biohydrogen (Guo et al. 2010; Tenca et al. 2011). Sugar-rich substrates disturb the optimum pH (5.0 to 6.0), which further affects the rate of production of biohydrogen by microbes. To overcome this problem, livestock manure can be a co-substrate with sugar-rich substrate, whose alkaline nature helps to maintain weak acidic pH that further sustains the progression of microbes for the production of biohydrogen (Tenca et al. 2011). Major microbes contributing to biohydrogen production from wastes include green algae, cyanobacteria and purple non-sulfur bacteria (Show et al. 2011). Vegetable wastes inoculated with sewage water produce 55% more hydrogen compared to non-sewage-supplemented waste (Mohanakrishna et al. 2010). Similarly, Tenca et al. (2011) co-fermented vegetable leftover with swine manure (an alkali-rich material) to achieve maximum biohydrogen production as it maintains a slightly acidic pH. Carrot, cabbage, lettuce and sugar beet wastes are also employed to produce biohydrogen. (Logan et al. 2002). To improve the rate of production and yield of biohydrogen, different types of bioreactors have been studied. Out of these, anaerobic dark fermenter bioreactors are found to be most effective (Jaitalee et al. 2010; Shi et al. 2009). So production of biohydrogen from vegetable/agri wastes leads us to the path of sustainable and clean energy generation.

11.3 THERMOCHEMICAL TRANSFORMATION PROCESSES (PYROLYSIS, GASIFICATION AND LIQUEFACTION)

Kitchen waste is a mixed biomass that consists of inedible parts of cereals, vegetables and fruits. Recycling and compressing kitchen waste was one of the methods to induce sustainable agriculture. There were thermochemical methods which include gasification, liquefaction and pyrolysis, to convert kitchen waste into usable product (fuel, electric energy, etc.). Gasification was involved with low-water content woody waste. For the thermochemical transformation processes, the extraction of fluid and pyrolysis may convert food waste into more beneficial products. The following thermochemical processes were discussed step by step:

a) **Pyrolysis** is one of the precise technologies used in the production of char, syngas and bio-oil. Temperature may vary from feedstock to feedstock, but low-temperature derived char and other by-products were qualitatively more enriched than high-temperature pyrolysis. Some of the reports stated the pyrolyzed char derived from kitchen waste enhanced the sustainability and yield of the crop and also recycled back the components to the environment (Peng et al. 2021). The pyrolysis process decreased the volume of feedstock (easy storage), handling and cost of transportation. The char thus produced is conserved for longer periods because of less moisture, and this results in minor damage to the char. Uneven heating, high energy efficiency and absence of fumes will facilitate lower ash content during pyrolysis. Some studies showed that the temperature affected the biochar composition and formation of product (Moral and Şensöz 2015; Chen et al. 2018; Mishra and Mohanty 2018). At 500°C, there was a formation of gaseous and liquid products due to a reduction in biochar yield. It was reported that the production of char increased due to the burning of biofuel. At higher temperatures, the biochar yield increased and the by-product formation decreased, and this inclined towards a constant rate (Chen at al. 2015).

b) **Gasification (Air or Steam)**

It was a valuable technique that converted the kitchen waste into important products (bio syngas) using various gases like air, CO_2 and steam. It had an elevated rate of conversion and the remaining materials after gasification were oils, tars and chars Morrin et al. 2012; Xiong et al. 2020). The configuration of these by-products depended on the retention rate, temperature, precursor, atmospheric condition and catalyst used (Jouhara et al. 2018). Gasification involved four steps (Saghir et al. 2018): a) feedstock drying; b) pyrolysis of feedstock; c) gasification or oxidation; d) combustion or reduction of feedstock. In this process, first there was incomplete oxidation of the precursor that extracted the energy from the feedstock and converted it into gaseous form (Elkhalifa et al. 2019). The mentioned reactions were categorized into homogenous and heterogeneous (Ram and Mondal 2019). Both the reactions take place as:

Homogenous reaction: $C + CO_2 \leftrightharpoons 2\ CO$; $C + (0.5)\ O_2 \rightarrow CO$; $C + 2\ H_2 \leftrightharpoons CH_4$; $C + H_2O \leftrightharpoons CO + H_2$

Heterogeneous reaction: $C + CO_2 \leftrightharpoons 2\ CO$; $C + (0.5)\ O_2 \rightarrow CO$; $C + 2\ H_2 \leftrightharpoons CH_4$; $C + H_2O \leftrightharpoons CO + H_2$

Catalysts used in gasification for alleviating the pollution of tar and also reducing the bio-oil oxygen content resulted in regulation of the mixture of gaseous by-products (Kim and Lee 2020). In recent studies, Ni/A12O3 was used widely due to its higher surface area and thermal stability (Andrew et al. 2016; Su et al. 2020; Farooq et al. 2021).

TABLE 11.1
Thermochemical Processes of Kitchen Waste

S. No.	Thermochemical processes	Temperature	Reference
1.	Gasification (air or steam)	700 °C	Valizadeh et al. 2021
2.	Steam gasification	230 °C, 260 °C, and 290 °C	Singh and Yadav 2021
3.	Steam gasification	850 °C	Zhang et al. 2021
4.	Gasification	400–600 °C	Shenbagaraj et al. 2021
5.	Pyrolysis	400–600 °C	Kumar et al. 2021
6.	Pyrolysis	200–300 °C	Peng et al. 2021
7.	Hydrothermal liquefaction	240 °C, 265 °C or 295 °C	Bayat et al. 2021
8.	Hydrothermal liquefaction	220 °C, 240 °C, 260 °C	Sharma et al. 2021
9.	Hydrothermal liquefaction	300 °C	Aierzhati et al. 2021
10.	Pyrolysis	400°C	Jia et al. 2021
11.	Steam gasification	923 K–1123 K	Xiong et al. 2020
12.	Hydrothermal liquefaction	<350 °C	Hongthong et al. 2020
13.	Hydrothermal liquefaction	-	Bayat et al. 2019
14.	Hydrolysis	55 °C	Li et al. 2017
15.	Hydrothermal liquefaction	250–300 °C	Déniel et al. 2016

c) Hydrothermal Liquefaction

Hydrothermal liquefaction (HTL) was a type of thermochemical processes that converted any type of biomass into crude oil and other co-products (char, gases, etc.). Char can be used as a crop yield enhancer, pollutant adsorbent and in wastewater management. It was also dependent upon temperature, retention time and biomass configuration (Qambrani et al. 2017; Gollakota et al. 2018; Lu et al. 2018; Madsen and Glasius 2019). The hostel/mess wastes were used as solid fuel for making hydrochar, which acts as pollutant remover. Some researchers reported that at 300°C, maximum crude oil was obtained via HTL of kitchen waste (Kostyukevich et al. 2018; Motavaf and Savage 2021). Some reports showed that the supplementation of catalysts boosted the oil production from 25–43% by weight (Posmanik et al. 2018; Cheng et al. 2020c). The oil yield depends upon the temperature and time. There were wide ranges of temperature to examine the yield and char structure (Chen et al. 2020). Sharma et al. (2020) has reported that the kitchen waste is carbonized during HTL via a) hydrolysis; b) dehydration; c) decarboxylation; d) aromatization; e) condensation; and f) polymerization. Hydrochar obtained from kitchen waste had increased hydrophobicity (Saqib et al. 2019). These methods of thermochemical conversion of kitchen/agri waste, i.e. pyrolysis, gasification and liquefaction produces various by-products which are shown in Figure 11.3.

In solid state fermentation, microorganisms grow on solid or insoluble substrate in the absence of free water (Bhargav et al. 2008). Lignocellulosic material such as wood chips, sawdust, straws, legume seeds and wheat bran are mainly used as substrate for fermentation. The field biomass or agri waste is used for the generation of important biofuels and other valuable products.

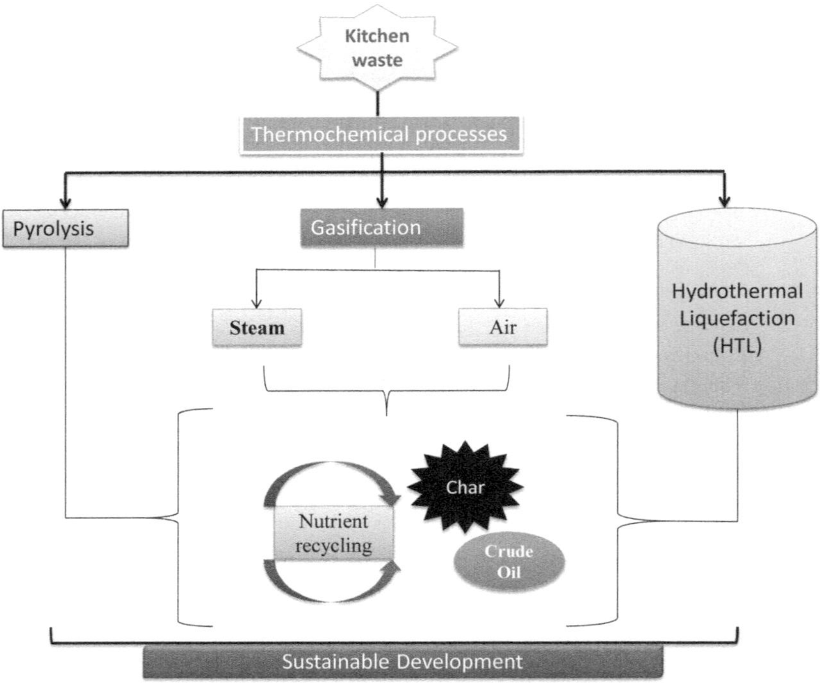

FIGURE 11.3 Thermochemical transformation processes.

11.4 UTILIZATION OF AGRI WASTES BY MEANS OF SOLID-STATE FERMENTATION

Kitchen and agricultural remains are good sources of bioactive components that can be recycled for the production of important products like biofuel, antibiotics, enzymes and mushrooms at an industrial scale. The use of kitchen and agri waste can be beneficial for the reduction of production costs and environmental contamination. These commercially important products can be obtained by a biotechnology solid-state fermentation (SSF), a process in which fermentation is carried out in the absence of free-flowing water and uses solid materials as substrates for further bioprocessing.

11.4.1 Biofuel Generation

The agri wastes are rich in polysaccharides like cellulose, hemicellulose and lignin, which support the growth of microorganisms and can be used as solid-state fermentation for the production of valuable products. Biofuels are important because they are used as substitutes for fossil fuels (Sadh et al. 2018). Production of biofuels has gained popularity worldwide. Production of ethanol from sugarcane and corn, biodiesel from rapeseed, soy and oil palm dominate the current market for biofuels. Second-generation biofuels produced from agri industrial waste is gaining attention. Table 11.2. shows various studies that have been conducted on the production of biofuels with the use of agri waste residues.

TABLE 11.2
Studies on Production of Biofuels with the Use of Agri Waste

S. No.	Microorganism	Substrate	Biofuel	References
1.	*Streptomyces fulvissimus* CKS7	Lignocellulosic materials including yellow gentian waste, horsetail waste, corn stover, cotton material	Bioethanol	Mihajlovski et al. 2020
2.	*Mogibacteriaceae and Ruminococcaceae*	Sugar beet pulp, fruit and vegetable waste and corn silage	Biohydrogen	Cieciura-Włoch et al. 2020
3.	*Fusarium oxysporum*	Olive mill waste	Bioethanol	M'barek et al. 2020
4.	*Clostridium beijerinckii*	Tomato pomace	Butanol and isopropanol	Hijosa-Valsero et al. 2019
5.	*Saccharomyces cerevisiae* TISTR 5020	Corn stalk juice	Bioethanol	Bautista et al. 2019
6.	*Saccharomyces cerevisiae* RK1.	Banana, grapes and mango	Bioethanol	Shah et al. 2019
7.	*Clostridium saccaroperbutylacetonicum*	Potato peel	Biobutanol	Hijosa-Valsero et al. 2018
8.	*Clostridium acetobutylicum*	Wheat starch wastewater	Biobutanol	Luo et al. 2018
9.	*Trametes versicolor*	Sugarcane bagasse and straw from rice, sweet sorghum and millet	Bioethanol	Harinikumar et al. 2017
10.	*Saccharomyces cerevisiae and Scheffersomyces stipites*	Apple pomace	Bioethanol	Pathania et al. 2017
11.	*Clostridium acetobutylicum* B 527	Pineapple	Biobutanol	Khedkar et al. 2017
12.	*Clostridium beijerinckii* CECT 508	Apple pomace	Biobutanol	Hijosa-Valsero et al. 2017
13.	*Aspergillus spp. and Saccharomyces cerevisiae*	Sweet potato	Bioethanol	Kumar et al. 2016
14.	*Saccharomyces cerevisiae*	Potato peel, carrot peel, and onion peel	Bioethanol	Mushimiyimana and Tallapragada 2016
15.	*Saccharomyces cerevisiae*	Sweet potato	Bioethanol	Kumar et al. 2014

11.4.2 Synthesis of Antibiotics from Various Agri Wastes

Antibiotics are the secondary metabolites which are synthesized by different microorganisms and selectively kill other microorganisms or prevent the growth of microorganisms (Tripathi 2008). Production of antibiotics using agricultural wastes as low-cost carbohydrate sources by employing solid-state fermentation has gained much recognition due to higher productivity, lower energy, shorter production period, and less investment cost (Mahalaxmi et al. 2010). In the solid-state fermentation process, the solid substrate facilitates the microbial growth by providing essential cofactors and nutrients. Various agricultural and kitchen wastes are used for the production of different antibiotics using solid-state fermentation. The production of oxytetracycline from cocoyam peels (a common household kitchen waste) by *Streptomyces speibonae* OXS1 in solid-state fermentation was documented by Ezejiofor et al. (2012). Recently, various studies were carried out using agri waste for the production of antibiotics. Paromomycin was produced through solid-state fermentation using *Streptomyces rimosus* subsp. *paromomycinus* NRRL 2455 with the cost-effective agro-industrial byproduct corn bran (El-Housseiny et al. 2021). The production of erythromycin by *Saccharopolyspora erythraea* NCIMB 12462 was investigated using sugar beet root and oatmeal (Shata et al. 2021). Potato waste was used as a sole substrate for producing antifungals using *Streptomyces* spp. (Schalchli et al. 2021). Zeng et al. (2019) reported that the *Streptomyces gilvosporeus* Z28 in combination with rapeseed cake, wheat bran, rice hull and crude glycerol led to the production of natamycin, which reduced 50.05% of the cost of raw materials and consumed less energy and wastewater release.

Onyegeme-Okerenta and Ebuehi (2017) studied that *Penicillium chrysogenum PCL501* fermented on cassava peels; a cellulosic waste material produced antibiotics which were effective against *Bacillus subtilis* and *Escherichia coli*, and their antibiotic activity was equivalent to commercial benzyl penicillin. Antibiotic production with *Streptomyces* sp. AS4 was carried out in solid-state fermentation using agri waste like orange peel, pineapple peel, pomegranate peel, banana peel, apple pomace, green gram husk, black gram husk, rice bran and wheat bran (Farraj et al. 2020).

11.4.3 Utilization of Agro Residue for the Production of Mushrooms

In order to utilize the residues obtained from the agro-forestry and agro-industries, mushroom cultivation acts as a prominent biotechnological process. For this process, the lignocellulosic agricultural crop residues and by-products generated by agro-industries that are rich in organic compounds are used and transformed into beneficial products. These residues are used as feedstock in a solid-state fermentation process in the presence of fungi basidiomycetes for the production of mushrooms.

Mushroom production is much expanded and also important economically, as it uses solid-state fermentation for the recovery of food protein from lignocellulosic materials. Commercial production of mushrooms is carried out at both large and small scales and is also an efficient and short biological process (Chiu and Moore 2008). In order to grow edible mushrooms, various pre-treatments are required to promote the development and growth of mushrooms by excluding competitor microorganisms (Gowda and Manvi 2019). Agricultural residues that are used for mushroom cultivation include banana leaves, coconut fiber pith, coconut husks, ground nutshells, maize straw, saw dust, barley straw, rice husk, cotton wastes, wheat bran, sunflower stalk and sugarcane bagasse. These are used for the cultivation of *Pleurotus* species, *Volvariella* species, *Calocybe indica* (Kwon and Kim 2004).

Several varieties of mushroom are cultivated using different residues generated from agro-industries, which are tabulated here:

TABLE 11.3
Production of Mushrooms by Using Agri Waste Residues

S. No.	Fungal species	Waste	References
1.	*Agaricus flocculosipes* R.L. Zhao, Desjardin, J. Guinberteau & K.D. Hyde	Horse manure + wheat straw	Thongklang et al. 2014
2.	*Agaricus subrufescens* Peck	Horse manure + wheat straw	Thongklang et al. 2014
3.	*Agrocybe cylindracea* (DC.) Maire	Rice bran; two-phase olive mill leftover; two-phase olive mill leftover with compost; wheat bran and wheat straw	(Zervakis et al. 2013)
4.	*Coprinus comatus* (O.F. Müll.) Pers	Spent of P. sajor-caju, P.ostreatus and P.florida mixed with 100 g of the different enrichment types (corn grit, rice grit and rice bran)	(Dulay et al. 2014)
5.	*Flammulina velutipes* (Curtis) Singer	Rice hull; palm unfilled fruit bunches; palmpressed residues	Harith et al. 2014
6.	*Pleurotus eryngii* (DC.) Quel	Wheat straw; raw two-phase olive mill leftover; two-phase olive mill leftover with compost	Zervakis et al. 2013)
7.	*Pleurotus floridanus* Singer	Wheat straw; barley straw; maize stem residue and lawn residue with supplementation of wheat bran and rice bran	Ng'etich et al. 2013
8.	*Pleurotus floridanus* Singer	Sorghum straw; paddy straw; sugarcane bagasse; banana leaves	Karuppuraj et al. 2014
9.	*Pleurotus ostreatus* (Jacq.) P. Kumm.	Coffee husk; cow dung; poultry manure; bone meal in different compositions	(Yang et al. 2016)
10.	*Pleurotus ostreatus* (Jacq.) P. Kumm.	Cotton waste and rice straw with rice bran additive at varying percentages	Sharma et al. 2013)

11.4.4 Production of Enzymes by Using Solid-State Fermentation

Kitchen and agri waste have carbon and nitrogen sources which support the growth of microorganisms and result in the production of many useful enzymes. The increased growth of fungi led to the conversion of lignocellulosic material into simple ones by the actions of numerous enzymes. Many enzymes such as cellulase, hemicellulase, laccase, lignin peroxidase (LiP) and manganese peroxidase (MnP) are used to break the lignocellulosic biomass of agri waste (Omar et al. 2017). It has been studied that the production of crude enzymes from solid state fermentation is useful for many industrial processes, such as hydrolysis of lignocellulosic materials, bio-bleaching of paper waste, bioethanol production, detergent formulation xylitol production and silver recovery from X-ray films. Table 11.4 shows the production of enzymes by microorganisms through fermentation using agri waste as a nutritional source.

TABLE 11.4
Production of Enzymes by Using Solid-State Fermentation

Sr. No.	Agriwaste derived media	Microorganisms for fermentation	Enzymes produced	References
1.	Corncob powder	*Trichoderma longibrachiatum, Aspergillus niger, A. flavus, A. fumigatus,* and *Botryodiplodia* spp.	Xylanases	Elegbede and Lateef 2018
2.	Wheat bran	*Trametes hirsuta*	Cellulase	Dave et al. 2021
3.	Corncob	*Saccharomyces cerevisiae*	Cellulase, xylanase and ligninase	Amadi et al. 2020
4.	*Luffa cylindrica* and *Litchi chinensis* peel	*Trichoderma reesei*	Cellulase	Verma et al. 2018
5	Rice straw, sugar cane bagasse, and solid tofu waste as additional carbon source to chicken feathers	*Bacillus* sp.	Keratinase	Nurkhasanah 2019
6.	Kitchen waste (discarding bones and eggshells, with only boneless meat, vegetables, rice and chicken meat remaining)	*Aspergillus niger*	Xylanase	Moid et al. 2021
7.	Wheat bran and corncobs	*Aspergillus niger*	Xylanase	Sharma Richhariya, and Dassani 2020
8.	Carrot pomace	*Penicillium oxalicum*	Inulinase	Singh et al. 2018
9.	Apple pomace	*Saccharomyces cerevisiae*	Cellulase, Xylanase and bioethanol	Kumar, Surya, and Verma 2021
10.	Sorghum stover	*Rhizopus oryzae*	Cellulase Xylanase	Pandey et al. 2016
11.	Wheat bran	*Trichoderma reesei* and *Neurospora crassa*	Cellulase	Verma and Kumar 2020
12.	Wheat straw	*Aspergillus flavus, A. niger, A. oryzae, Penicillium chrysogenum, Penicillium sp* and *Trichoderma harzianum*	Cellulase	Toscano-Palomar et al. 2015
13.	Wheat straw, wheat bran, rice straw and rice husk	*Trichoderma reesei*	Cellulase and xylanase	Taherzadeh-Ghahfarokhi et al. 2019
14.	Wheat bran	*Aspergillus Niger*	Pectinase	Janveja and Soni 2016

Sr. No.	Agriwaste derived media	Microorganisms for fermentation	Enzymes produced	References
15.	Wheat bran	*Aspergillus niger*	Cellulase and xylanase	Kumar et al. 2018
16.	Empty fruit bunches (EFB), rice straw, corncob and rice husk	*Marasmius* sp, *Trametes hirsuta, Trametes versicolor and Phanerochaete crysosporium.*	Laccase	Risdianto et al. 2012
17.	Rice straw	*Aspergillus niger*	Cellulases and hemicellulases (*endo*-β-1,4-glucanase, *exo*-β-1,4-glucanase, β-1,4-glucosidase, endo-β-1,4-xylanase and endo-β-1,4-mannanase)	Kaur et al. 2020
18.	Wheat straw	*Aspergillus niger*	β-1,4-Endoxylanase	Azzouz et al. 2021
19.	Bamboo (*Bambusa tulda*) pulp	*Bacillus australimaris*	Xylanase	Dutta et al. 2020

11.5 CONCLUSION

Kitchen and agri wastes contain many nutritional components and bioactive compounds. These wastes are rich in variable composition of mineral elements, proteins and polysaccharides, so they should be considered as raw material for industrial processes. Kitchen and agri wastes are easily accessible in huge amounts at no cost, hence decreasing the production costs of biofuels. Conversion of this renewable biomass into bioenergy via anaerobic digestion is both economically and ecologically viable. Many valuable products like biofuels, antibiotics, enzymes and mushrooms are produced through solid state fermentation. To conclude, biofuel generation from waste biomass through biochemical processes will provide clean, renewable, techno-economic and alternative solutions for reprocessing of biowaste, which can be further utilized as value-added products.

11.6 REFERENCES

Agyeman, F.O. and Tao, W., (2014). Anaerobic co-digestion of food waste and dairy manure: Effects of food waste particle size and organic loading rate. *Journal of Environmental Management*, *133*, pp. 268–274.

Aierzhati, A., Watson, J., Si, B., Stablein, M., Wang, T. and Zhang, Y., (2021). Development of a mobile, pilot scale hydrothermal liquefaction reactor: Food waste conversion product analysis and techno-economic assessment. *Energy Conversion and Management: X*, *10*, p. 100076.

Al Farraj, D.A., Varghese, R., Vágvölgyi, C., Elshikh, M.S., Alokda, A.M. and Mahmoud, A.H., 2020. Antibiotics production in optimized culture condition using low cost substrates from Streptomyces sp. AS4 isolated from mangrove soil sediment. *Journal of King Saud University-Science*, *32*(2), pp. 1528–1535.

Alves, H.J., Junior, C.B., Niklevicz, R.R., Frigo, E.P., Frigo, M.S. and Coimbra-Araújo, C.H., (2013). Overview of hydrogen production technologies from biogas and the applications in fuel cells. *International Journal of Hydrogen Energy*, *38*(13), pp. 5215–5225.

Amadi, O.C., Egong, E.J., Nwagu, T.N., Okpala, G., Onwosi, C.O., Chukwu, G.C., Okolo, B.N., Agu, R.C. and Moneke, A.N., (2020). Process optimization for simultaneous production of cellulase, xylanase and ligninase by Saccharomyces cerevisiae SCPW 17 under solid state fermentation using Box-Behnken experimental design. *Heliyon*, *6*(7), p.e04566.

Andrew, R., Gokak, D.T., Sharma, P. and Gupta, S., (2016). Novel hydrogen-rich gas production by steam gasification of biomass in a research-scale rotary tubular helical coil gasifier. *International Journal of Energy Research*, *40*(13), pp. 1788–1799.

Azzouz, Z., Bettache, A., Boucherba, N., Prieto, A., Martinez, M.J., Benallaoua, S. and de Eugenio, L.I., (2021). Optimization of β-1, 4-endoxylanase production by an aspergillus niger strain growing on wheat straw and application in Xylooligosaccharides production. *Molecules*, *26*(9), p. 2527.

Balan, V., Rogers, C.A., Chundawat, S.P., da Costa Sousa, L., Slininger, P.J., Gupta, R. and Dale, B.E., (2009). Conversion of extracted oil cake fibers into bioethanol including DDGS, canola, sunflower, sesame, soy, and peanut for integrated biodiesel processing. *Journal of the American Oil Chemists' Society*, *86*(2), pp. 157–165.

Balat, M., (2007). Global bio-fuel processing and production trends. *Energy Exploration & Exploitation*, *25*(3), pp. 195–218.

Bautista, K., Unpaprom, Y. and Ramaraj, R., (2019). Bioethanol production from corn stalk juice using Saccharomyces cerevisiae TISTR 5020. *Energy Sources, Part A: Recovery, Utilization, and Environmental Effects*, *41*(13), pp. 1615–1621.

Bayat, H., Cheng, F., Dehghanizadeh, M., Soliz, N., Brewer, C.E. and Jena, U., (2019). Hydrothermal liquefaction of food waste: Bio-crude oil characterization, mass and energy balance. In *2019 ASABE Annual International Meeting* (p. 1). American Society of Agricultural and Biological Engineers.

Bayat, H., Dehghanizadeh, M., Jarvis, J.M., Brewer, C. and Jena, U., (2021). Hydrothermal liquefaction of food waste: Effect of process parameters on product yields and chemistry. *Frontiers in Sustainable Food Systems*, *5*, p. 160.

Bhargav, S., Panda, B.P., Ali, M. and Javed, S., (2008). Solid-state fermentation: An overview. *Chemical and Biochemical Engineering Quarterly*, *22*(1), pp. 49–70.

Bhuvaneswari, M. and Sivakumar, N., (2019). Bioethanol production from fruit and vegetable wastes. *Bioprocessing for Biomolecules Production*, pp. 417–427.

Binod, P., Sindhu, R., Singhania, R.R., Vikram, S., Devi, L., Nagalakshmi, S., Kurien, N., Sukumaran, R.K. and Pandey, A., (2010). Bioethanol production from rice straw: An overview. *Bioresource Technology*, *101*(13), pp. 4767–4774.

Bouallagui, H., Haouari, O., Touhami, Y., Cheikh, R.B., Marouani, L. and Hamdi, M., (2004). Effect of temperature on the performance of an anaerobic tubular reactor treating fruit and vegetable waste. *Process Biochemistry*, *39*(12), pp. 2143–2148.

Bouallagui, H., Touhami, Y., Cheikh, R.B. and Hamdi, M., (2005). Bioreactor performance in anaerobic digestion of fruit and vegetable wastes. *Process Biochemistry*, *40*(3–4), pp. 989–995.

Bryant, M.P., (1979). Microbial methane production—theoretical aspects. *Journal of Animal Science*, *48*(1), pp. 193–201.

Câmara-Salim, I., González-García, S., Feijoo, G. and Moreira, M.T., (2021). Screening the environmental sustainability of microbial production of butyric acid produced from lignocellulosic waste streams. *Industrial Crops and Products*, *162*, p. 113280.

Cardona, C.A., Quintero, J.A. and Paz, I.C., (2010). Production of bioethanol from sugarcane bagasse: Status and perspectives. *Bioresource Technology*, *101*(13), pp. 4754–4766.

Carlsson, M., Lagerkvist, A. and Ecke, H., (2008). Electroporation for enhanced methane yield from municipal solid waste. In *ORBIT 2008: Moving Organic Waste Recycling Towards Resource Management and Biobased Economy 13/10/2008–15/10/2008*. Wageningen, The Netherlands.

Cavinato, C., Giuliano, A., Bolzonella, D., Pavan, P. and Cecchi, F., (2012). Bio-hythane production from food waste by dark fermentation coupled with anaerobic digestion process: A long-term pilot scale experience. *International Journal of Hydrogen Energy*, *37*(15), pp. 11549–11555.

Chatanta, D.K., Attri, C., Gopal, K., Devi, M., Gupta, G. and Bhalla, T.C., (2008). Bioethanol production from apple pomace left after juice extraction. *Internet Journal of Microbiology*, *5*(2).

Chen, D., Yin, L., Wang, H. and He, P., (2015). Reprint of: Pyrolysis technologies for municipal solid waste: A review. *Waste Management*, *37*, pp. 116–136.

Chen, F. and Dixon, R.A., (2007). Lignin modification improves fermentable sugar yields for biofuel production. *Nature Biotechnology*, *25*(7), pp. 759–761.

Chen, L., Yu, Z., Fang, S., Dai, M. and Ma, X., (2018). Co-pyrolysis kinetics and behaviors of kitchen waste and chlorella vulgaris using thermogravimetric analyzer and fixed bed reactor. *Energy Conversion and Management*, *165*, pp. 45–52.

Chen, W.H., Lin, Y.Y., Liu, H.C. and Baroutian, S., (2020). Optimization of food waste hydrothermal liquefaction by a two-step process in association with a double analysis. *Energy*, *199*, p. 117438.

Cheng, F., Tompsett, G.A., Murphy, C.M., Maag, A.R., Carabillo, N., Bailey, M., Hemingway, J.J., Romo, C.I., Paulsen, A.D., Yelvington, P.E. and Timko, M.T., (2020). Synergistic effects of inexpensive mixed metal oxides for catalytic hydrothermal liquefaction of food wastes. *ACS Sustainable Chemistry & Engineering*, *8*(17), pp. 6877–6886.

Chiu, S. and Moore, D., (2008). Threats to biodiversity caused by traditional mushroom cultivation technology in China. *Fungal Conservation*, p. 111.

Cieciura-Włoch, W., Borowski, S. and Otlewska, A., (2020). Biohydrogen production from fruit and vegetable waste, sugar beet pulp and corn silage via dark fermentation. *Renewable Energy*, *153*, pp. 1226–1237.

das Neves, M.A., Kimura, T., Shimizu, N. and Nakajima, M., (2007). State of the art and future trends of bioethanol production. *Dynamic Biochemistry, Process Biotechnology and Molecular Biology*, *1*(1), pp. 1–14.

Das, D. and Veziroglu, T.N., (2008). Advances in biological hydrogen production processes. *International Journal of Hydrogen Energy*, *33*(21), pp. 6046–6057.

Dave, B.R., Parmar, P., Sudhir, A., Singal, N. and Subramanian, R.B., (2021). Cellulases production under solid state fermentation using agro waste as a substrate and its application in saccharification by Trametes hirsuta NCIM. *Journal of Microbiology, Biotechnology and Food Sciences*, *2021*, pp. 203–208.

Déniel, M., Haarlemmer, G., Roubaud, A., Weiss-Hortala, E. and Fages, J., (2016). Energy valorisation of food processing residues and model compounds by hydrothermal liquefaction. *Renewable and Sustainable Energy Reviews*, *54*, pp. 1632–1652.

Dulay, R.M.R., Gagarin, W.S., Abella, E.A., Kalaw, S.P. and Reyes, R.G., (2014). Aseptic cultivation and nutrient compositions of Coprinus comatus (OF Müll.) Pers. on Pleurotus mushroom spent. *Journal of Microbiology and Biotechnology*, *4*(3), pp. 1–7.

Dutta, P.D., Neog, B. and Goswami, T., (2020). Xylanase enzyme production from Bacillus australimaris P5 for prebleaching of bamboo (Bambusa tulda) pulp. *Materials Chemistry and Physics*, *243*, p. 122227.

Elegbede, J.A. and Lateef, A., (2018). Valorization of corn-cob by fungal isolates for production of xylanase in submerged and solid state fermentation media and potential biotechnological applications. *Waste and Biomass Valorization*, *9*(8), pp. 1273–1287.

El-Housseiny, G.S., Ibrahim, A.A., Yassien, M.A. and Aboshanab, K.M., (2021). Production and statistical optimization of Paromomycin by Streptomyces rimosus NRRL 2455 in solid state fermentation. *BMC Microbiology*, *21*(1), pp. 1–13.

Elkhalifa, S., Al-Ansari, T., Mackey, H.R. and McKay, G., (2019). Food waste to biochars through pyrolysis: A review. *Resources, Conservation and Recycling*, *144*, pp. 310–320.

Ezejiofor, T.I.N., Duru, C.I., Asagbra, A.E., Ezejiofor, A.N., Orisakwe, O.E., Afonne, J.O. and Obi, E., (2012). Waste to wealth: Production of oxytetracycline using streptomyces species from household kitchen wastes of agricultural produce. *African Journal of Biotechnology*, *11*(43), pp. 10115–10124.

Facchin, V., Cavinato, C., Fatone, F., Pavan, P., Cecchi, F. and Bolzonella, D., (2013). Effect of trace element supplementation on the mesophilic anaerobic digestion of foodwaste in batch trials: The influence of inoculum origin. *Biochemical Engineering Journal*, *70*, pp. 71–77.

Farooq, A., Song, H., Park, Y.K. and Rhee, G.H., (2021). Effects of different Al2O3 support on HDPE gasification for enhanced hydrogen generation using Ni-based catalysts. *International Journal of Hydrogen Energy*, *46*(34), pp. 18085–18092.

Ferry, J.G., (2010). The chemical biology of methanogenesis. *Planetary and Space Science*, *58*(14–15), pp. 1775–1783.

Fivga, A., Speranza, L.G., Branco, C.M., Ouadi, M. and Hornung, A., (2019). A review on the current state of the art for the production of advanced liquid biofuels. *Aims Energy*, *7*(1), pp. 46–76.

Ghosh, S.K., (2016). Biomass & bio-waste supply chain sustainability for bio-energy and biofuel production. *Procedia Environmental Sciences*, *31*, pp. 31–39.

Gollakota, A.R.K., Kishore, N. and Gu, S., (2018). A review on hydrothermal liquefaction of biomass. *Renewable and Sustainable Energy Reviews*, *81*, pp. 1378–1392.

Gómez, X., Fernández, C., Fierro, J., Sánchez, M.E., Escapa, A. and Morán, A., (2011). Hydrogen production: Two stage processes for waste degradation. *Bioresource Technology*, *102*(18), pp. 8621–8627.

Gou, C., Yang, Z., Huang, J., Wang, H., Xu, H. and Wang, L., (2014). Effects of temperature and organic loading rate on the performance and microbial community of anaerobic co-digestion of waste activated sludge and food waste. *Chemosphere*, *105*, pp. 146–151.

Gowda, N.A. and Manvi, D., (2019). Agro-residues Disinfection Methods for Mushroom Cultivation. *Agricultural Reviews*, *40*(2), pp. 93–103.

Günerken, E., D'Hondt, E., Eppink, M.H.M., Garcia-Gonzalez, L., Elst, K. and Wijffels, R.H., (2015). Cell disruption for microalgae biorefineries. *Biotechnology Advances*, *33*(2), pp. 243–260.

Guo, X.M., Trably, E., Latrille, E., Carrere, H. and Steyer, J.P., (2010). Hydrogen production from agricultural waste by dark fermentation: A review. *International Journal of Hydrogen Energy*, *35*(19), pp. 10660–10673.

Gupta, A. and Verma, J.P., (2015). Sustainable bio-ethanol production from agro-residues: A review. *Renewable and Sustainable Energy Reviews*, *41*, pp. 550–567.

Harinikumar, K.M., Kudahettige-Nilsson, R., Devadas, A., Holmgren, M. and Sellstedt, A., (2017). Bioethanol production from four abundant Indian agricultural wastes. *Biofuels*, *11*(5), pp. 607–613.

Harith, N., Abdullah, N. and Sabaratnam, V., (2014). Cultivation of Flammulina velutipes mushroom using various agro-residues as a fruiting substrate. *Pesquisa Agropecuária Brasileira*, *49*, pp. 181–188.

Hijosa-Valsero, M., Garita-Cambronero, J., Paniagua-García, A.I. and Díez-Antolínez, R., (2019). Tomato waste from processing industries as a feedstock for biofuel production. *BioEnergy Research, 12*(4), pp. 1000–1011.

Hijosa-Valsero, M., Paniagua-García, A.I. and Díez-Antolínez, R., (2017). Biobutanol production from apple pomace: The importance of pretreatment methods on the fermentability of lignocellulosic agro-food wastes. *Applied Microbiology and Biotechnology, 101*(21), pp. 8041–8052.

Hijosa-Valsero, M., Paniagua-García, A.I. and Díez-Antolínez, R., (2018). Industrial potato peel as a feedstock for biobutanol production. *New Biotechnology, 46*, pp. 54–60.

Hongthong, S., Raikova, S., Leese, H.S. and Chuck, C.J., (2020). Co-processing of common plastics with pistachio hulls via hydrothermal liquefaction. *Waste Management, 102*, pp. 351–361.

Ivanova, G., Rákhely, G. and Kovács, K.L., (2009). Thermophilic biohydrogen production from energy plants by Caldicellulosiruptor saccharolyticus and comparison with related studies. *International Journal of Hydrogen Energy, 34*(9), pp. 3659–3670.

Jaitalee, L., Dararat, S. and Chavalparit, O., (2010). Bio-hydrogen production potential from market waste. *Environment Asia, 3*(2), pp. 115–122.

Janveja, C. and Soni, S.K., (2016). A combinatorial statistical approach for hyperproduction of pectinase from aspergillus Niger C-5 via solid state fermentation of wheat bran and its kinetic characterization. *Journal of Multidisciplinary Engineering Science Studies (JMESS), 2* (11), pp. 1097–1110.

Jia, H., Ben, H. and Wu, F., (2021). Effect of biochar prepared from food waste through different thermal treatment processes on crop growth. *Processes, 9*(2), p. 276.

Jørgensen, H., Kristensen, J.B. and Felby, C., (2007). Enzymatic conversion of lignocellulose into fermentable sugars: Challenges and opportunities. *Biofuels, Bioproducts and Biorefining, 1*(2), pp. 119–134.

Jouhara, H., Ahmad, D., van den Boogaert, I., Katsou, E., Simons, S. and Spencer, N., (2018). Pyrolysis of domestic based feedstock at temperatures up to 300 C. *Thermal Science and Engineering Progress, 5*, pp. 117–143.

Kapdan, I.K. and Kargi, F., (2006). Bio-hydrogen production from waste materials. *Enzyme and Microbial Technology, 38*(5), pp. 569–582.

Karthikeyan, O.P. and Visvanathan, C., (2012). Effect of C/N ratio and ammonia-N accumulation in a pilot-scale thermophilic dry anaerobic digester. *Bioresource Technology, 113*, pp. 294–302.

Karuppuraj, V., Chandra Sekarenthiran, S. and Perumal, K., (2014). Continuous production of Pleurotus florida and Calocybe indica by utilizing locally available lignocellulosic substrates for additional income generation in rural area. *International Journal of Pharmaceutical Sciences Review and Research, 29*(1), pp. 196–199.

Kaur, J., Chugh, P., Soni, R. and Soni, S.K., (2020). A low-cost approach for the generation of enhanced sugars and ethanol from rice straw using in-house produced cellulase-hemicellulase consortium from A. niger P-19. *Bioresource Technology Reports, 11*, p. 100469.

Khedkar, M.A., Nimbalkar, P.R., Gaikwad, S.G., Chavan, P.V. and Bankar, S.B., (2017). Sustainable biobutanol production from pineapple waste by using Clostridium acetobutylicum B 527: Drying kinetics study. *Bioresource Technology, 225*, pp. 359–366.

Kim, J.K., Oh, B.R., Chun, Y.N. and Kim, S.W., (2006). Effects of temperature and hydraulic retention time on anaerobic digestion of food waste. *Journal of Bioscience and Bioengineering, 102*(4), pp. 328–332.

Kim, S. and Lee, J., (2020). Pyrolysis of food waste over a Pt catalyst in CO2 atmosphere. *Journal of Hazardous Materials, 393*, p. 122449.

Kostyukevich, Y., Vlaskin, M., Borisova, L., Zherebker, A., Perminova, I., Kononikhin, A., Popov, I. and Nikolaev, E., (2018). Investigation of bio-oil produced by hydrothermal liquefaction of food waste using ultrahigh resolution Fourier transform ion cyclotron resonance mass spectrometry. *European Journal of Mass Spectrometry*, *24*(1), pp. 116–123.

Krylova, A.Y., Kozyukov, E.A. and Lapidus, A.L., (2008). Ethanol and diesel fuel from plant raw materials: A review. *Solid Fuel Chemistry*, *42*(6), pp. 358–364.

Kumar, A., Duhan, J.S., Gahlawat, S. and Gahlawat, S.K., (2014). Production of ethanol from tuberous plant (sweet potato) using Saccharomyces cerevisiae MTCC-170. *African Journal of Biotechnology*, *13*(28).

Kumar, A., Sadh, P.K., Kha, S.U.R.E. and Dhuan, J.S., (2016). Bio-ethanol production from sweet potato using co-culture of saccharolytic molds (Aspergillus spp.) and Saccharomyces cerevisiae MTCC170. *Journal of Advanced Biotechnology*, *6*(1), pp. 822–827.

Kumar, B.A., Amit, K., Alok, K. and Dharm, D., (2018). Wheat bran fermentation for the production of cellulase and xylanase by Aspergillus niger NFCCI 4113. *Research Journal of Biotechnology*, *13*, p. 5.

Kumar, D., Surya, K. and Verma, R., (2021). Bioethanol production from apple pomace using co-cultures with saccharomyces cerevisiae in solid-state fermentation. *Journal of Microbiology, Biotechnology and Food Sciences*, *2021*, pp. 742–745.

Kumar, M., Srivastava, N., Upadhyay, S.N. and Mishra, P.K., (2021). Thermal degradation of dry kitchen waste: Kinetics and pyrolysis products. *Biomass Conversion and Biorefinery*, pp. 1–18.

Labatut, R.A., Angenent, L.T. and Scott, N.R., (2011). Biochemical methane potential and biodegradability of complex organic substrates. *Bioresource Technology*, *102*(3), pp. 2255–2264.

Laufenberg, G., Kunz, B. and Nystroem, M., (2003). Transformation of vegetable waste into value added products: (A) the upgrading concept;(B) practical implementations. *Bioresource Technology*, *87*(2), pp. 167–198.

Lee, D.H., Behera, S.K., Kim, J.W. and Park, H.S., (2009). Methane production potential of leachate generated from Korean food waste recycling facilities: A lab-scale study. *Waste Management*, *29*(2), pp. 876–882.

Lee, S.Y., Sankaran, R., Chew, K.W., Tan, C.H., Krishnamoorthy, R., Chu, D.T. and Show, P.L., (2019). Waste to bioenergy: A review on the recent conversion technologies. *BMC Energy*, *1*(1), pp. 1–22.

Li, P., Xie, Y., Zeng, Y., Hu, W., Kang, Y., Li, X., Wang, Y., Xie, T. and Zhang, Y., (2017). Bioconversion of welan gum from kitchen waste by a two-step enzymatic hydrolysis pretreatment. *Applied Biochemistry and Biotechnology*, *183*(3), pp. 820–832.

Lin, Y. and Tanaka, S., (2006). Ethanol fermentation from biomass resources: Current state and prospects. *Applied Microbiology and Biotechnology*, *69*(6), pp. 627–642.

Logan, B.E., Oh, S.E., Kim, I.S. and Van Ginkel, S., (2002). Biological hydrogen production measured in batch anaerobic respirometers. *Environmental Science & Technology*, *36*(11), pp. 2530–2535.

Lu, J., Li, H., Zhang, Y. and Liu, Z., (2018). Nitrogen migration and transformation during hydrothermal liquefaction of livestock manures. *ACS Sustainable Chemistry & Engineering*, *6*(10), pp. 13570–13578.

Luo, W., Zhao, Z., Pan, H., Zhao, L., Xu, C. and Yu, X., (2018). Feasibility of butanol production from wheat starch wastewater by Clostridium acetobutylicum. *Energy*, *154*, pp. 240–248.

M'barek, H.N., Arif, S., Taidi, B. and Hajjaj, H., (2020). Consolidated bioethanol production from olive mill waste: Wood-decay fungi from central Morocco as promising decomposition and fermentation biocatalysts. *Biotechnology Reports*, *28*, p. e00541.

Madsen, R.B. and Glasius, M., (2019). How do hydrothermal liquefaction conditions and feedstock type influence product distribution and elemental composition? *Industrial & Engineering Chemistry Research*, *58*(37), pp. 17583–17600.

Mahalaxmi, Y., Sathish, T., Rao, C.S. and Prakasham, R.S., (2010). Corn husk as a novel substrate for the production of rifamycin B by isolated Amycolatopsis sp. RSP 3 under SSF. *Process Biochemistry*, *45*(1), pp. 47–53.

Mihajlovski, K., Buntić, A., Milić, M., Rajilić-Stojanović, M. and Dimitrijević-Branković, S., (2020). From Agricultural waste to biofuel: Enzymatic potential of a bacterial isolate Streptomyces fulvissimus CKS7 for bioethanol production. *Waste and Biomass Valorization*, pp. 1–10.

Mishra, R.K. and Mohanty, K., (2018). Pyrolysis kinetics and thermal behavior of waste sawdust biomass using thermogravimetric analysis. *Bioresource Technology*, *251*, pp. 63–74.

Mohanakrishna, G., Goud, R.K., Mohan, S.V. and Sarma, P.N., (2010). Enhancing biohydrogen production through sewage supplementation of composite vegetable based market waste. *International Journal of Hydrogen Energy*, *35*(2), pp. 533–541.

Moid, M.M., Idris, N., Othman, R. and Wahid, D.A., (2021), March. Development of xylanase as detergent additive to improve laundry application. In *IOP Conference Series: Materials Science and Engineering* (Vol. 1092, No. 1, p. 012053). IOP Publishing.

Moralı, U. and Şensöz, S., (2015). Pyrolysis of hornbeam shell (Carpinus betulus L.) in a fixed bed reactor: Characterization of bio-oil and bio-char. *Fuel*, *150*, pp. 672–678.

Morita, M. and Sasaki, K., (2012). Factors influencing the degradation of garbage in methanogenic bioreactors and impacts on biogas formation. *Applied Microbiology and Biotechnology*, *94*(3), pp. 575–582.

Morrin, S., Lettieri, P., Chapman, C. and Mazzei, L., (2012). Two stage fluid bed-plasma gasification process for solid waste valorisation: Technical review and preliminary thermodynamic modelling of sulphur emissions. *Waste Management*, *32*(4), pp. 676–684.

Motavaf, B. and Savage, P.E., (2021). Effect of process variables on food waste valorization via hydrothermal liquefaction. *ACS ES&T Engineering*, *1*(3), pp. 363–374.

Mushimiyimana, I. and Tallapragada, P., (2016). Bioethanol production from agro wastes by acid hydrolysis and fermentation process. *Journal of Scientific & Industrial Research*, *75*, pp. 383–388.

Nagao, N., Tajima, N., Kawai, M., Niwa, C., Kurosawa, N., Matsuyama, T., Yusoff, F.M. and Toda, T., (2012). Maximum organic loading rate for the single-stage wet anaerobic digestion of food waste. *Bioresource Technology*, *118*, pp. 210–218.

Nasir, I.M., Ghazi, T.I.M. and Omar, R., (2012). Production of biogas from solid organic wastes through anaerobic digestion: A review. *Applied Microbiology and Biotechnology*, *95*(2), pp. 321–329.

Ng'etich, O.K., Nyamangyoku, O.I., Rono, J.J., Niyokuri, A.N. and Izamuhaye, J.C. (2013). Relative performance of Oyster mushroom (*Pleurotus florida*) on agro industrial and agricultural substrate. *International Journal of Agronomy and Plant Production, 4*(1), 109–116.

Nurkhasanah, U., (2019), May. Preliminary study on keratinase fermentation by Bacillus sp. MD24 under solid state fermentation. In *IOP Conference Series: Earth and Environmental Science* (Vol. 276, No. 1, p. 012016). IOP Publishing.

Omar, F.N., Hafid, H.S., Baharuddin, A.S., Mohammed, M.A.P. and Abdullah, J., (2017). Oil palm fiber biodegradation: Physico-chemical and structural relationships. *Planta*, *246*(3), pp. 567–577.

Onyegeme-Okerenta, B.M. and Ebuehi, O.A.T., (2017). Physiological effect of natural penicillin extract on cassava peel media by Penicillium chrysogenum PCL501 on E. coli infected wistar rats. *IOSR Journal of Pharmacy*, *7*(4), pp. 34–42.

Pandey, A.K., Edgard, G. and Negi, S., (2016). Optimization of concomitant production of cellulase and xylanase from Rhizopus oryzae SN5 through EVOP-factorial design technique and application in Sorghum Stover based bioethanol production. *Renewable Energy*, *98*, pp. 51–56.

Paritosh, K., Kushwaha, S.K., Yadav, M., Pareek, N., Chawade, A. and Vivekanand, V., (2017). Food waste to energy: An overview of sustainable approaches for food waste management and nutrient recycling. *BioMed Research International*, *2017*.

Pathania, S., Sharma, N. and Handa, S., (2017). Immobilization of co-culture of Saccharomyces cerevisiae and Scheffersomyces stipitis in sodium alginate for bioethanol production using hydrolysate of apple pomace under separate hydrolysis and fermentation. *Biocatalysis and Biotransformation*, *35*(6), pp. 450–459.

Pattanaik, L., Pattnaik, F., Saxena, D.K. and Naik, S.N., (2019). Biofuels from agricultural wastes. In *Second and Third Generation of Feedstocks* (pp. 103–142). Elsevier.

Peng, C., Feng, W., Zhang, Y., Guo, S., Yang, Z., Liu, X., Wang, T. and Zhai, Y., (2021). Low temperature co-pyrolysis of food waste with PVC-derived char: Products distributions, char properties and mechanism of bio-oil upgrading. *Energy*, *219*, p. 119670.

Posmanik, R., Martinez, C.M., Cantero-Tubilla, B., Cantero, D.A., Sills, D.L., Cocero, M.J. and Tester, J.W., (2018). Acid and alkali catalyzed hydrothermal liquefaction of dairy manure digestate and food waste. *ACS Sustainable Chemistry & Engineering*, *6*(2), pp. 2724–2732.

Prabakar, D., Manimudi, V.T., Mathimani, T., Kumar, G., Rene, E.R. and Pugazhendhi, A., (2018). Pretreatment technologies for industrial effluents: Critical review on bioenergy production and environmental concerns. *Journal of Environmental Management*, *218*, pp. 165–180.

Qambrani, N.A., Rahman, M.M., Won, S., Shim, S. and Ra, C., (2017). Biochar properties and eco-friendly applications for climate change mitigation, waste management, and wastewater treatment: A review. *Renewable and Sustainable Energy Reviews*, *79*, pp. 255–273.

Ram, M. and Mondal, M.K., (2019). Investigation on fuel gas production from pulp and paper waste water impregnated coconut husk in fluidized bed gasifier via humidified air and CO2 gasification. *Energy*, *178*, pp. 522–529.

Raposo, F., Fernández-Cegrí, V., De la Rubia, M.A., Borja, R., Béline, F., Cavinato, C., Demirer, G.Ö.K.S.E.L., Fernández, B., Fernández-Polanco, M., Frigon, J.C. and Ganesh, R., (2011). Biochemical methane potential (BMP) of solid organic substrates: Evaluation of anaerobic biodegradability using data from an international interlaboratory study. *Journal of Chemical Technology & Biotechnology*, *86*(8), pp. 1088–1098.

Renewable fuel association, (2017). www.ethanolrfa.org/resources/industry/statistics/.

Risdianto, H., Sofianti, E., Suhardi, S.H. and Setiadi, T., (2012). Optimisation of laccase production using white rot fungi and agriculture wastes in solid state fermentation. *ITB Journal of Engineering Science*, *44*(2), pp. 93–105.

Sadh, P.K., Duhan, S. and Duhan, J.S., (2018). Agro-industrial wastes and their utilization using solid state fermentation: A review. *Bioresources and Bioprocessing*, *5*(1), pp. 1–15.

Saghir, M., Rehan, M. and Nizami, A.S., (2018). Recent trends in gasification based waste-to-energy. *Gasification for Low-grade Feedstock*, pp. 97–113.

Saini, J.K., Saini, R. and Tewari, L., (2015). Lignocellulosic agriculture wastes as biomass feedstocks for second-generation bioethanol production: Concepts and recent developments. *3 Biotech*, *5*(4), pp. 337–353.

Saqib, N.U., Sarmah, A.K. and Baroutian, S., (2019). Effect of temperature on the fuel properties of food waste and coal blend treated under co-hydrothermal carbonization. *Waste Management*, *89*, pp. 236–246.

Sarkar, N., Ghosh, S.K., Bannerjee, S. and Aikat, K., (2012). Bioethanol production from agricultural wastes: An overview. *Renewable Energy*, *37*(1), pp. 19–27.

Schalchli, H., Hormazábal, E., Astudillo, Á., Briceño, G., Rubilar, O. and Diez, M.C., (2021). Bioconversion of potato solid waste into antifungals and biopigments using Streptomyces spp. *Plos One*, *16*(5), p. e0252113.

Schattauer, A., Abdoun, E., Weiland, P., Plöchl, M. and Heiermann, M., (2011). Abundance of trace elements in demonstration biogas plants. *Biosystems Engineering*, *108*(1), pp. 57–65.

Shah, K., Vyas, R. and Patel, G., (2019). Bioethanol production from pulp of fruits. *Bioscience Biotechnology Research Communications*, *12*(2), pp. 464–471.

Sharma, H.B., Panigrahi, S. and Dubey, B.K., (2021). Food waste hydrothermal carbonization: Study on the effects of reaction severities, pelletization and framework development using approaches of the circular economy. *Bioresource Technology*, *333*, p. 125187.

Sharma, S., Yadav, R.K.P. and Pokhrel, C.P., (2013). Growth and yield of oyster mushroom (Pleurotus ostreatus) on different substrates. *Journal on New Biological Reports*, *2*(1), pp. 03–08.

Sharma, T.K., Richhariya, J. and Dassani, S., (2020). Production and optimization of enzyme xylanase from Aspergillus niger isolated from decaying litter of Orchna wildlife sanctuary, mp, india. *Journal of Advanced Scientific Research*, *11*(2).

Shata, H., Deen, A.M., Abdelwahed, N.A. and Farid, M.A., (2021). Statistical optimization of erythromycin production by Saccharopolyspora erythraea under solid state fermentation of agro-industrial materials using response surface methodology. *Journal of Microbiology, Biotechnology and Food Sciences*, *2021*, pp. 692–697.

Shenbagaraj, S., Sharma, P.K., Sharma, A.K., Raghav, G., Kota, K.B. and Ashokkumar, V., (2021). Gasification of food waste in supercritical water: An innovative synthesis gas composition prediction model based on artificial neural networks. *International Journal of Hydrogen Energy*, *46*(24), pp. 12739–12757.

Shi, Y., Gai, G.S., Zhao, X.T. and Hu, Y.Y., (2009). Influence and simulation model of operational parameters on hydrogen bio-production through anaerobic microorganism fermentation using two kinds of wastes. In *WCECS 2009. Proceedings of the World Congress on Engineering and Computer Science, vol. II*. San Francisco, USA.

Show, K.Y., Lee, D.J. and Chang, J.S., (2011). Bioreactor and process design for biohydrogen production. *Bioresource Technology*, *102*(18), pp. 8524–8533.

Siegert, I. and Banks, C., (2005). The effect of volatile fatty acid additions on the anaerobic digestion of cellulose and glucose in batch reactors. *Process Biochemistry*, *40*(11), pp. 3412–3418.

Singh, A., Kuila, A., Adak, S., Bishai, M. and Banerjee, R., (2012). Utilization of vegetable wastes for bioenergy generation. *Agricultural Research*, *1*(3), pp. 213–222.

Singh, D. and Yadav, S., (2021). Steam gasification with torrefaction as pretreatment to enhance syngas production from mixed food waste. *Journal of Environmental Chemical Engineering*, *9*(1), p. 104722.

Singh, R.S., Chauhan, K., Singh, J., Pandey, A. and Larroche, C., (2018). Solid-state fermentation of carrot pomace for the production of inulinase by Penicillium oxalicum BGPUP-4. *Food Technology and Biotechnology*, *56*(1), pp. 31–39.

Sivamani, S., Chandrasekaran, A.P., Balajii, M., Shanmugaprakash, M., Hosseini-Bandegharaei, A. and Baskar, R., (2018). Evaluation of the potential of cassava-based residues for biofuels production. *Reviews in Environmental Science and Bio/Technology*, *17*(3), pp. 553–570.

Su, H., Kanchanatip, E., Wang, D., Zhang, H., Mubeen, I., Huang, Z. and Yan, M., (2020). Catalytic gasification of food waste in supercritical water over La promoted Ni/Al2O3 catalysts for enhancing H2 production. *International Journal of Hydrogen Energy*, *45*(1), pp. 553–564.

Taherzadeh-Ghahfarokhi, M., Panahi, R. and Mokhtarani, B., (2019). Optimizing the combination of conventional carbonaceous additives of culture media to produce lignocellulose-degrading enzymes by Trichoderma reesei in solid state fermentation of agricultural residues. *Renewable Energy*, *131*, pp. 946–955.

Talebnia, F., Karakashev, D. and Angelidaki, I., (2010). Production of bioethanol from wheat straw: An overview on pretreatment, hydrolysis and fermentation. *Bioresource Technology*, *101*(13), pp. 4744–4753.

Tampio, E., Ervasti, S., Paavola, T., Heaven, S., Banks, C. and Rintala, J., (2014). Anaerobic digestion of autoclaved and untreated food waste. *Waste Management*, *34*(2), pp. 370–377.

Tang, Y.Q., Koike, Y., Liu, K., An, M.Z., Morimura, S., Wu, X.L. and Kida, K., (2008). Ethanol production from kitchen waste using the flocculating yeast Saccharomyces cerevisiae strain KF-7. *Biomass and Bioenergy*, *32*(11), pp. 1037–1045.

Tayeh, H.A., Najami, N., Dosoretz, C., Tafesh, A. and Azaizeh, H., (2014). Potential of bioethanol production from olive mill solid wastes. *Bioresource Technology*, *152*, pp. 24–30.

Tenca, A., Schievano, A., Perazzolo, F., Adani, F. and Oberti, R., (2011). Biohydrogen from thermophilic co-fermentation of swine manure with fruit and vegetable waste: Maximizing stable production without pH control. *Bioresource Technology*, *102*(18), pp. 8582–8588.

Thongklang, N., Sysouphanthong, P., Callac, P. and Hyde, K.D., (2014). First cultivation of Agaricus flocculosipes and a novel Thai strain of A. subrufescens. *Mycosphere*, *5*(6), pp. 814–820.

Toscano-Palomar, L., Montero-Alpirez, G., Stilianova-Stoytcheva, M. and Vertiz-Pelaez, E., (2015). Cellulase Production from Filamentous Fungi for Its Application in the Hydrolysis of Wheat Straw. *MRS Online Proceedings Library (OPL)*, *1763*.

Tripathi, K.D., (2008). *Antimicrobial Drugs: Essentials of Medical Pharmacology* (p. 733). JP Brothers Medical Publishers (p) Ltd.

Valizadeh, S., Lam, S.S., Ko, C.H., Lee, S.H., Farooq, A., Yu, Y.J., Jeon, J.K., Jung, S.C., Rhee, G.H. and Park, Y.K., (2021). Biohydrogen production from catalytic conversion of food waste via steam and air gasification using eggshell-and homo-type Ni/Al2O3 catalysts. *Bioresource Technology*, *320*, p. 124313.

Van Haandel, A. C., & Lettinga, G. (1994). *Anaerobic sewage treatment*. A practical guide for regions with a hot climate. John Whiley and sons. Gran Bretaña. https://doi.org/10.2166/9781780409627

Van Lier, J.B., Mahmoud, N. and Zeeman, G., (2008). Anaerobic wastewater treatment. *Biological Wastewater Treatment: Principles, Modelling and Design*, pp. 415–456.

Verma, N., Bansal, M.C. and Kumar, V., (2011). Pea peel waste: A lignocellulosic waste and its utility in cellulase production by Trichoderma reesei under solid state cultivation. *Bioresources*, *6*(2), pp. 1505–1519.

Verma, N. and Kumar, V., (2020). Impact of process parameters and plant polysaccharide hydrolysates in cellulase production by Trichoderma reesei and Neurospora crassa under wheat bran based solid state fermentation. *Biotechnology Reports*, *25*, p.e00416.

Verma, N., Kumar, V. and Bansal, M.C., (2018). Utility of Luffa cylindrica and Litchi chinensis peel, an agricultural waste biomass in cellulase production by Trichoderma reesei under solid state cultivation. *Biocatalysis and Agricultural Biotechnology*, *16*, pp. 483–492.

Vijayaraghavan, K., Varma, V.S. and Nalini, S.K., (2012). Hydrogen generation from biological solid waste of milk processing effluent treatment plan. *International Journal of Current Trends and Resources*, *1*, pp. 17–23.

Walker, M., Iyer, K., Heaven, S. and Banks, C.J., (2011). Ammonia removal in anaerobic digestion by biogas stripping: An evaluation of process alternatives using a first order rate model based on experimental findings. *Chemical Engineering Journal, 178*, pp. 138–145.

Wang, X., Yang, G., Feng, Y., Ren, G. and Han, X., (2012). Optimizing feeding composition and carbon—nitrogen ratios for improved methane yield during anaerobic co-digestion of dairy, chicken manure and wheat straw. *Bioresource Technology, 120*, pp. 78–83.

Whelan, M.J., Everitt, T. and Villa, R., (2010). A mass transfer model of ammonia volatilisation from anaerobic digestate. *Waste Management, 30*(10), pp. 1808–1812.

Wolfe, R.S., (2011). Techniques for cultivating methanogens. *Methods in Enzymology, 494*, pp. 1–22.

Wu, X., Zhu, J., Dong, C., Miller, C., Li, Y., Wang, L. and Yao, W., (2009). Continuous biohydrogen production from liquid swine manure supplemented with glucose using an anaerobic sequencing batch reactor. *International Journal of Hydrogen Energy, 34*(16), pp. 6636–6645.

Xiong, S., He, J., Yang, Z., Guo, M., Yan, Y. and Ran, J., (2020). Thermodynamic analysis of CaO enhanced steam gasification process of food waste with high moisture and low moisture. *Energy, 194*, p. 116831.

Xu, F., Li, Y., Ge, X., Yang, L. and Li, Y., (2018). Anaerobic digestion of food waste—Challenges and opportunities. *Bioresource Technology, 247*, pp. 1047–1058.

Yang, D., Liang, J., Wang, Y., Sun, F., Tao, H., Xu, Q., Zhang, L., Zhang, Z., Ho, C.T. and Wan, X., (2016). Tea waste: An effective and economic substrate for oyster mushroom cultivation. *Journal of the Science of Food and Agriculture, 96*(2), pp. 680–684.

Yao, C., Yang, X., Roy Raine, R., Cheng, C., Tian, Z. and Li, Y., (2009). The effects of MTBE/ethanol additives on toxic species concentration in gasoline flame. *Energy & Fuels, 23*(7), pp. 3543–3548.

Yenigün, O. and Demirel, B., (2013). Ammonia inhibition in anaerobic digestion: A review. *Process Biochemistry, 48*(5–6), pp. 901–911.

Yuan, W., Gong, Z., Wang, G., Zhou, W., Liu, Y., Wang, X. and Zhao, M., (2018). Alkaline organosolv pretreatment of corn stover for enhancing the enzymatic digestibility. *Bioresource Technology, 265*, pp. 464–470.

Zabed, H., Faruq, G., Sahu, J.N., Boyce, A.N. and Ganesan, P., (2016). A comparative study on normal and high sugary corn genotypes for evaluating enzyme consumption during dry-grind ethanol production. *Chemical Engineering Journal, 287*, pp. 691–703.

Zeng, X., Miao, W., Zeng, H., Zhao, K., Zhou, Y., Zhang, J., Zhao, Q., Tursun, D., Xu, D. and Li, F., (2019). Production of natamycin by Streptomyces gilvosporeus Z28 through solid-state fermentation using agro-industrial residues. *Bioresource Technology, 273*, pp. 377–385.

Zervakis, G.I., Koutrotsios, G. and Katsaris, P., (2013). Composted versus raw olive mill waste as substrates for the production of medicinal mushrooms: An assessment of selected cultivation and quality parameters. *BioMed Research International, 2013*, pp. 1–13.

Zhang, J., Cui, Y., Zhang, T., Hu, Q., Tong, Y.W., He, Y., Dai, Y., Wang, C.H. and Peng, Y., (2021). Food waste treating by biochar-assisted high-solid anaerobic digestion coupled with steam gasification: Enhanced bioenergy generation and porous biochar production. *Bioresource Technology, 331*, p. 125051.

12 Various Value-Added Products from Agricultural and Bio-Waste

Amit Kumar Tiwari, Dan Bahadur Pal, Vikash Shende, Vinay Raj, Anjali Prasad and Sunder Lal Pal

CONTENTS

12.1 INTRODUCTION

Agricultural products such as oilseeds are good sources of oils and fats, which are widely used for the preparation of food items, the production of medicine, etc. The other sources of these oils and fats are animal or synthetic mediums. Most of the edible oils and fats are utilized in frying, cooking and baking of food items. Oils and fats are usually found in a liquid form at ambient temperature, although some plant oils (e.g. coconut oil, palm oil and groundnut) that contain high amounts of saturated fat are found in a solid state. A wide range of edible oils are from plant origin materials, like palm seed, sunflower seed, rice bran, soybean seed, olive seed, corn seed, groundnut seed and other vegetable oils, as well as animal origin fat and oils. A huge quantity of used vegetable oil waste is generated and recycled by oil refineries, potato chip processing units, snacks industries, restaurants and domestic kitchens. Deep-frying is a process where high heat is applied to cook the food which is submerged in boiling fat/oil (Goswami et al., 2015). Waste oil should not be released or dumped into drainage systems or sewage systems because it creates

DOI: 10.1201/9781003245773-12

several problems in the cleaning, operation and maintenance of these systems (Said et al., 2016). Waste oil should not be dumped in open areas because it may lead to the development of bad smell, soil and water pollution (Alberto et al., 2020). Recycling of waste oil can be used for the purpose of fuel directly, biodiesel production, soap manufacturing, production of feed for animals, food for pets, cosmetics and detergents. Actually, the transfer of intellectual and knowledgeable information is generally hampered by the limited information about the available new technologies. In this regard, the evolution of improved and new technologies and raw materials for the conversion of biowaste materials into value-added products describes a basic target to reveal the transformations that occurs at the laboratory scale (De Almeida and Borsato, 2019). Within the confined list of available potential biowaste materials, waste cooking oil is one of the most important materials for conversion into value-added products (Borrello et al., 2017). Around 20–30% of the edible oil and fats are treated as waste after cooking globally; therefore, the management and disposal of the huge amount of this biowaste becomes a big issue worldwide, especially for highly populated and dense urban areas. Nowadays, this waste and similar types of waste received from biorefineries have been utilized as raw materials for production of biofuels, soaps, etc.

In the concepts for new product development (NPD), safety, performance and economic criteria are extremely important. To develop new and economic products from oil-based waste materials, oleo chemistry was introduced in the 1950s, and research in this technical area was also started in different R&D institutions and production industries. A huge variety of value-added products from fats and oil-based materials have been developed for different applications like biodiesel, surfactants, lubricants, polymer applications, emollients for home and personal-care units, pesticides, insecticide and minerals.

12.2 PRODUCTION OF OLEO-CHEMICALS FROM WASTE OIL OF BIOREFINERY

A huge number of edible oils extracted from vegetables like palm, oil seeds and coconut are available freely in tropical and moderate climates, while animal fats are obtained from beef tallow and fish oil in cold-climate countries. Biodiesel, lubricants, surfactants, surface coating, polymeric material, pharmaceuticals and cosmetics can be produced from these. The set of products that are available are shown in Figure 12.1. The most critical procedure to convert both vegetable and animal oil/fat into oleo chemicals is hydrolysis, and chemicals such as glycerols and mixed fatty acids are produced at high temperatures. The reaction of oils can also occur with pure water; however, at low temperatures, the rates are very small because oil is less soluble in water. To increase the speed of chemical reactions, a specific and suitable catalyst is required; during commercial production of biodiesels, these specific catalysts are always required (Demirbas, 2008c). For instance, a reaction could be carried out under reflux with dilute HCl, H_2SO_4 or aqueous NaOH solution as catalysts, and the products are glycerols and sodium salt of fatty acids. Glycerol is a sensitive molecule which can undergo several reactions (Zheng, et al., 2008).

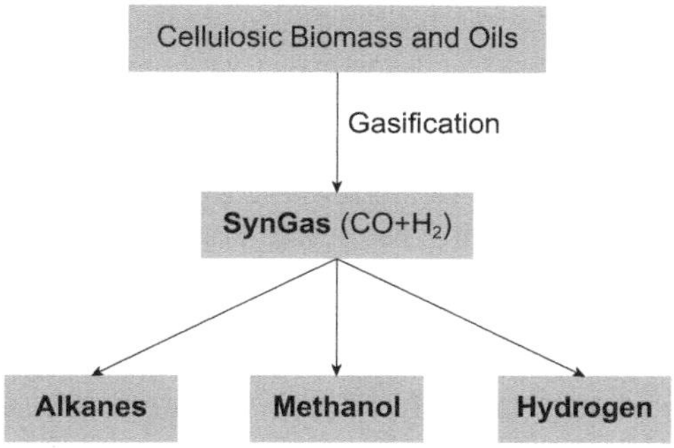

FIGURE 12.1 Production of oleo-chemicals.

At a higher temperature and pressure, the oil becomes more soluble and the rate of hydrolysis increases rapidly, even in the absence of a catalyst. Fats can be broken down by an enzyme such as lipase; commercially, there are different methods for splitting up oil to produce glycerol and fatty acid at low temperatures and medium pressures. In commercial processes for oil splitting, normally one uses acid catalysts with a concentration of sulfuric acid in the range of 1–2%. The process is attractive due to being cost-effective and simple, but the main drawback is the requirement of large reactor vessels. Medium to high pressures don't need continuous production of glycerol, and a high percentage of hydrolysis is achieved in autoclaves with 6–10 hours of residence time. Continuous counter-current processes with continuous removal of glycerol by water stream are usually operated at high pressures. The hydrolysis occurring is 80–90%, and constitutes the minimum requirement for industrial reactions. However, in the case of unsaturated fat (such as castor oil), hydrolytic methods involving lower temperature and pressure are required. Industrial interest is in low temperatures and pressures, and it is easy to implement in an industrial process.

Fatty amines also form an important class of basic oleo-chemicals, which are derived from fatty acid alkyl esters or fatty acids by reaction with ammonia. The preparation is carried out in two steps: first, forming nitrile, followed by hydrogenation to the desired product. The most efficient dehydrating catalysts are cerium, osmium oxide, aluminium, iron and thorium, and the hydrogenation of nitrile leads to the production of mixtures of tertiary, secondary and primary amines; process selectivity depends upon the choice of the catalyst. The choice of Ni and Co catalysts favour the production of 1° amines, and promoters such as boron and chromium could be added to enhance the activity and selectivity. Secondary amines are easily produced by catalytic hydrogenation of nitriles in two steps. First, the nitrile is hydrogenated to 1° amine in the presence of ammonia, and then, it is converted to 2° amine in the presence of hydrogen It is then used as an intermediate for producing di-fatty methyl quaternary ammonium compounds.

3° amines are differentiated into asymmetrical methyl di-fatty amines and symmetrical tri-fatty amines. The symmetrical products can be directly obtained from nitriles and form 2° amines by catalytic hydrogenation. The most significant tertiary amines are the asymmetric ones like di-methyl fatty amines and methyl di-fatty amines, which are procured by reductive alkylation of 2° or 1° amines. Tertiary amines are used as cosmetic ingredients, fuel additives, bactericides, emulsifiers, flotation agents, foaming agents, corrosion inhibitors, fungicides and intermediates for preparing quaternary ammonium salt. The most important quaternary ammonium salts are di-fatty methyl ammonium salt, which is different from ammonium salts. The oleo chemicals are used as surface active compounds (sulfonic derivatives, alkylamines, alkanoamides), coatings and cosmetics, detergents (mono- and di-aliphatic acid glycerides, sulfonated monoglycerides), textiles, pharmaceutical industries, soap, derivatives such as epoxidized plant oils used as stabilizers and plasticizers for producing PVC, and also used for producing dispersants, fluid polyurethane resin and foams. The chemicals are prepared by reactions such as epoxy ring-opening, epoxidation, oxidation and dimerization, hydroformylation and hydrocarboxylation, ozonolysis, etc.

Epoxidation of double bonds of the oleo-acids can be carried out using peracids (such as performic acids and peracetic acids). People also carry out epoxidation in the presence of nitrogen-containing compounds or ammonia. There is an inclination to use heterogeneous catalysts for these reactions; for this, the solid catalysts used are Amberlite IR 120 or other resins, zeolites with Tin metal, Ti-β Zeolites, Ti-MCM41 catalyst derived from hydrotalcite clay having magnesium with heteropolyanions and MoO_2 and phase transfer catalysts such as tungsten peroxo complexes.

There are three ways to add a carbonyl group to double bonds of oleo-acids: hydrofomylation, the Koch reaction and hydrocarboxylation. Hydroformylation is the most significant and has been broadly studied. Cobalt carbonyl is the most commonly used catalyst, but one can also use Ruthenium or Co as catalysts. The Rh catalyst is used for hydroformylation at low pressure and temperature, and the product formed is pure aldehyde; as opposed to this, a Co catalyst leads to the formation of the mixture of aldehydes and mono aldehydes, oleo acids, oleo esters and amides, which are used as lubricants and in latexes. The aldehydes can be utilized for manufacture of polyesters, poly-amides, polyols and polyacids, and the triglycerides and polyols can be reacted to benzaldehyde diisocyanate. Dimerization of fatty acids is a vital reaction for the synthesis of dicarboxylic acid, which is used for the production of polyamides or polyesters, as well as printing inks, hot melt adhesives and epoxy coatings. Dimers can further be converted into fatty alcohols after hydrogenation by a suitable catalyst. Dimerization can follow various mechanisms and give different mixtures of oleo chemicals, which can be used to form diols and oleic acid. In heterogeneous catalysis, layered clay, montmorillonite, has shown good properties for di-merization of acids.

12.3 UTILIZATION OF BYPRODUCT (GLYCERINE) OBTAINED FROM BIODIESEL PRODUCTION

The vegetable oils are the fatty acid esters of glycerol. These oils are derived from various nuts, like soyabean, rapeseed, jetropha oil, etc. The inedible oils are neem, mahua, jetropha, karanja rubber seeds, micro algae, silk cotton, nagchampa and rubber seed tree, and these have been characterized by high free fatty acid (FFA) content. These oils have different chemical structures compared to petrol-based diesel fuels and have high viscosity, low energy, higher engine wear and tear and cloud points, lower engine speed and power, higher engine knocking rate, injector cocking, engine compatibility and higher pour. The main commodity sources from biodiesel production and from inedible oil like jetropha, karanja, nagchampa, rubber seed tress, neem etc. are available cheaply, particularly in India. The crude vegetable oils produced by oil seed processes have undesirable substances giving colour, gums and unwanted products. In view of this, before preparing biodiesel, a physical refining process including degumming, bleaching, winterization and deodorization must be done.

There has been an attempt to utilize vegetable oil directly as fuel in place of diesel, but in view of high viscosity, the direct use or in blends is unsatisfactory and impractical because of cocking, oil ring sticking and thickening, carbon deposits, contamination by vegetable oils and gelling of the lubricating oil. People have attempted to reduce viscosity of oil by microemulsification, dilution, catalytic cracking, pyrolysis and transesterification. The method based upon pyrolysis and microemulsion has been studied in the past and has not been found satisfactory. These two methods use ethanol as diluent (4%), which lowers the viscosity of the vegetable oil and increases the break torque, break thermal efficiency and break power. Since the boiling point of vegetable oil is higher than ethanol, it leads to the formation of unburned blends. About 25% of sunflower oil has been blended with about 75% diesel; in the end, it works satisfactorily for short times, but for long-term use, it produces problems for the direct injection engine. The high viscosity of oil can also be reduced by preparing microemulsion of vegetable oil with immiscible fluids such as methanol, ethanol and 1-butanol and nonionic amphiphiles. The microemulsion has been prepared with insoluble butanol, hexanol and octanol.

12.4 PRODUCTION OF BIODIESEL THROUGH TRANSESTERIFICATION OF WASTE OIL

Among all options, transesterification has emerged as the chosen one because the physical properties of fatty acid ester (which is biodiesel) are very similar to diesel fuel. The process of making it is simple and depends heavily upon the choice of catalyst. In the esterification of an acid, alcohol is used as the nucleophilic agent; in the hydrolysis of an ester, an alcohol displaces the glycerin. This alcoholysis of an ester is called transesterification and the overall reaction is as follows:

$$\text{Oils}\left(\text{triglycerides}\right) + \text{methanol} \rightleftharpoons \text{Glycerine} + \text{Monoalkyl ester} \tag{2}$$

This reaction (2) is catalyzed either by base or acid catalyst, and the reaction consists of many continuously reversible reactions producing monoglycerides and diglycerides as intermediates. The first step involves producing alkyl esters and diglycerides, followed by alkyl esters, monoglycerides and finally glycerol. Excess alcohol is utilized to carry the reaction to completion and alkyl esters are separated from alcohol and catalyst phase during the reaction, which leads to a further fall in reverse rates of reaction. The first step with a base catalytic reaction involves a base and methanol to give an alkoxide species. This leads to an intermediate which disintegrates to form a corresponding anion of the diglyceride and alkyl ester. This step is later followed by the deprotonation of methanol by the diglyceride anion, regenerating the alkoxide species.

The oils, which are triglycerides, have limited solubility in methanol, and as a result of this, it is tough to measure the rate of reaction due to the presence of multiple phases. The mechanical mixing of the two phases is vital for enhancing the mass transfer between the different phases. Adding a suitable solvent such as tetrahydrofuran (THF) increases the solubility of the alcohol and improves the rate of reaction. Water has a negative impact on the reaction because monoglycerides, diglycerides, triglycerides and alkyl esters could react with water to give free fatty acids (FFA) by the following reaction:

$$R-COOCH_3 + H_2O \rightarrow R-COOH(FFA) + CH_3OH \quad (1.3)$$

Experiments have shown that basic catalysts are considerably more reactive and result in fewer corrosion problems than an acidic catalyst. Most of the biodiesel procedures use mineral acids for esterification of an alkali base catalyst and FFA to transesterification. One can use sodium methoxide, sodium hydroxide and potassium hydroxide for the production of biodiesel. This is economical at low pressure and temperature, with high conversion occurring in a single step. However, these catalysts have a tendency of emulsifying, as a result of which interest in solid catalysts has gained considerable attention. A variety of new catalysts have been explored to avoid use of costly material and to facilitate the recovery of catalyst, including (a) sulfonated amorphous carbon, (b) the ion exchange resin, (c) sodium ethoxide, (d) solid acid catalyst such as ZnO (Mok et al., 1992). The requirement of catalysts can be minimized with high temperature and pressure to generate supercritical fluid conditions where alcohols can react with triglycerides of fatty acid.

12.5 PRODUCTION OF ETHANOL AND FUELS FROM BIOMASS

The process of converting H_2 and CO mixture into liquid hydrocarbon is known as Fischer Tropsch synthesis (FTS). The first FTS plant began in 1938 in Germany and was later shut down due to safety reasons. In 1955, Sasol started a commercial production of liquid chemicals and fuels from coal in Sasolburg, South Africa. The FTS is a crucial step in converting carbon-containing feedstock to liquid fuels such as diesel. The advantages are (i) flexibility in feedstocks (coal, natural gas, biomass), (ii) the huge availability of raw materials involved in the procedure, (iii) the ultraclean (low sulfur-containing) products that are formed, and (iv) aptness for converting difficult-to-process resources. A major disadvantage of the FTS is the polymerization-like features of the process, producing wide product spectrum. The product can have from low molecular weight, like methane, to very high molecular weight, like waxes.

The selectivity of lesser molecular weight hydrocarbons increases, and the olefin to paraffin ratio in products shows the increase in Al_2O_3/SiO_2 ratio.

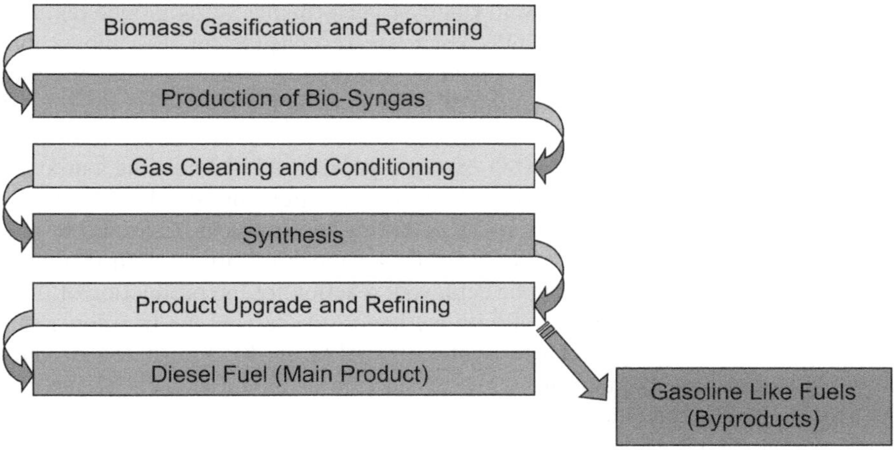

FIGURE 12.2 Production of diesel fuel from syngas.

The bio-syngas is obtained from a biomass (Figure 12.2), and it consists of mainly H_2O, CO, CO_2 and CH_4. The composition of bio-syngas can be changed by CH_4 refer water-syngas and CO_2 removal. The basic reactions in FTS are as follows:

$$nCO+2nH_2O \rightarrow (CH_2) + nH_2O \quad \text{(i)}$$
$$nCO + (2n+1)H_2 \rightarrow C_2H_{2n+1} + nH_2O \quad \text{(ii)}$$
$$nCO + (2n+m/2)H_2 \rightarrow C_nH_{2n} + nH_2O \quad \text{(iii)}$$

In this case, M is the number of hydrogen atoms per carbon atom and n is the average length of a hydrocarbon chain. The product is mixture of paraffin and olefin, and all reactions are exothermic. To prevent poising of FTS catalyst tar, hydrogen cyanide, alkalines and dust particles, ammonia, hydrogen sulfide and carbonyl sulfide must be removed thoroughly. The process of diesel from bio-syngas can be represented by the following scheme.

12.6 APPLICATION OF REACTIVE CATALYSTS AND TRANSESTERIFICATION REACTIONS

H3PW12O40 /Ta2O5: 1.7.mmol TaCl3 in 85.9mmol ethanol solution and aq. H3PW12O40 is added drop wise and stirred at room temp for 1h. The resulting clear sol was heated at 353K for 48h in vacuum and product was centrifuged to separate; finally, ethanol was removed by drying at 318K and final product is formed. When lauric acid reacts with ethanol at 80^{0}C/99, the molar ratio of ethanol to lauric acid with a constant total volume from 1:1 to 7:1, the yield of laurate increases from 12.7% to above 99.9%, reaction was performed at 351 ± 2 K for 3 h. It is due to the Implicit Acid Catalyst Performance of its unique textural properties H3PW12O40 and Ta2O5, which are responsible for this impeccable acid catalytic property. The HPA is held by the support either by ionic bonding or simple adsorption and the leaching of these may be the reason for the loss of activity (Xu et al., 2008).

WO_3/ZrO_2: sulfated zirconia (SO_4/ZrO_2) was prepared by kneading a mixture of 1860 g of hydrated zirconia, and 575 g of ammonium sulfate, 1120 g of alumina together with deionized water; by calcining in air at 675 ^{0}C for 1.5 h, sulfated zirconia-alumina (SZA) was obtained. Soyabean oil mixed with methanol at 20^0–30 ^{0}C /85% conversion can be achieved; reaction for the production of fatty acid methyl ester from soybean oil with methanol was carried out in a flow reactor with 4.0 g of catalyst: molar ratio of methanol to oil 40. Tungstated zirconia alumina is a promising catalyst for the production of biodiesel fuels because of its activity for the esterification as well as the transesterification. The catalyst is in powder form and gets carried away in every run through entrapment (Furuta et al., 2004).

$Zn_{1.2}H_{0.6}PW_{12}O_{40}$: $H_3PW_{12}O_{40}$ (8g) was dissolved in water (20ml) with stirring; a solution of zinc acetate in ethanol is added to this and stirred for 30 min. Finally, these ZnPW complexes were calcined at 400^{0}C. Waste cooking oil and palmitic acid 65/95, an increase in the methanol/acid molar ratio from 1:1 to 1:5 had a notable effect on the yield of ester. The yield of methyl palmitate increased from 59.9% to 98.1%, to a free fatty acid content of 1 wt % after 10 h reaction at 65 ^{0}C. The catalyst can be reused and recycled with negligible loss in activity over five cycles. After long uses, the losses in the activity can be attributed to the fact that the HPA is held by ionic forces (E. Li et al., 2009).

HPA with Ionic liquid: methylimidazole (9.02 g, 0.11 mol) and 1,3-propanesultone (12.21 g, 0.10 mol) were dissolved in toluene (100 mL) and stirred at 50 °C for 24 h under nitrogen atmosphere. A white precipitate is (MIMPS) formed, washed, and filtered; finally, added $H_3PW_{12}O_{40}$ and stirred for 24h at room temp, and water is removed by vacuum. Final product is solid acid acetic with n-butanol. 110/92, acetic acid (30 mmol), n-butanol (36 mmol), and catalyst (0.06 mmol) in a round-bottomed flask with a water segregator and refluxed at 110 °C for 1.5h with stirring. The highly efficient performance of this family of catalysts is due to the pseudo-liquid behaviour of HPA salts that allows PS acidic species in the bulk of the heteropoly compound to act fully as active centers for esterification. Ionic liquid will have a tendency to associate with the hydrophilic phase and tends to leach with every run (Leng et al., 2009a).

HPA with organic cation: Leng et al. (2009b) reported that Reactant is Citric acid with n-butanol130/94; the reaction was carried out at 130^{0}C, and 80.2 wt% of the catalyst was recovered and 84.5% yield was obtained. The catalytic activity of HPA salts is because of its pseudo-liquid phase behaviour. In this case, the HPA catalyst over organic support could not resist its leaching because it may be held by ionic bond. Work of various other researchers is also reviewed, and a few examples of reactive catalysts for the transesterification reactions and reactants are shown in Table 12.1.

TABLE 12.1
Reactive Catalysts for the Transesterification Reactions and Reactants

Researchers	Catalyst	Reactants
Leng et al., 2009b	Solid organic HPA	Acetic Acid + N-Butanol
Alsalme et al., 2008	HPA	Hexanoic Acid + Methanol
Bokade and Yadav, 2007	HPA supported with Clay (K-10)	Sunflower Oil + Methanol
Narasimharao et al., 2007	Cs doped heteropolytungstate (H3PW12O40 HPW) salts	Palmitic Acid + Methanol
Yan et al., 2008	CaO/MgO	Rapeseed Oil + Methanol
Li and Rudolph, 2008; Fujie Lu et al., 2015	Silica loaded with MgO as solid base catalyst	Blended Vegetable Oil + Ethanol
Pena et al., 2009	CH_3ONa, NaOH, and KOH (with Co solvent)	Castor Oil + Hexane + Methanol
Dalai et al., 2006	CaO, MgO, $Ba(OH)_2$, Li/CaO and Zeotile	Canola Oil + Methanol
Kiss et al., 2008	Sulfated zirconia, Titania and tin oxide	Dodecanoic Acid + 2-Ethylhexanol
Singh and Fernando, 2007	MgO, CaO, BaO, PbO, and MnO2	Rapeseed Oil + Methanol
Kim et al., 2004	Na/ $NaOH/Al_2O_3$	Soybean Oil + N-Hexane + Methanol
Bancquart et al., 2001	MgO, CeO_2, $La2O_3$ and ZnO	Glycerol + Methyl Strearate (Under N_2 Atmosphere)
Einolft et al., 2008	DBTDL, H_2SO_4 and Tin chloride dihydrate	Refined Rice Bran Oil + Methanol
Mcneff et al., 2008	Porous zirconia, titania and alumina micro-particulate	Waste Cooking Oil + Methanol
Xie et al., 2007	Lithium-Doped ZnO Catalysts	Soybean Oil + Methanol
Yan et al., 2009	$ZnO-La_2O_3$ catalyst	Crude Palm Oil + Methanol
Yang and Xie, 2007	Modified zinc oxide with alkali earth metals	Edible Grade Soybean Oil + Methanol
Benjapornkulaphong et al., 2009	$LiNO_3/Al_2O_3$, $NaNO_3/Al_2O_3$ and KNO_3/Al_2O_3	Palm Kernel Oil + Methanol
Corma et al., 2005	Calcined Li-Al hydrotalcites and MgO or Al-Mg hydrotalcites	Glycerol + Fatty Acid + Methyl Ester
Mazzocchia et al., 2004	Montmorillonite KSF, Montmorillonite K-10 and Zeolites	Rape Oil + Methanol
Kawashima et al., 2009	Calcium oxide	Rapeseed Oil + Methanol
Cho et al., 2009	Calcium oxide catalysts	Tributyrin + Methanol
Umdu et al., 2009	Al2O3 supported CaO and MgO catalysts	Lipid + Methanol
Caballero et al., 2009	Calcium Zincate	Edible Sunflower Oil + Methanol
Carmo et al., 2009	Mesoporous Aluminosilicate Al-MCM-41	Palmitic Acid + Methanol + Ethanol
Dıaz et al., 2000	MCM-41-Type Silica	Sulphonic Acid
Bossaert et al., 1999	Zeolites, Sulfonic Resins, and Sulfonic Mesoporous	Glycerol + Lauric Acid
Dıaz et al., 2001	MCM-41-type mesoporous silicas functionalized with sulfonic acid	Glycerol + Lauric Acid
Rhijn et al., 1998	Sulfonic acid functionalized ordered mesoporous materials	D-Sorbitol + Lauric Acid

(Continued)

TABLE 12.1
Continued

Researchers	Catalyst	Reactants
Cauvel et al., 1997	MCM-41 Type Silicas (Functionalized with Amino Groups)	Lauric Acid+ Glycidol+ Toluene
Corma et al., 1998	Cs-MCM-41, Cs-Sepiolite, MgO, and calcined hydrotalcites with different Al/Mg ratios	Rapeseed Oil + Glycerol
Lin et al., 1999	Base functionalized MCM-41	Fatty Acids +Glycidol
Vyas et al., 2009	KNO_3/Al_2O_3	Jatropha Oil + Methanol
J. Li et al., 2009	MgCoAl-LDH	Canola Oil + Methanol
Dizge et al., 2009	Immobilized lipase	Canola Oil +Methanol
Georgogianni et al., 2009	NaOH and Mg MCM-41, Mg-Al Hydrotalcite, and K+ impregnated zirconia	Rapeseed Oil + Methanol
Chuayplod and Trakarnpruk, 2009	Mg(Al)La Hydrotalcites and Metal/ MgAl Oxides	Rice Bran + Methanol
Zeng et al., 2008	Mg-Al hydrotalcite catalysts	Refined Rape Oil + Methanol
Xie et al., 2006	Mg-Al hydrotalcites	Soybean Oil + Methanol
Yakovlev et al., 2009	Ni-Cu Catalysts	Methyl Oleate + H_2
Marciniuk et al., 2009	Acidic lanthanum and aluminium diphosphates	Oleic Acid + Methanol
Suwannakarn et al., 2009	Tungstated zirconia (WZ)	Lauric Acid + Methanol
Chen et al., 2007	Mesoporous Sulfated Silica-Zirconia	Lauric Acid + Methanol
Ramos et al., 2008	Zeolites using different metal loading	Refined Sunflower Oil + Methanol
Suppes et al., 2004	Zeolite and metal catalysts	Refined Soybean Oil + Methanol
Jaenicke et al., 2000	K_2O, BaO and K_2O/La_2O_3 into MCM-41	Lauric Acid + Glycidol
Iso et al., 2001	Immobilized lipase	Triolein + Methanol
Abreu et al., 2003	Tin, Lead, Mercury, Zinc, 3-hydroxy-2-methyl-4-pyrone)$_2$(H2O)$_2$	Soybean Oil + Methanol
Santacesaria et al., 2004	V_2O_5 on SiO_2, TiO_2-SiO_2 as support	Butane + Butene
Corma et al., 2005	Li-Al hydrotalcites, MgO or Al-Mg hydrotalcites	Methyl Oleate + Methanol
Peters et al., 2006	Amberlyst 15, Smopex-101, zeolites, sulphated zirconia niobium acid, sulphuric acid, etc.	Acetic Acid + Butanol
Jitputti et al., 2006	ZrO_2, ZnO, SO_4/SnO_2, SO_4–/ZrO_2, KNO_3/KL, zeolite and KNO_3/ZrO_2	Palm Kernel Oil + Methanol
Takagaki et al., 2006	Carbon material	Ethanol + Oleic Acid
Liu et al., 2006	Nafion/silica nanocomposite catalyst (SAC-13) and H_2SO_4	Acetic Acid + Methanol
Shah and Gupta, 2007	Lipases	Jatropha Oil + Methanol
Yang and Xie, 2007	Alkaline earth metal-doped zinc oxide	Soybean Oil + Methanol
Monteiroa et al., 2008	KOH, NaOH, and $NaOCH_3$	Soybean Oil + Methanol
Silva et al., 2008	Cu(II), Co(II)	Soybean And Babassu Oils + Methanol
Marchetti and Errazu, 2008	Enzyme Lipozyme CALB	1-Propanol, 2-Propanol + Butanol, Pure Oleic Acid
Ebiura et al., 2005	K_2CO_3-loaded alumina catalyst	Trioleoyl Glycerol

Researchers	Catalyst	Reactants
Mcneff et al., 2008	Zirconia, titania and alumina	Soybean Oil + Acidulated Soapstock
Kansedo et al., 2009	Montmorillonite KSF	Palm Oil + Methanol
Sharma and Singh, 2008	NaOH/KOH	Karanja Oil + Methanol
Garcia et al., 2008	S-ZrO_2	Soyabean Oil + Methanol
Soriano et al., 2009	$AlCl_3$ or $ZnCl_2$	Canola Oil + Stearic Acid
Murugesan et al., 2009	NaOH, NaOMe, KOH, H_2SO_4, H_3PO_4 HCl	All Vegetable Oil + Methanol
Demirbas, 2008a	Alkali catalyst	Vegetable Oil + Methanol
Demirbas, 2008b	Alkali catalyst	Vegetable Oil + Alcohols
Yakovlev et al., 2009	Ni-Cu Catalyst	Rapeseed Oil + Anisole
Chongkhong et al., 2009	Sulfuric acid catalyst	Palm Fatty Acid Distillate + Methanol
Demirbas, 2009	Acid and base catalyst	Rapeseed Oil, Canola Oil + Methanol
Kolaczkowski et al., 2009	Acid or alkaline catalysts, zinc amino acid complex [znl_2]	Methanol + Rapeseed Oil
Kansedo et al., 2009	Montmorillonite KSF	Palm Oil + Methanol
Sakai et al., 2009	KOH, CaO	Vegetable Oil
Patil et al., 2009	BaO, SrO, MgO, and CaO	Jatropha, Camelina Oil + Methanol
Melero et al., 2009	Sulfonic acid-modified mesostructured catalyst	Refined Soybean Oil + Methanol
Baig and Ng, 2010	Heteropolyacid catalyst (PSA)	Soybean Oil + Methanol
Sharma et al., 2010	Eggshells after calcination (as a nonconventional heterogeneous catalyst)	Karanja Oil + Methanol
Achtena et al., 2008	NaOH, KOH	Kernel Oil + Methanol
Guzman et al., 2010	NiMo/Al_2O_3	Crude Palm Oil + Methanol
Šimácek et al., 2010	Ni-Mo/alumina	Rapeseed Oil + Methanol
Toda et al., 2005	Sugar Catalyst	Vegetable Oil + Methanol
Brahmkhatri and Patel, 2011	Acid catalyst comprising 12-tungstophosphoric acid (30%) and MCM-41	Palmitic Acid + Methanol
Santos et al., 2011	Sodium ethoxide	Waste Frying Oil + Methanol
Martínez et al., 2011	CaO Nanoparticles/NaX Zeolite	Sunflower Oil + Methanol
Sánchez-Cantú et al., 2011	Commercial hydrated lime	Soybean Oil + Methanol
Borges et al., 2011	Natural porous silica, pumice	Sunflower Oil + Frying Oil + Methanol
Fang et al., 2011	Dicationic Ionic Liquids	Naoh+ Oleic Acid
Camara and Aranda, 2011	Niobium oxide catalyst	Stearic, Palmitic, + Lauric Acid
Thitsartarn, and Kawi, 2011	Sulfated Zr-Supported Mesoporous Silica	Palm Oil + Methanol
Lerkkasemsan et al., 2011	SO_4/ZrO, $AcAl_2O_3$	Palmitic Acid + Methanol
Morales et al., 2011	Sulfonic Acid-Functionalized	Glycerol
Z. Yuan et al., 2011	Tungstophosphoric Acid	Glycerol + Ethanol
Das et al., 2011	Mesoporous Zirconium Oxophosphates	Acetic Acid + Alcohol
Kapil et al., 2011	Hyrotalcite Catalysts	Vegetable Oils + Methanol

12.7 CONCLUSION

In this chapter, on the basis of review of the available literature, a few processes and methods are discussed to develop sustainable and value-added materials from unusable fats and oil received from various sources like refineries, potato processing units, snacks plants, restaurants and kitchens. The use of biobased raw materials for the manufacturing of several useful products may be a great perspective for green and eco-friendly chemistry. Production and utilization of safe and green products can help us to keep our surroundings clean and healthy and also resolve the scarcity of petroleum fuel. As too much pollution is created by uncontrolled use of petroleum products and dumping of inedible cooked oil and its derivatives, the utilization of biowaste as raw materials is the best solution to this problem. Several synthetic chemicals are utilized in different industries to produce or synthesize many useful materials that otherwise have a bad effect on human health and the environment. These materials can be produced from biowaste materials (especially from fat and oil waste). Various processes like pyrolysis, hydrogeneration and oxidation are useful methods for preparation of lubricants, oleo-chemicals and biofuels. A lot of research has been done in this field, but still more work is required in this area worldwide. To provide a better livelihood to everyone in the coming era, it is important to utilize these resources properly. Biorefinery waste can be easily converted into value-added products like ecofriendly chemicals, biofuel, and energy by using the promising technologies discussed in this chapter. To achieve the target of proper integration, an optimization, separation and conversion process is required. To achieve the final goal, several optimized processes and treatments are developed by researchers. Each technique has its own benefit, which is completely dependent on the properties of the biowaste used and processing parameters.

12.8 REFERENCES

F.R. Abreu, D.G. Lima, E.H. Hamú, S. Einloft, J.C. Rubimand and P.A.Z. Suarez, New metal catalysts for soybean oil transesterification, *JAOCS*, 80 (6) (2003).

W.M.J. Achtena, L. Verchotb, Y.J. Frankenc, E. Mathijsd, V.P. Singhe, R. Aertsa and B. Muysa, Jatropha bio-diesel production and use, *Biomass and Bioenergy*, 32 (2008), 1063–1084.

A. Alsalme, E.F. Kozhevnikova and I.V. Kozhevnikov, Heteropoly acids as catalysts for liquid-phase esterification and transesterification, *Applied Catalysis A: General*, 349 (2008), 170–176.

A. Baig and F.T. Ng, A single-step solid acid-catalyzed process for the production of biodiesel from high free fatty acid feedstocks. *Energy & Fuels*, 24(9) (2010), 4712–4720.

S. Bancquart, C. Vanhove, Y. Pouilloux and J. Barrault, Glycerol transesterification with methyl stearate over solid basic catalysts-I. Relationship between activity and basicity, *Applied Catalysis A: General*, 218 (2001), 1–11.

S. Benjapornkulaphong, C. Ngamcharussrivichai and K. Bunyakiat, Al_2O_3-supported alkali and alkali earth metal oxides for transesterification of palm kernel oil and coconut oil, *Chemical Engineering Journal*, 145 (2009), 468–474.

V.V. Bokade and G.D. Yadav, Synthesis of biodiesel and bio-lubricant by transesterification of vegetable oil with lower and higher alcohols over heteropolyacids by clay (K-10), *Industrial Chemistry Environmental*, 85 (B5) (2007), 372–377.

M.E. Borges, L. Díaz, M.C. Alvarez-Galván and A. Brito, High performance heterogeneous catalyst for biodiesel production from vegetal and waste oil at low temperature. *Applied Catalysis B: Environmental*, 102(1–2) (2011), 310–315.

M. Borrello, F. Caracciolo, A. Lombardi, S. Pascucci and L. Cembalo, Consumers' perspective on circular economy strategy for reducing food waste, *Sustainability*, 9 (2017), 141.

W.D. Bossaert, D.E.D. Vos, W.M.V. Rhijn, J.B.P.J. Grobet and P.A. Jacobs, Mesoporous sulfonic acids as selective heterogeneous catalysts for the synthesis of monoglycerides, *Journal of Catalysis*, 182 (1999), 156–164.

V. Brahmkhatri and A. Patel, Biodiesel production by esterification of free fatty acids overl2-tungstophosphoric acid anchored to MCM-41, *Industrial & Engineering Chemistry Research*, 50 (2011), 6620–6628.

J.M.R. Caballero, J.S.G. Lez, J.M. Robles, R.M. Tost, A.J. Lopez and P.M. Torres, Calcium Zincate as precursor of active catalysts for biodiesel production under mild conditions, *Applied Catalysis B: Environmental*, 91 (2009), 339–346.

L.D.T. Camara and D.A.G, Aranda, reaction kinetic study of biodiesel production from fatty acids esterification with ethanol, *Industrial & Engineering Chemistry Research*, 50 (2011), 2544–2547.

A.C. Carmo Jr., L.K.C. deSouza, C.E.F. daCosta, E. Longo, J.R. Zamian and G.N.D. Filho, Production of biodiesel by esterification of palmitic acid over mesoporous aluminosilicate Al-MCM-41, *Fuel*, 88 (2009), 461–468.

A. Cauvel, G. Renard and D. Brunel, Monoglyceride synthesis by heterogeneous catalysis using MCM-41 type silicas functionalized with amino groups, *Journal of Organic Chemistry*, 62 (1997), 749–751.

X.R. Chen, Y.H. Ju and C.Y. Mou, Direct synthesis of mesoporous sulfated silica-zirconia catalysts with high catalytic activity for biodiesel via esterification. *The Journal of Physical Chemistry C*, 111(50) (2007), 18731–18737.

Y.B. Cho, G. Seo and D.R. Chang, Transesterification of tributyrin with methanol over calcium oxide catalysts prepared from various precursors, *Fuel Processing Technology*, 90 (2009), 1252–1258.

S. Chongkhong, C. Tongurai and P. Chetpattananondh, Continuous esterification for biodiesel production from palm fatty acid distillate using economical process, *Renewable Energy*, 34 (2009), 1059–1063.

P. Chuayplod and W. Trakarnpruk, Transesterification of rice bran oil with methanol catalyzed by Mg (Al)La hydrotalcites and metal/MgAl oxides, *Industrial & Engineering Chemistry Research*, 48 (2009), 4177–4183.

A. Corma, S. Bee, A. Hamid, S. Iborra and A. Vel, Lewis and brönsted basic active sites on solid catalysts and their role in the synthesis of monoglycerides, *Journal of Catalysis*, 234 (2005), 340–347.

A. Corma, S. Iborra, S. Miquel and J. Primo, Catalysts for the production of fine chemicals: Production of food emulsifiers, monoglycerides, by glycerolysis of fats with solid base catalysts, *Journal of Catalysis*, 173 (1998), 315–321.

A.K. Dalai, M.G. Kulkarni and L.C. Meher, Biodiesel productions from vegetable oils using heterogeneous catalysts and their applications as lubricity additives, *IEEE*, 20 (2006), 1–8.

S.K. Das, M.K. Bhunia, A.K. Sinha and A. Bhaumik, Synthesis, characterization and biofuel application of mesoporous zirconium oxophosphates, *ACS Catalysis*, 1 (2011), 493–501.

S.T. De Almeida and M. Borsato, Assessing the efficiency of End of Life technology in waste treatment—A bibliometric literature review, *Resources, Conservation and Recycling*, 140 (2019), 189–208.

A. Demirbas, Relationships derived from physical properties of vegetable oil and biodiesel fuels, *Fuel*, 87 (2008a), 1743–1748.

A. Demirbas, Biofuels sources, biofuel policy, biofuel economy and global biofuel projections, *Energy Conversion and Management*, 49 (2008b), 2106–2116.

A. Demirbas, *Biodiesel a Realistic Fuel Alternative for Diesel Engines*, Springer-Verlag London Limited, (2008c).

A. Demirbas, Progress and recent trends in biodiesel fuels, *Energy Conversion and Management*, 50 (2009), 14–34.

I. Dıaz, C.M. Alvarez, F. Mohino, J.P. Pariente and E. Sastre, Combined alkyl and sulfonic acid functionalization of MCM-41-type silica, *Journal of Catalysis*, 193 (2000), 283–294.

I. Dıaz, F. Mohino, J.P. Pariente and E. Sastre, Synthesis, characterization and catalytic activity of MCM-41-type mesoporous silicas functionalized with sulfonic acid, *Applied Catalysis A: General*, 205 (2001), 19–30.

N. Dizgea, B. Keskinlera and A. Tanrisevenb, Biodiesel production from canola oil by using lipase immobilized onto hydrophobic microporous styrene—divinylbenzene copolymer, *Biochemical Engineering Journal*, 44 (2009), 220–225.

T. Ebiura, T. Echizen, A. Ishikawa, K. Murai and T. Baba, Selective transesterification of triolein with methanol to methyl oleate and glycerol using alumina loaded with alkali metal salt as a solid-base catalyst. *Applied Catalysis A: General*, 283(1–2) (2005), 111–116.

S. Einloft, T.O. Magalhaes, A. Donato, J. Dullius and R. Ligabue, Biodiesel from rice bran oil: Transesterification by tin compounds, *Energy & Fuel*, 22 (2008), 671–674.

D. Fang, J. Yang and C. Jiao, Dicationic ionic liquids as environmentally benign catalysts for biodiesel synthesis, *ACS Catalysis*, 1 (2011), 42–47.

S. Furuta, H. Matsuhashi and K. Arata, Biodiesel fuel production with solid superacid catalysis in fixed bed reactor under atmospheric pressure, *Catalysis Communications*, 5 (2004), 721–723.

C.M. Garcia, S. Teixeira, L.L. Marciniuk and U. Schuchardt, Transesterification of soybean oil catalyzed by sulfated zirconia, *Bioresource Technology*, 99 (2008), 6608–6613.

K.G. Georgogianni, A.K. Katsoulidis, P.J. Pomonis, G. Manos and M.G. Kontom, Fuel, transesterification of soybean frying oil to biodiesel using heterogeneous catalysts, *Processing Technology*, 90 (2009), 1016–1022.

G. Goswami, R. Bora and M. Rathore, Oxidation of cooking oils due to repeated frying and human health. Conference: 2nd International Conference on Science, Technology and Management (ICSTM-15) At: Delhi University, Conference Centre, New Delhi Volume: 4, Sept issue, 2015.

A. Guzman, J.E. Torres, L.P. Prada and M.L. Nunez, Hydroprocessing of crude palm oil at pilot plant scale, *Catalysis Today*, 156 (2010), 38–43.

M. Iso, B. Chenb, M. Eguchi, T. Kudo and S. Shrestha, Production of biodiesel fuel from triglycerides and alcohol using immobilized lipase, *Journal of Molecular Catalysis B: Enzymatic*, 16 (2001), 53–58.

S. Jaenicke, G.K. Chuah, X.H. Lin and X.C. Hu, Organic—inorganic hybrid catalysts for acid- and base-catalyzed reactions, *Microporous and Mesoporous Materials*, 35–36 (2000), 143–153.

B. Jitputti, P. Kitiyanan, K. Rangsunvigit, L. Bunyakiat, L. Attanatho and P. Jenvanitpanjakul, Transesterification of crude palm kernel oil and crude coconut oil by different solid catalysts, *Chemical Engineering Journal*, 116 (2006), 61–66.

J. Kansedo, K.T. Lee and S. Bhatia, Biodiesel production from palm oil via heterogeneous transesterification, *Biomass and Bioenergy*, 33 (2009), 271–276.

A. Kapil, K. Wilson, A. Lee and J. Sadhukhan, Kinetic modeling studies of heterogeneously catalyzed biodiesel synthesis reactions, *Industrial & Engineering Chemistry Research*, 50 (2011), 4818–4830.

A. Kawashima, K. Matsubara and K. Honda, Acceleration of catalytic activity of calcium oxide for biodiesel production, *Bioresource Technology*, 100 (2009), 696–700.

H.J. Kima, B.S. Kanga, M.J. Kima, Y.M. Park, D.K. Kimb, J.S. Lee and K.Y. Lee, Transesterification of vegetable oil to biodiesel using heterogeneous base catalyst, *Catalysis Today*, 93–95 (2004), 315–320.

A. Kiss, A.C. Dimian and G. Rothenberg, Biodiesel by catalytic reactive distillation powered by Metal Oxides, *Energy & Fuel*, 22 (2008), 598–604.

S.T. Kolaczkowski, U.A. Asli and M.G. Davidson, A new heterogeneous ZnL_2 catalyst on a structured support for biodiesel production, *Catalysis Today*, 147S (2009), S220–S224.

Y. Leng, J. Wang, D. Zhu, X. Ren, H. Ge and L. Shen, Heteropolyanion-based ionic liquids: Reaction-induced self-separation catalysts for esterification. *Angewandte Chemie*, 121(1) (2009a), 174–177.

Y. Leng, J. Wang, D. Zhu, Y. Wu and P. Zhao, Sulfonated organic heteropolyacid salts: Recyclable green solid catalysts for esterifications. *Journal of Molecular Catalysis A: Chemical*, 313(1–2) (2009b), 1–6.

N. Lerkkasemsan, N. Abdoulmoumine, L. Achenie and F. Agblevor, Mechanistic modeling of palmitic acid esterification via heterogeneous catalysis, *Industrial & Engineering Chemistry Research*, 50 (2011), 1177–1186.

E. Li and V. Rudolph, Transesterification of vegetable oil to biodiesel over MgO-functionalized mesoporous catalysts, *Energy & Fuel*, 22 (2008), 45–149.

J. Li, X. Wang, W. Zhu and F. Cao, $Zn_{1.2}H_{0.6}PW_{12}O_{40}$ Nanotubes with double acid sites as heterogeneous catalysts for the production of biodiesel from waste cooking oil, *Chem Sus Chem*, 2 (2009), 177–183.

E. Li, Z.P. Xu, V. Rudolph, Synthesis of biodiesel via acid catalysis, *Applied Catalysis B: Environmental*, 88 (2009), 42–49.

X. Lin, G.K. Chuah and S. Jaenicke, Base-functionalized MCM-41 as catalysts for the synthesis of Monoglycerides, *Journal of Molecular Catalysis A: Chemical*, 150 (1999), 287–294.

Y. Liu, E. Lotero and J.G. Goodwin Jr, A comparison of the esterification of acetic acid with methanol using heterogeneous versus homogeneous acid catalysis. *Journal of Catalysis*, 242(2) (2006), 278–286.

F. Lua, Wei Yu, Xinhai Yu a, b *and Shan-Tung Tu, Transesterification of vegetable oil to biodiesel over MgoLi2O catalysts templated by a PDMS-PEO comb-like copolymer, *Energy Procedia*, 75 (2015), 72–77.

A. Mannu, S. Garroni, J. Ibanez and A. Mele, Available technologies and materials for waste cooking oil recycling, *Processes*, 8 (2020), 366. doi:10.3390/pr8030366.

J.M. Marchetti and A.F. Errazu, Comparison of different heterogeneous catalysts and different alcohols for the esterification reaction of oleic acid. *Fuel*, 87(15–16) (2008), 3477–3480.

L.L. Marciniuk, C.M. Garcia, R.B. Muterle and U. Schuchardt, Bioenergy II: Acid diphosphates as catalysts for the production of methyl and ethyl esters from vegetable oils, *International Journal of Chemical Reactor*, 7 (2009), A44.

S.L. Martínez, R. Romero, J.C. Lopez, A. Romero, V.S. Mendieta and R. Natividad, preparation and characterization of CaO nanoparticles/NaX zeolite catalysts for the transesterification of sunflower oil, *Industrial & Engineering Chemistry Research*, 50 (2011), 2665–2670.

C. Mazzocchia, G. Modica, A. Kaddouri and R. Nannicini, Fatty acid methyl esters synthesis from triglycerides over heterogeneous catalysts in the presence of microwaves, *Comptes Rendus Chimie*, 7 (2004), 601–605.

C.V. Mcneff, L.C. Mcneff, B. Yan, D.T. Nowlan, M. Rasmussen, A.E. Gyberg, B.J. Krohn, R.L. Fedie and T.R. Hoye, A continuous catalytic system for biodiesel production, *Applied Catalysis A: General*, 343 (2008), 39–48.

J.A. Melero, L.F. Bautista, G. Morales, J. Iglesias and D. Briones, Biodiesel production with heterogeneous sulfonic acid-functionalized mesostructured catalysts, *Energy & Fuels*, 23 (2009), 539–547.

W.S. Mok, M.J. Antal Jr and G. Varhegyi, Productive and parasitic pathways in dilute acid-catalyzed hydrolysis of cellulose, *Industrial & Engineering Chemistry Research*, 31(1) (1992), 94–100.

M.R. Monteiro, A.R.P. Ambrozin, L.M. Liao and A.G. Ferreira, Critical review on analytical methods for biodiesel characterization, *Talanta*, 77 (2008), 593–605.

G. Morales, M. Paniagua, J.A. Melero, G. Vicente and C. Ochoa, Sulfonic acid-functionalized catalysts for the valorization of glycerol via transesterification with methyl acetate, *Industrial & Engineering Chemistry Research*, 50 (2011), 5898–5906.

A. Murugesan, C. Umarani, T.R. Chinnusamy, M. Krishnan, R. Subramanian and N. Neduzchezhain, Production and analysis of bio-diesel from non-edible oils—A review, *Renewable and Sustainable Energy Reviews*, 13 (2009), 825–834.

K. Narasimharao, D.R. Brownb, A.F. Leea, A.D. Newmana, P.F. Siril b, S.J. Tavener and Wilsona, Structure—activity relations in Cs-doped heteropolyacid catalysts for biodiesel production, *Journal of Catalysis*, 248 (2007), 226–234.

S. Nurdin, R. Yunus, A. Nour, J. Gimbun, N. Azman and M. Sivaguru, Restoration of waste cooking oil (WCO) using alkaline hydrolysis technique (ALHYT) for future, *Biodetergent*, 11 (2016), 6405–6410.

P.D. Patil, V.G. Gude and S. Deng, Biodiesel production from Jatropha curcas, waste cooking, and Camelina sativa oils. *Industrial & Engineering Chemistry Research*, 48(24) (2009), 10850–10856.

R. Pena, R. Romero, S.L. Martınez, M.J. Ramos, A. Martınez and R. Natividad, Transesterification of castor oil: Effect of catalyst and co-solvent, *Industrial & Engineering Chemistry Research*, 48 (2009), 1186–1189.

T.A. Peters, N.E. Benes, A. Holmen and J.T.F. Keurentjes, Comparison of commercial solid acid catalysts for the esterification of acetic acid with butanol, *Applied Catalysis A: General*, 297 (2006), 182–188.

M.J. Ramos, A. Casas, L. Rodrıguez, R. Romero and A. Perez, Transesterification of sunflower oil over zeolites using different metal loading: A case of leaching and agglomeration studies, *Applied Catalysis A: General*, 346 (2008), 79–85.

W.M.V. Rhijn, D.E.D. Vos, B.F. Sels, W.D. Bossaert and P.A. Jacobs, Sulfonic acid functionalized ordered mesoporous materials as catalysts for condensation and esterification reactions, *Jacobs, Chemical Communications* (1998), 317–378.

T. Sakai, A. Kawashima and T. Koshikawa, Economic assessment of batch biodiesel production processes using homogeneous and heterogeneous alkali catalysts, *Bioresource Technology*, 100 (2009), 3268–3276.

M. Sánchez-Cantú, L.M. Pérez-Díaz, R.M. Rosales, E. Ramírez, A. Apreza-Sies, I. Pala-Rosas, E. Rubio-Rosas, M. Aguilar-Franco and J.S. Valente, Commercial hydrated lime as a cost-effective solid base for the transesterification of wasted soybean oil with methanol for biodiesel production, *Energy & Fuels*, 25 (2011), 3275–3282.

E. Santacesaria, M. Cozzolino, M.D. Serio, A.M. Venezia and R. Tesser, Vanadium based catalysts prepared by grafting: Preparation, properties and performances in the ODH of butane, *Applied Catalysis A: General*, 270 (2004), 177–192.

R.O. Santos, I.G.C. Andreia, A.M. Giannetti and R.B. Torres, Optimization of the transesterification reaction in biodiesel production and determination of density and viscosity of biodiesel/diesel blends at several temperatures, *Journal of Chemical & Engineering Data*, 56 (2011), 2030–2038.

S. Shah and M.N. Gupta, Lipase catalyzed preparation of biodiesel from Jatropha oil in a solvent free system. *Process Biochemistry*, 42(3) (2007), 409–414.

Y.C. Sharma and B. Singh, Development of biodiesel from Karanja, a tree found in rural India, *Fuel*, 87 (2008), 1740–1742.

Y.C. Sharma, B. Singh and J. Korstad, Application of an efficient nonconventional heterogeneous catalyst for biodiesel synthesis from pongamia pinnata oil, *Energy Fuels*, 24 (2010), 3223–3231.

R.B.D. Silva, A.F.L. Neto, L.S.S.D. Santos, J.R.D.O. Lima, M.H. Chaves, J.R.D. Santos Jr, G.M.D. Lima, E.M.D. Moura and C.V.R.D. Moura, Catalysts of Cu(II) and Co(II) ions adsorbed in chitosan used in transesterification of soy bean and babassu oils—A new route for biodiesel syntheses, *Bioresource Technology*, 99 (2008), 6793–6798.

P. Simacek, D. Kubicka, G.A Sebor and M. Pospisil, Fuel properties of hydroprocessed rapeseed oil, *Fuel*, 89 (2010), 611–615.

A.K. Singh and S.D. Fernando, Reaction kinetics of soybean oil transesterification using heterogeneous metal oxide catalysts, *Chemical Engineering & Technology*, 30 (12) (2007), 1716–1720.

N.U. Soriano Jr., R.V. Ditti and D.S. Argyropoulos, Biodiesel synthesis via homogeneous lewis acid-catalyzed transesterification, *Fuel*, 88 (2009), 560–565.

G.J. Suppes, M.A. Dasari, E.J. Doskocil, P.J. Mankidy and M.J. Goff, Transesterification of soybean oil with zeolite and metal catalysts, *Applied Catalysis A: General*, 257 (2004), 213–223.

K. Suwannakarn, E. Lotero, K. Ngaosuwan and J.G. Goodwin Jr, Simultaneous free fatty acid esterification and triglyceride transesterification using a solid acid catalyst with in situ removal of water and unreacted methanol, *Industrial & Engineering Chemistry Research*, 48 (2009), 2810–2818.

A. Takagaki, M. Toda, M. Okamura, J.N. Kondo, S. Hayashi, K. Domen and M. Hara Esterification of higher fatty acids by a novel strong solid acid, *Catalysis Today*, 116 (2006), 157–161.

W. Thitsartarn and S. Kawi, Transesterification of oil by sulfated Zr-supported mesoporous silica, *Industrial and Engineering Chemistry Research*, 50 (13) (2011), 7857–7865. ScholarBank@NUS Repository.

M. Toda, A. Takagaki, M. Okamura, J.N. Kondo, S. Hayashi, K. Domen and M. Hara, Biodiesel made with sugar catalyst, *Nature*, 438 (2005), 178.

E.S. Umdu, M. Tuncer and E. Seker, Transesterification of Nannochloropsisoculata microalga's lipid to biodiesel on Al2O3 supported CaO and MgO catalysts, *Bioresource Technology*, 100 (2009), 2828–2831.

A.P. Vyas, N. Subrahmanyam and P.A. Patel, Production of biodiesel through transesterification of Jetropha oil using KNO_3/Al_2O_3 solid catalyst, *Fuel*, 88 (2009), 625–628.

W. Xie, H. Peng and L. Chen, Calcined Mg—Al hydrotalcites as solid base catalysts for methanolysis of soybean oil, *Journal of Molecular Catalysis A: Chemical*, 246 (2006), 24–32.

W. Xie, Z. Yang and H. Chun, Catalytic properties of lithium-doped ZnO catalysts used for biodiesel preparations, *Industrial & Engineering Research*, 46 (2007), 7942–7949.

L. Xu, X. Yang, X. Yu and Y. Guo, Preparation of mesoporous polyoxometalate tantalum pentoxide composite catalyst for efficient esterification of fatty a Maynurkader, *Catalysis Communications*, 9 (2008), 1607–1611.

V.A. Yakovlev, S.A. Khromova, O.V. Sherstyuk, V.O. Dundich, D.Yu. Ermakov, V.M. Novopashina, M.Yu. Lebedev, O. Bulavchenko and V.N. Parmon, Development of new catalytic systems for upgraded bio-fuels production from bio-crude-oil and biodiesel, *Catalysis Today*, 144 (2009), 362–366.

S. Yan, H. Lu and B. Liang, Supported CaO catalysts used in the transesterification of rapeseed oil for the purpose of biodiesel production, *Energy & Fuel*, 22 (2008), 646–651.

S. Yan, S.O. Salley and K.Y. Simon Ng, Simultaneous transesterification and esterification of unrefined or waste oils over ZnO-La2O3 catalysts, *Applied Catalysis A: General*, 353 (2009), 203–212.

Z. Yang and W. Xie, Soybean oil transesterification over zinc oxide modified with alkali earth metals, *Fuel Processing Technology*, 88 (2007), 631–638.

Z. Yuan, S. Xia, P. Chen, Z. Hou and X. Zheng, Etherification of biodiesel-based glycerol with bioethanol over tungstophosphoric acid to synthesize glyceryl ethers, *Energy & Fuels*, 25 (2011). doi:10.1021/ef200366q.

H.Y. Zeng, Z. Feng, X. Deng and Y. Li, Activation of Mg—Al hydrotalcite catalysts for transesterification of rape oil, *Fuel*, 87 (2008), 3071–3076.

Y. Zheng, X. Chen and Y. Shen, Commodity chemicals derived from glycerol, an important biorefinery feedstock, *Chemical Reviews*, 108 (12) (2008), 5253–5277. https://doi.org/10.1021/cr068216s (Retraction published Chem Rev. 2010 Mar 10;110(3):1807).

Index